The principles and practice of electron microscopy
Second Edition

The first edition of this book was widely praised as an excellent introduction to electron microscopy in journals of physics, materials science, earth science and biological science. In this completely revised new edition the author has taken the opportunity to expand the coverage of certain existing topics and to add much new material.

As with the first edition, the principal aim of the book is to present the subject of electron microscopy in a readable way, both to readers who may never see an electron microscope but who need to know what these instruments are and how they perform, and also to practising electron microscopists by providing an up-to-date picture of the changes taking place in fields other than their own.

Compared with the first edition, the number of chapters has been increased from 5 to 8 (with 4 Appendixes). One of the new chapters explains the interactions taking place between an electron beam and a sample, which results in signals that can be used for imaging and analysis. Another deals with microanalysis in electron microscopes, the invaluable added bonus of using electrons instead of light for microscopy. The third new chapter results from the logical separation of specimen preparation techniques and the interpretation of micrographs into separate chapters. Throughout the book the coverage has been brought completely up to date, whilst retaining techniques devised in the early days of electron microscopy which are still relevant and widely used. Current live topics such as computer control of microscopes, energy-filtered imaging, cryo- and environmental microscopy, digital imaging and high-resolution scanning and transmission microscopy are all described and illustrated, with reference to recent work in these fields. The case studies in the first edition were very highly praised in the reviews, and some interesting new examples are included in this new edition. The bibliography has been enlarged and updated to include references to current work as well as retaining classical references to the basic principles underlying the subject.

This guide to electron microscopy, written by an author with thirty years practical experience of the technique, will be invaluable to new and experienced users of electron microscopes in any area of science and technology.

The principles and practice of electron microscopy

Second Edition

Ian M. Watt

CAMBRIDGE
UNIVERSITY PRESS

Published by the Press Syndicate of the University of Cambridge
The Pitt Building, Trumpington Street, Cambridge CB2 1RP
40 West 20th Street, New York, NY 10011–4211, USA
10 Stamford Road, Oakleigh, Melbourne 3166, Australia

First published 1985
Second edition 1997

Printed in Great Britain at the University Press, Cambridge

A catalogue record for this book is available from the British Library

Library of Congress cataloguing in publication data

Watt, Ian M., 1926–
The principles and practice of electron microscopy / Ian M. Watt.
– 2nd ed.
     p.   cm.
Includes bibliographical references (p.   ) and indexes.
ISBN 0 521 43456 4 (hc). – ISBN 0 521 43591 9 (pb)
1. Electron microscopy.   I. Title.
QH212.E4W38   1996
502′.8′25–dc20   95-37567 CIP

ISBN 0 521 43456 4 hardback
ISBN 0 521 43591 9 paperback

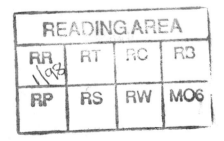

Contents

Preface to the first edition

Electron microscopes have been available commercially for 45 years now, and several thousands are in use throughout the world, operating in a wide range of fields. Although they started out as transmission instruments, with consequent limitations for certain types of examination, the introduction of the scanning microscope only 20 years ago revolutionised electron microscopy and broadened its scope very considerably. More recently the availability of a range of analytical techniques and the development of both hybrid and more specialised models of microscope have made it necessary to consider which member of the e.m. family is most appropriate to a particular examination, and how it can best be employed. This book has been written for anyone whose work or study encounters the results of electron microscopy, whether in biology, medicine or technology. Its purpose is to describe in simple terms the range of instruments and techniques now available and to illustrate how their capabilities are employed. The book mentions some of the factors which have to be considered and optimised by the microscopist, in the hope that the user of the results of microscopy will feel he understands more of what is going on, even though he may never see an electron microscope in operation. If this results in a closer collaboration between the 'customer' and the microscopist it will be to the advantage of both.

In attempting to give an easily understood survey of the broad field of electron microscopy the author has been helped by his background as a general-purpose microscopist providing a service in electron microscopy to 'customers' from a range of disciplines; initially at Sira (formerly the British Scientific Instrument Research Association) and more recently at the Research Centre of Johnson Matthey p.l.c. He was already immersed in microscopy when the scanning microscope was introduced (initially as a technique looking for applications – how the scene has changed!) and, having had the use of both types of microscope, is able to appreciate the strong and weak points of both. This balanced viewpoint was helped by regular association over a number of years with other practising microscopists who formed the steering committee of a research group on electron microscope techniques, based on the EM Unit at Sira.

It is only possible to mention a few of the many people to whom the author is grateful for support and stimulation in his first 25 years in electron microscopy. Without the forethought of the late Dr John Thomson and Dr K. M. Greenland, who arranged for an unwanted electron microscope to be transferred to Sira in 1958 and encouraged the author to become an expert in using it, the episode would not have begun. A series of colleagues in the EM Unit at Sira, notably Ann Allnutt and the late Noel Wraight, shared and nourished his enthusiasm for the subject. The supporters of the Electron Microscope Techniques Group have already been mentioned. Thanks are due to the present Management of Sira Ltd for permission to reproduce micrographs and results obtained at Chislehurst. Similarly, the author is grateful for the help and encouragement received at Johnson Matthey Research Centre from numerous colleagues, but particularly from the Group Research Director, Dr G. J. K. Acres and the Research Manager, Dr M. J. Cleare. The facilities and staff provided for word processing and copying have helped considerably in the preparation of the final version of this book.

Finally, I must apologise to my family for the disruption of family life which this book has caused over a number of years.

Ian M. Watt
Reading, 1983

Preface to the second edition

Since publication of the previous edition the pace of computerisation and automation has been relentless, and in much current equipment the mouse reigns supreme! But at the same time as making the new edition necessary to encompass the new technologies, the potential readership has been broadened to include those new users who could be helped with some of the background to their 'point and shoot' microscopes.

Since electron microscopes seldom wear out, and tight economic conditions worldwide have led users to delay replacing their instrument until design changes have slowed down, the book has to be updated by adding information: little of the old technology is not still used by someone, somewhere (one only has to survey the current literature to see how much work is still being reported using JSM 35s and EM 301s, for example). The newer instruments work using the same principles as they did twenty years ago, but the electron optics and the results are better understood than they were. Hence we have new chapters on processes occurring between beam and specimen, and the analytical capabilities of electron microscopes. The digital imaging revolution is described, and the developments in dealing with moist specimens, from cryo-preparation to environmental microscopy.

Since the first edition the writer has retired from practical electron microscopy, but has been encouraged to keep abreast of the latest developments in microscopy as a whole as Technical Editor of *Microscopy and Analysis* magazine. He hopes that this book will help microscopists and their customers to get a feeling for what is going on in the still expanding field of microscopy.

I should like to thank the microscopists and suppliers who have willingly provided illustrations and case studies for this new edition.

Reading, 1995

Microscopy with light and electrons **1**

As the unaided human eye moves closer to an object it produces in the brain of the observer a progressively larger and more detailed image until the eye is about 25 cm from the object, at the so-called *near point* of the eye. The observer's eye cannot focus on objects closer than this and the image rapidly becomes blurred as the distance decreases further. What is now needed is a device to give the effect of moving closer but with the enlarged image still appearing no closer than the near point; that is, to help the human eye to see smaller detail than it could unaided. This device is a magnifying system or microscope which may take one of a variety of forms to be described in this book. It may produce an aerial image in space, in which case only one person at a time can see it, or it may form the image on a screen or photographic material so that several people can view it or it can be reproduced elsewhere. Before looking into the types of microscope which are available we can consider the very wide range of objects which are to be examined, a range from postage stamps and fabrics down to the individual atoms of which matter is composed. This range of dimensions covers six orders of magnitude, requiring enlargements up to millions of times greater than life-size. Such enlargements are beyond the capability of traditional microscopes and this book endeavours to explain the new technology which makes it possible to meet these requirements.

1.1 Microscope systems using light

1.1.1 The simple microscope

The simplest of all microscopes is the hand lens or magnifying glass which is a single biconvex lens of glass or plastic. An object placed close behind the lens appears to be bigger when seen through it, so that more detail can be seen in it than without the use of the lens. Figure 1.1 illustrates diagrammatically and photographically the enlargement produced by a simple hand lens. A wider field of view is seen when the eye is placed close to the lens.

The magnification produced by a single lens is approximately $25/f$ where f is the focal length of the lens in centimetres. To increase the magnification the focal length must be reduced, which in practice requires the lens to have more sharply curved surfaces. A $\times 10$ magnifier is already a bulbous piece of glass and shows various defects in image formation. These occur to varying extents in all lens systems and will be described in more detail in the next section.

Defects of lenses

There are several image defects which result from the shape and size of the lens and one which depends on the material of which it is made. Figure 1.2 shows a magnified image of black printing on white paper as seen through the single bicon-vex lens of a ×5 watchmaker's eyeglass, photographed on panchromatic film. The original image was tinged with colour, particularly away from the centre of the field of view, and this has been recorded as an apparent smearing of the image. This is because the single lens suffers from *chromatic aberration* and results from the material of the lens having a different light-bending ability (refractive index) for light of different colours. This is the same effect which enables us to split up white light into a spectrum of colours by passing it through a prism. In a simple lens the effect of

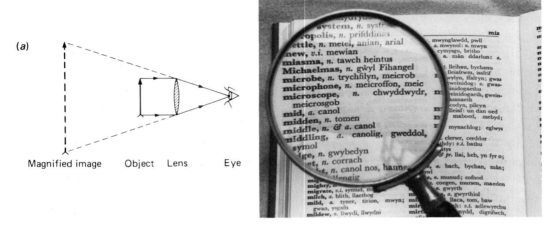

(*a*)

Magnified image   Object   Lens   Eye

Figure 1.1. The magnifying glass. (*a*) Ray diagram showing how the lens converges light from the object so that it appears to come from a larger 'image' which is sufficiently far away for the eye to focus it comfortably. (*b*) Photograph through a magnifying glass.

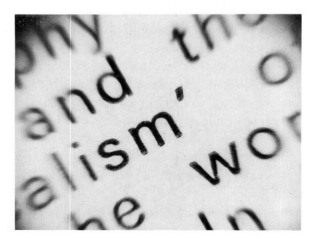

Figure 1.2. Black printing on white paper seen through a single biconvex lens.

chromatic aberration is to form a succession of coloured images in different planes along the lens axis, the blue one being closer to the lens than the red one (Figure 1.3(*a*)). A screen placed at B will show a focused blue image ringed with red, and vice versa when placed at R. (There are, of course, the other spectral colours in between but only these two are illustrated, for clarity.)

Fortunately the degree of *dispersion* of coloured rays, i.e. the difference between the refractive index for the two extremes of the spectrum compared to the average refractive index, varies between different types of glass, 'flint' glass having a higher dispersion than 'crown' glass. It is therefore possible to replace a single 'chromatic' lens by an *achromatic doublet* which has the same focal length for light of two different colours, e.g. red and blue (Figure 1.3(*b*)). If a third glass type is used the correction can be made for three colours, and the lens combination is called *apochromatic*. A simpler way of avoiding chromatic aberration is to form the image using light of only one colour, i.e. monochromatic illumination, in which case only one focal length is involved. Figure 1.4 is the same as Figure 1.2 but illuminated with sodium yellow light. Chromatic aberration has now been eliminated and the image is cleaner but still imperfect. The remaining faults are caused by the shape and size of the lens and are: spherical aberration, distortion, field curvature, astigmatism and coma. For the present purpose it is not necessary to understand these in detail, but their effects will be described briefly in connection with our simple microscope.

Spherical aberration, perhaps more aptly described by the German term

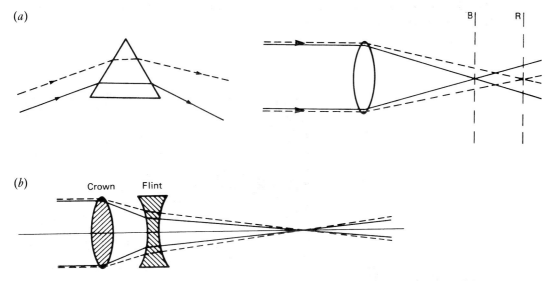

Figure 1.3. (*a*) Formation of chromatic image. Solid line represents blue light, broken line red.
(*b*) Achromatic doublet with same focal length for blue and red light.

öffnungsfehler or aperture defect, becomes troublesome when image-forming rays make more than a small angle with the axis of the lens or, looking at it in another way, when light rays make large angles of incidence with the surfaces of the lens. The resultant effect is that central rays are brought to a focus further from the lens than rays passing through the periphery of the lens. The degree of spherical aberration is measured by the axial separation of the foci for axial and marginal rays. A lens suffering from this defect will not have a unique focus, but only a 'best' focus or 'circle of least confusion' in the image. This is shown diagrammatically in Figure 1.5. It is this focus, not the ideal or *Gaussian* focus, which is used in practice. One way of minimising the aberration is by inserting an opaque 'stop' so that only the central part of the lens is used, but this is wasteful of light and darkens the image. The aberration may also be minimised by splitting the refraction between several surfaces so that no large angles of refraction are required at any single surface. Fortunately it is possible to eliminate the defect completely by replacing a single lens by several with appropriate surface curvatures. If different types of glass are used for the separate lenses the combination may be made achromatic at the same time.

A lens or lens combination which has been corrected for spherical and chromatic aberrations may still show *distortion*, a defect in which the magnification varies across the image field. Figure 1.6 shows the effects of the two types of distortion, a square object having its sides apparently sucked in by *pincushion distortion* and blown outwards by *barrel distortion*. It should be noted that in this particular defect the sharpness of the image is not affected, but the magnification varies with distance from the lens axis.

Spherical and chromatic aberrations both influence the quality of image detail on the lens axis. Images of off-axis points are affected by astigmatism, field curvature and coma. In the first, illustrated in Figure 1.7, an off-axis point is imaged as a line

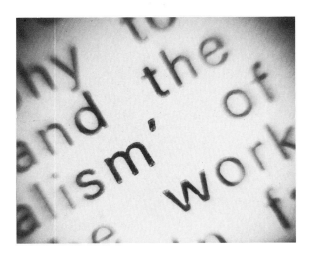

Figure 1.4. As Figure 1.2 but photographed using monochromatic light from a sodium vapour lamp. Principal defect is now spherical aberration.

in two orthogonal planes. Again there is only a 'best' image between the two line images, but the effect can be eliminated by careful design of the lens components. Field curvature is an effect by which a plane object is reproduced as a curved image. Lenses in which astigmatism and field curvature are both corrected are called *anastigmats*. Photographic lenses are examples of this type and the designs are frequently complex, involving four, five or six separate lenses (elements) of carefully calculated shape and separations to eliminate or at the least minimise the effects of the aberrations which have been mentioned. Because of the sometimes conflicting requirements of each defect the final design is a compromise to fit a particular set of imaging conditions. Hence, if the lens is used for a different purpose, e.g. a lens corrected for parallel light from distant objects used as a close-up lens with strongly divergent light, certain aberrations will no longer be eliminated and the image quality will deteriorate.

Returning once more to our simple microscope, we can see that it is possible to improve the performance of a single lens by making it of several components, at the very least an achromatic doublet corrected for spherical aberration; better quality magnifying glasses are improved in this way. Even so, if magnifications of more than about ×10 are required, and particularly if a magnified image is required on a screen

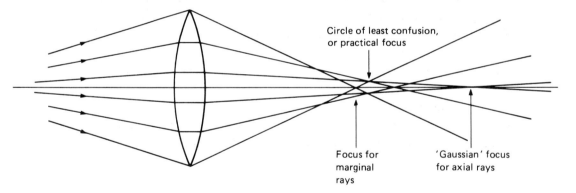

Figure 1.5. Imaging of a point object by a lens with spherical aberration. The practical focus will appear to be where the imaging rays occupy the smallest diameter, i.e. at the plane of the circle of least confusion.

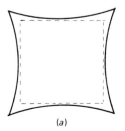

(a) (b)

Figure 1.6. The image of a square object formed by a lens with (a) pincushion and (b) barrel distortion.

or photographic film, the simple microscope is preferably replaced by a two-lens system, the second magnifying the image formed by the first. This system is called a *compound microscope,* and will now be described.

1.1.2 The compound microscope

Most simply (Figure 1.8) this comprises two convex lenses appropriately spaced from each other and the object. The object is placed just beyond the focus of the first or *objective* lens, which forms an enlarged real *intermediate* image between the two lenses. The second lens or *eyepiece* can be used as a magnifying glass to form a magnified virtual image of the intermediate image in a plane convenient for the observer to see. Alternatively the second lens can be used to project the second image on to a screen, (and is then called the *projector* lens).

If the individual magnifications produced by the two lenses are ×10 the overall magnification of the microscope will be ×100. With a more powerful (i.e. shorter focal length) objective lens, e.g. ×40, and the same eyepiece the overall magnification of the microscope will be ×400. Practical microscope systems have a choice of highly corrected objective lenses, usually mounted on a revolving nosepiece so that the overall magnification can be rapidly changed between several values (e.g. ×50, ×100, ×400, and ×1000) by changing objectives. The eyepiece magnification can also be altered by substituting different lenses, but this should only be done with discretion (see later, *empty magnification*). A practical compound microscope will of course be an enhancement of the simple two-lens theoretical system shown in the figure; an important component is the *condenser* lens or system of lenses which focuses illumination on the specimen; this must be free from aberrations and fulfil

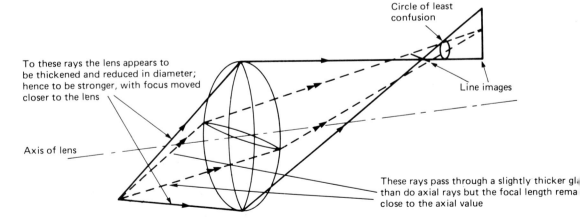

Figure 1.7. Diagrammatic representation of the defect of astigmatism. This is a defect whereby the image of an off-axis point becomes two lines at right angles to one another, in different planes. The separation between the planes is a measure of the degree of astigmatism in the lens. The best circular rendering of the point is the circle of least confusion between the two line images.

certain requirements to match the illumination to the objective lens being used. The positioning of the condenser determines whether the compound microscope is a *transmission* instrument (biological or research microscope), for looking **through** specimens at their internal structure or operates in *reflection* (dissecting or metallurgical microscope) to look **at** the surface of the specimen. There are significant differences in the construction of these two types, but the magnifying roles of the objective and eyepiece are the same.

It might be expected that by increasing the number of magnifying stages above two, or making each one more powerful, an overall magnification of tens- or even hundreds of thousands might be obtained, enabling very fine detail to be examined with compound light microscopes. Unfortunately the nature of light itself makes this last achievement impossible, for reasons which will now be described.

1.2 The diffraction barrier

Light is a form of electromagnetic energy which is transmited as a wave motion. We can visualise many of its effects if we picture it as a set of ripples moving away from its source, e.g. from a lamp filament. Whenever the waves impinge on an obstacle, secondary wavelets are set up centred on the obstacle and these can interact or interfere with the main wave-front to produce local increases or decreases of illumination. Thus, passing light through a hole in an opaque screen

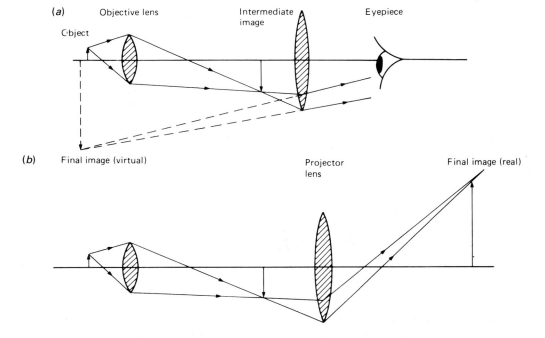

Figure 1.8. Diagrammatic representation of the principle of the compound microscope. (*a*) Virtual image viewed directly by the observer. (*b*) Real image projected onto a screen or recording medium.

results in a bright patch of light with fringe (*diffraction*) effects intruding into the geometrical shadow, because of the effect of the secondary wavelets from the edges of the hole.

The action of an optically denser medium is to slow down a wave-front of light. A biconvex lens converts a plane wave-front from a distant source into a contracting one which converges to the focus of the lens. The effect of diffraction, however, is to make the image of a point source not a second point but a small bright spot surrounded by less intense light and dark haloes. The spot is called the '*Airy disc*' by opticians and it can be shown by calculation to contain 84% of the light in the image plane. The intensity distribution in and around the disc is shown approximately in Figure 1.9(*a*). The radius of the disc or, more precisely, of the dark ring surrounding it is $1.22F\lambda/d$ where F is the focal length of the lens, d its diameter and λ the wavelength of light. It should be noted that this is the smallest point which can be projected by a perfect lens; the effect of lens aberrations on the image may be to make the central maximum wider than the value given above.

If two points are imaged by the lens, each will be represented by an Airy disc and haloes. Moving the object points together will move the intensity patterns together until they overlap (Figure 1.9(*b*)). As the overlap increases there will be a point at which the elongated patch of light can just be identified as coming from two points close together. According to Rayleigh this situation where the two individual points can just be discerned separately, or resolved, occurs when the intensity maximum of one pattern falls in the first intensity minimum of the other, i.e. the separation between the two peaks is $1.22F\lambda/d$ (Figure 1.9(*c*)). Any lesser separation would result in formation of a continuous blur with no discernible minimum in it (Figure 1.9(*d*)). The criterion of the *resolving power* of an imaging system as the separation between two object points where images can just be seen as separate is common to all microscopy, although a little difficult to apply in certain cases.

1.2.1 Resolving power of microscopes

In 1873 the German optician Ernst Abbe published his theory of the formation of images by microscopes. In it he related the diffraction-limited *resolving power*, defined by the Rayleigh criterion described above, to the angle subtended by the objective lens at a point object. Abbe's theoretical relationship, borne out in practical observations, stated that the minimum resolvable separation d_0 in the object was given by

$$d_0 = 0.61\lambda/n\sin\alpha$$

where λ=wavelength of the illuminant, α=half-angle subtended by the objective at the object, and n=refractive index in the space between object and objective lens.

This relationship can be used to work out the theoretical limit of resolution of any microscope system. The product $n\sin\alpha$ is called the *numerical aperture* (NA) of the objective lens, and is usually engraved by the manufacturer on the lens mount. Since α can never exceed 90° an objective in air cannot resolve distances smaller than 0.6λ; an oil immersion system will be limited to a resolving power a little less than 0.5λ. In round numbers the resolving power of the light microscope is limited by diffraction to about 0.3 µm for white light and about one-half this using UV light and quartz lenses.

A typical laboratory microscope with ×40 objective of NA 0.7 or thereabouts has its resolution limit at about 0.5 µm with white light. The relationship between NA, angular aperture and resolving power is shown diagrammatically for several microscope objectives in Figure 1.10.

Useful and empty magnification

We saw earlier that the purpose of a microscope is to magnify specimen detail and present it at a distance where the eye can see it. The useful magnification of a

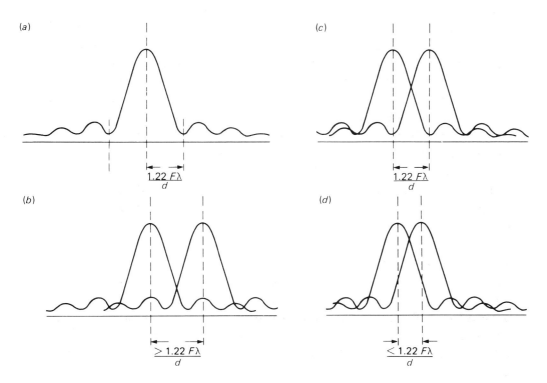

Figure 1.9 (*a*) Intensity distribution in image of point object formed by perfect lens. (*b*) Overlapping patterns of two point images sufficiently far apart to be recognised clearly as separate points. (*c*) 'Rayleigh criterion': two points are just 'resolved' when the separation of the peaks equals the radius of the Airy disc. (*d*) Image points are not resolved when the separation of peak intensities is less than the radius of the Airy disc.

microscope is therefore one which will increase the size of the minimum resolvable distance until it is large enough for the eye to see. The normal eye is just able to resolve two points separated by 0.1 mm (100 μm) at its nearest distance of distinct vision. The least magnification necessary to make use of a microscope's resolution d_0 is thus $100/d_0$ times (when d_0 is expressed in μm). For white light this figure approximates to 300 times the NA of the objective. In practice this figure should be increased, since it is tiring to work at the extreme of visual resolution for any length of time, but a magnification of 1000 times the NA is the maximum which can usefully be employed. Higher enlargements than this, although permissible in large display micrographs which will not be viewed closely, only enlarge the unresolvable blurs in the image and do not introduce any further detail into the micrograph. This over-enlargement is said to introduce *empty magnification* into the image.

Taking as an example the laboratory microscope with ×40 objective of NA 0.7 this should be used with a stronger eyepiece than ×5 to enable its potential resolving power to be utilised; ×10 would be suitable, or ×15, but anything in excess of this will only contribute empty magnification.

Depth of field and focus

We have seen how the diffraction of light places an upper limit of about ×1000 on the useful magnification of most light microscopes. A further impediment in the way of examining fine detail of practical three-dimensional objects is the fact that this detail is only seen in a very shallow depth of the specimen. The *depth of field* , or thickness of specimen which will be seen acceptably sharply in the magnified image, may be only a fraction of a micrometre in some cases and comparable with the resolv-

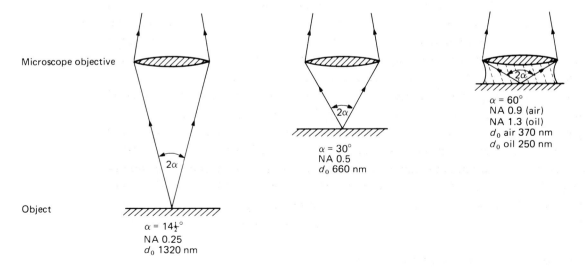

Figure 1.10. Relationship between angular and numerical apertures of microscope objectives and the diffraction-limited resolving power for λ = 550 nm.

ing power. Specimen detail outside this zone will be enlarged into an out-of-focus blur, which confuses the overall image. The *depth of focus* is the range of axial distances in the image space over which the image remains acceptably sharp. Figure 1.11(*a–d*) shows diagrammatically the concepts of depth of focus and depth of field in lens systems and illustrates how these depths are less for a wide aperture lens than for one with a narrower aperture. The microscopist can produce a mental impression of fine detail over a wider range than the depth of field of his instrument if he racks the focused plane up and down within the object. However, this impression of detail in depth cannot normally be recorded in a single photograph to show to another person.

1.3 The confocal microscope

The principal drawback of imaging an extended object with a light microscope is due not so much to the shallowness of the zone of sharp focus but to the fact that the sharp image detail may be obscured by blurred out-of-focus detail from the specimen planes on either side of it. If these could be removed one of the major limitations of light microscopy would have been eliminated.

Looking back at Figure 1.11(*c*) it can be seen that light from a point outside the focused image plane, whilst focused either closer or further away than the chosen image plane, is spread over a disc in the focused image plane. Suppose a small pinhole diaphragm is positioned in the focused plane, this will intercept nearly all of the out-of-focus light, but transmit all the sharply focused image of the object point. A normal field of view can be imaged sharply if it is sampled as a succession of focused points covering the field by translating (*scanning*) either the specimen or the illuminating beam of the microscope. There are several ways in which this can be achieved in practical microscopes, and the resulting scanned images may be either digitised and stored for reconstruction, or seen in original colours as live images, depending on the system used. The resulting images have the common characteristic of being clear and sharp, with an ultimate resolution slightly better than the compound microscope. The principle of using a point object and point image at **conj**ugate **foci** of the objective lens gives the *confocal microscope* its name. Confocal microscopes are generally used in reflection, when the same pinhole may be used for illumination and imaging. The imaging beam may be reflected light or the fluorescence stimulated by the incident radiation. For comparison between the various types of confocal microscope the reader can refer to a publication such as Wilson (1990). Boyde (1994) presents a bibliography covering the instrumentation and applications up to March 1993. Figure 1.12 shows a comparison between conventional and confocal images.

1.4 Beating the wavelength restriction

Microscopy would have remained a long way from the goal of seeing matter on an atomic scale were it not for a combination of discoveries which in a period of three

(a)

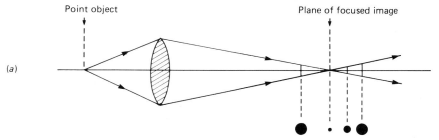

Point object

Plane of focused image

Out of focus images form circular discs of diameter
depending on distance from plane of focus

(b)

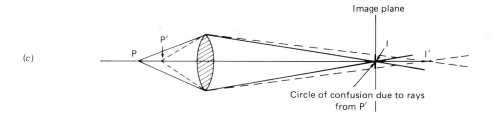

Depth of focus

Depth of focus

(c)

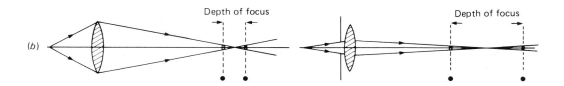

Image plane

Circle of confusion due to rays
from P′

(d)

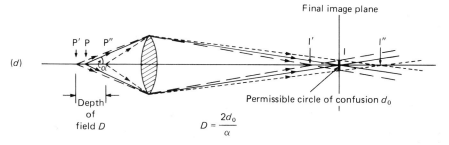

Final image plane

Permissible circle of confusion d_0

Depth
of
field D

$$D = \frac{2d_0}{\alpha}$$

Figure 1.11. Depth of focus and depth of field in microscope systems. Depth of focus is the latitude in positioning the
screen or photographic material before the image sharpness deteriorates beyond a chosen amount. (*a*) The focused
image of a point is a disc, because of diffraction, plus any aberration effects, so there is a finite depth of focus even at
the highest resolving power of the system. (*b*) The depth of focus can be increased by reducing the angular aperture of
the lens.

Conversely, (*c*) if the image plane is fixed it can be seen that the image of an object point such as P′, situated closer or
further than P along the lens axis, will be a disc or circle of confusion whose size depends on the amount of defocus
PP′ and the angular aperture of the imaging rays (not illustrated). (*d*) The *depth of field* P′P″ is the axial distance in the
object space between points whose discs in the final image plane have diameters up to a permissible maximum value
d_0.

years showed the existence of suitable radiation with shorter wavelength and the necessary lens systems for focusing it.

1.4.1 Electronic illumination

A glow discharge in a gas at low pressure and an incandescent metallic filament in a vacuum are both sources of electrons (or cathode rays as they were originally known), demonstrated by J. J. Thomson in 1897 to be negatively charged particles of very small mass. Electrons from such sources may be attracted towards and through a hole in a positively charged plate, called the anode, and may be formed into a beam by suitable magnetic and electrical fields. The cathode ray tube, fore-runner of today's television tubes and video monitors, used these principles without, however, going beyond the basic knowledge given above. An electron which has been accelerated by an anode potential of V volts is called a 'V volt electron' and has an energy of V electron volts (eV); the higher its energy the faster it will be travelling and, by the principle of Relativity, the heavier it becomes.

The utilisation of the electron to form the basis of a new optics – electron optics – was made possible by combining the discoveries of de Broglie (1924) and Busch

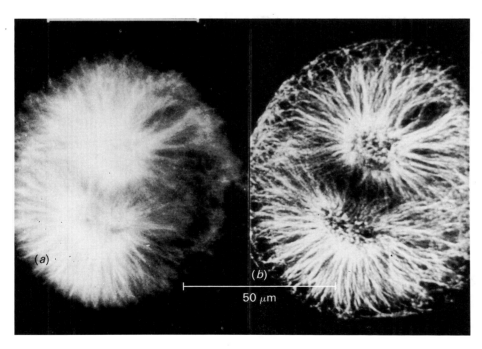

Figure 1.12. Fertilised egg of a sea urchin (*Psammechinus*) stained with antitubulin. Fluorescence micrographs made by (*a*) conventional microscopy and (*b*) scanning confocal microscopy. Elimination of out-of-focus fluorescence results in improved detail in the confocal image at the periphery and in the mitotic centres. (Reproduced from White, Amos & Fordham (1987) by copyright permission of the Rockefeller University Press).

Table 1.1. *Variation of electronic velocity, mass and wavelength with energy*

V (volts)	Velocity (km h^{-1})	Velocity, relative to velocity of light	Mass, relative to rest mass	λ (nm)
1	593	0.0020	1.000 002 0	1.226
10	1 876	0.0063	1.000 019 6	0.387 8
10^2	5 932	0.0198	1.000 195 7	0.122 6
10^3	18 730	0.0625	1.001 957	0.038 76
10^4	58 460	0.1949	1.019 57	0.012 20
10^5	164 400	0.5482	1.195 7	0.003 70 (3.7 pm)
10^6	282 200	0.9411	2.956 9	0.000 87 (0.87 pm)
3×10^6	296 700	0.989	6.870	0.000 36 (0.36 pm)

(1926). De Broglie showed that a moving electron has a dual personality. It can be regarded either as a moving charged particle or as a radiation with associated wavelength. The key relationship deduced by de Broglie was expressed by $\lambda = h/mv$, where λ is the de Broglie wavelength of the particle, h is a constant (equal to Planck's constant), and m, v are the mass and velocity of the electron. Working from this concept we see that a beam of moving electrons can be regarded as a beam of invisible radiation which, given the means of focusing it, could be used to form an image. The wavelength of the radiation varies inversely with the square root of the accelerating voltage (subject to relativistic correction at higher energies). The real advance is apparent when we put numbers into the equation and see that the electron wavelength is in fact many orders of magnitude shorter than that of light. Typical values are listed in Table 1.1.

Thus, a 100 keV electron has a wavelength of 0.0037 nm. The significance of this in Abbe's resolving power relationship $d_0 = 0.6\lambda/n\sin\alpha$ is obvious. If an electron objective lens of NA 1.4 were available then a 100 kV electron microscope would be capable of resolving detail as fine as 0.0016 nm, or about a hundredth of the size of an atom. All this presupposes firstly that the electron lenses are available and can be used in a compound microscope, and secondly that the lenses can be made as perfect as glass lenses in light-optical microscopy. So let us now look at the nature and characteristics of electron lenses.

1.5 Microscopy with electrons

From here on it must be remembered that the generation, manipulation and use of electron beams can only take place in a vacuum. How this is done in practice is dealt with in later chapters.

1.5.1 *Electron lenses*

The direction of travel of a moving electron will be altered either by applying a magnetic field or an electric field. This makes possible two different types of lens in electron-optical systems.

Magnetic lenses

An electron moving in a magnetic field experiences a force tending to change its direction of motion except when it is travelling parallel to the magnetic lines of force. If the magnetic field is suitably shaped a beam of electrons diverging from a point at one end of the field can be made to converge to a second point at the other end; i.e. the field has a focusing action on electrons in a similar way to a convex glass lens and light rays. There is, however, no magnetic equivalent of the diverging, concave, lens. Magnetic lenses can be made to give magnified or diminished images, and sequences of lenses can be used to form compound microscope systems. The focusing properties of magnetic fields were worked out in theory by Busch in 1926, although it remained for others, e.g. Knoll and Ruska, to exploit the principles by constructing practical lenses.

The principle of a magnetic lens is illustrated in Figure 1.13(*a*). This shows an electron entering the magnetic field in the gap between two hollow cylindrical magnets arranged with unlike poles facing one another. The focusing action, resulting in the electron path cutting the axis, is accompanied by a circumferential motion, illustrated in Figure 1.13(*b*), which is a view back along the axis of the lens. In practice the complete figure would be rotated about the axis of symmetry xx′ so that a parallel beam of electrons entering the lens would be focused to a point on the axis of the field. Electrons travelling along the axis would be moving parallel to the magnetic lines of force and would be undeflected in their passage through the lens field.

In electron microscopes the magnetic lenses may use permanent magnets, as assumed in Figure 1.13, or (more usually) may use the electromagnetic effect

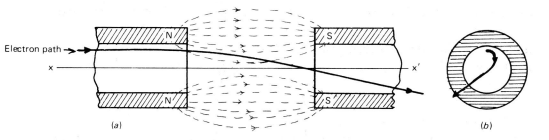

Figure 1.13. (*a*) Focusing action of a magnetic field on a moving electron. (*b*) Axial view back from right to left, showing the combination of rotation with the focusing action of the field.

associated with a current-bearing wire. A coil of many turns of wire wound on a hollow cylindrical former constitutes a *solenoid*, and has a magnetic field along its axis which acts as a lens to an electron beam passing through it. Moreover, the strength of the magnetic field and hence the focal length of the lens can be varied by altering the current in the coil. A very strong localised magnetic field is obtained if the solenoid is given a core consisting of two soft iron tubes with a short gap between their ends. When the outer ends of the tubes are joined by an iron shroud or sleeve outside the coil we have the basic lens form used in electron microscopes. The inner ends of the iron tubes have now become the *pole-pieces* of an *electromagnetic lens*; their shapes, internal diameters and separation are important parameters in lens design, as also is the strength of the magnetising field, specified by the number of ampere turns (current $I \times$ number of turns N in the coil) applied to magnetise the iron. The progression from solenoid to practical electromagnetic lenses is shown schematically in Figure 1.14.

Since the magnification of an electromagnetic lens is varied merely by altering the current through the coil there is the possibility of making a magnifying system of several lenses in series whose magnification can be varied either continuously or in fixed steps over a wide range, without physically moving any of the components. A less desirable feature, however, is the rotation of image which occurs as the magnification is varied. This can be reduced in several ways and we will return to this point in Chapter 3.

Electrostatic lenses

As well as introducing the concept of electron focusing by magnetic fields Busch also calculated the effect on an electron beam of electrostatic forces between axially symmetrical electrodes at different potentials. Electrode configurations of several types have lens properties and show some advantages over magnetic lenses, such as rotation-free imaging and the ability to work with simpler and less highly stabilised power supplies. However, they require a high precision in construction and alignment and extreme cleanliness in operation to avoid voltage breakdown and the formation of surface insulating films. Without any clear advantages and with certain operational disadvantages the purely electrostatic electron microscope has had only a limited success commercially. Those instruments which have been constructed use *unipotential* or *Einzel* lenses for objective and projector. This three-electrode lens, shown in principle in Figure 1.15, consists of three coaxial annular electrodes, the outer two being held at zero (earth) potential whilst the central one is at the high negative potential of the electron source. The focal length and aberrations of the lens are determined by the inter-electrode separation and the diameter of the bore. The focal length can be varied by altering the potential of the middle electrode.

Although electrostatic lenses are seldom used as part of the magnifying lens system in electron microscopes the illuminating systems invariably contain an

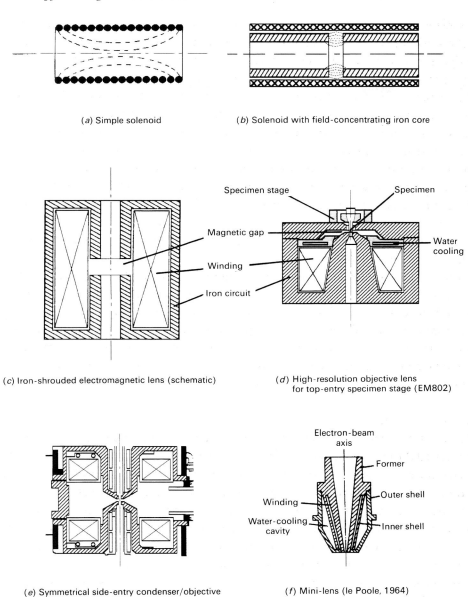

(*a*) Simple solenoid

(*b*) Solenoid with field-concentrating iron core

Specimen stage

Specimen

Magnetic gap

Winding

Iron circuit

Water cooling

(*c*) Iron-shrouded electromagnetic lens (schematic)

(*d*) High-resolution objective lens for top-entry specimen stage (EM802)

Electron-beam axis

Former

Winding

Outer shell

Water-cooling cavity

Inner shell

(*e*) Symmetrical side-entry condenser/objective lens of Riecke–Ruska type (EM400)

(*f*) Mini-lens (le Poole, 1964) N.B. at about 4 x the scale of the adjacent lenses

Figure 1.14. (*a*–*c*) Simplified diagrams of the progression from a focusing solenoid to an iron-shrouded electromagnetic lens similar in form to a weak lens used in electron microscopes. (*d*, *e*) Typical objective lenses for top-entry and side-entry specimen stages. The mini-lens (*f*) dispenses with iron and uses very high currents in a water-cooled coil.
(Diagrams (*d*), (*f*) by courtesy of Kratos Ltd; (*e*) by courtesy of Philips Electron Optics.)

electrostatic lens to extract and accelerate the electrons. This latter application is described more fully in Appendix 4.

1.5.2 Aberrations and defects of electron lenses

As with their glass counterparts, electron lenses suffer from defects such as spherical and chromatic aberrations, as well as astigmatism and distortion. Unfortunately in the electron-optical case it is not possible to cancel out or correct aberrations simply by combining positive and negative elements of different refractive index. Instead it is necessary to choose the operating conditions so that the aberrations have the least damaging effect on performance, i.e. on resolving power.

Spherical aberration

The two main factors determining the resolving power of a microscope system are spherical aberration and diffraction in the first, or objective, lens. Both of these cause a point object to be imaged as a disc instead of a point, the diameter of the disc being dependent on the angular aperture of the lens.

The effect of spherical aberration is shown diagrammatically in Figure 1.16. Extreme rays making an angle α to the lens axis are focused closer to the lens than the paraxial rays, which meet at the ideal or Gaussian focus. The marginal rays form a disc of radius Δr_i in this plane, centred on the Gaussian image point P′. In the object plane the equivalent disc would have a radius $\Delta r = \Delta r_i / M$, where M is the magnification produced by the lens. The nature of the aberration is such that $\Delta r = C_s \alpha^3$, where

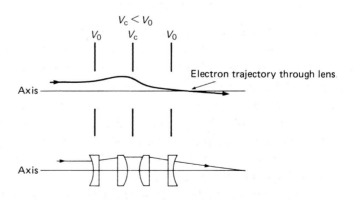

Figure 1.15. Electrostatic Einzel lens and its light-optical equivalent.

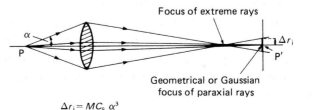

Figure 1.16. Formation of disc of confusion because of spherical aberration of lens.

$$\Delta r_i = M C_s \alpha^3$$

α is measured in radians and C_s is a constant for a particular lens, and is called its *spherical aberration coefficient*.

Using the Rayleigh criterion the resolving power of the lens as determined by spherical aberration will have the same magnitude as Δr: $d_0 = C_s \alpha^3$; i.e. for best resolving power α should be **small**.

We saw earlier (page 0) that the resolving power of a lens was limited by diffraction to a value $d_0 = 0.6\lambda / n\sin\alpha$. In an electron lens $n = 1$ and α is small enough to be equated to $\sin\alpha$. The diffraction limited resolving power is thus $d_0 = 0.6\lambda / \alpha$, and for the best resolving power α should therefore be **large**.

In order to balance the requirements of the two opposing effects it is necessary to choose an optimum angular aperture α_{opt} for the objective lens which gives the same limiting resolution by the two criteria. It follows directly from the two previous equations that

$$\alpha_{opt} = (0.6\lambda / C_s)^{1/4}$$
and the minimum $d_0 = 0.7 C_s^{1/4} \lambda^{3/4}$.

Values for C_s have been calculated for a variety of configurations of magnetic lens. C_s depends on the focal length of the lens, its bore and the separation between the pole-pieces. Numerically its value is approximately equal to the focal length. For example, consider a commercial objective lens with focal length 4.0 mm and C_s 3.5 mm. Putting $C_s = 3.5$ mm into the equations above:

$$\alpha_{opt} = 5 \times 10^{-3} \text{ radians } (= 0.29 \text{ degrees})$$
and $d_0 = 0.45$ nm at an electron energy of 100 keV.

So we see that because of spherical aberration the angular aperture of objective lenses in electron microscopy has to be very narrow, usually between 10^{-3} and 10^{-2} radians, which is hundreds of times smaller than in the light microscope (e.g. 0.5 radian for NA 0.5). However, since the wavelength is more than 10^5 times shorter, a sacrifice of a factor of 10^2 in aperture will still leave us 10^3 better off in resolving power. For an account of the alternatives which have been considered for reducing the spherical aberration coefficient the reader is referred to Septier (1966). The association of lower C_s values with shorter focal lengths has been an important influence in the design of electron microscopes.

The smaller aperture gives the electron lens a depth of field which is hundreds of times greater than that of the light objective *at the same resolution*. At highest resolution the depth of field of the electron lens is shallower, e.g. 0.2 μm for the lens quoted above, but this is still greater than the thickness of specimen which can be examined at the high resolution. Depth of field is therefore not the same problem in electron optics as in light microscopy, although there are certain specialised cases where it becomes a restriction.

The approximate equality between C_s value and focal length applies only to electromagnetic lenses; C_s is higher for electrostatic lenses which would have a resolution limit nearly ten times larger than the best magnetic designs.

Chromatic aberration

The other aberration affecting the performance of the electron microscope is chromatic aberration. In electron optics the equivalent of colour is electron energy, and chromatic aberration is the lens defect which degrades the image whenever electrons in the beam cease to be monoenergetic. This may be the result of electrons starting from the gun with a spread of energies, or of the accelerating voltage fluctuating with time, or the electron beam losing energy through collisions in passing through the specimen. Chromatic aberration is not a defect which can be simply corrected as in a glass achromat or apochromat, but one which can only be minimised.

If we consider what is involved in an electron microscope we see that for it to have a constant Gaussian focus the focal length of the lenses must remain constant. This will depend on the stability of (i) the electron energy and (ii) the current in the lens coil in the case of a magnetic lens or the electrode potentials in electrostatic instruments.

The chromatic aberration constant C_c of a magnetic objective lens is usually slightly smaller numerically than the focal length. Chromatic effects produce a broadening of the Gaussian focus into a disc of confusion in the same way as with spherical aberration, and the radius of the disc, Δr_c, is given by

$$\Delta r_c = C_c \alpha E / E$$

where ΔE is the deviation of the electron energy from its mean value E.

To find the order of stability required we put $C_c = 3.4$ mm for the lens we considered earlier, and find that for Δr_c 0.45 nm and $\alpha = 5 \times 10^{-3}$ radians, $\Delta E = 2.6$ in 10^5 V. This figure assumes that all the instability is in the electron energy and none in the lens current. Allowing for some fluctuation in this the practical stability required in the electron energy will be nearer 1 in 10^5 V. The focal length of a magnetic lens is influenced by the square of the lens current, so the sensitivity to current fluctuations is twice as stringent as that to energy variations. Thus lens supplies must also be stabilised for minutes at a time to better than 1 part in 10^5 for 0.5 nm resolution with the lens quoted. The requirements for stability of supplies at very high resolutions are thus very stringent, since the practical performance of a microscope is that permitted by the combined effects of the aberrations and diffraction.

We shall see in Chapter 2 that there are several ways in which the electron beam can become 'coloured' in passing through the specimen. The effects of this will degrade the microscope image more or less according to the C_c of the objective lens. We shall also see later that removing the differently coloured electrons by inserting an electron energy filter into the microscope can have a very beneficial effect on image clarity.

Astigmatism

Astigmatism is the other defect which can degrade the resolution of an electron lens, as in light optics. In electron optics it is caused by the electron-deflecting fields not being perfectly symmetrical about the lens axis, so that the lens has a different focal length in different orientations. In practice, however carefully the pole-pieces of a magnetic lens are machined there will always be a slight departure from circularity in the bore. There may also be slight inhomogeneity in the iron used for the pole-pieces. These defects result in a lens whose focal length varies slightly around the lens axis. The result is similar to the astigmatism illustrated earlier (Figure 1.7), two separate line foci being obtained at right angles to each other. The degree of astigmatism is measured by the separation between the two foci. The permissible maximum value of astigmatism varies as the square of the required resolving power; for a resolution of 1 nm at 100 kV it must be smaller than 0.4 μm.

Fortunately this is one of the few defects in electron lenses which can be corrected, and every electron microscope has an *astigmatism corrector* or *stigmator* in at least one lens and sometimes in several. An astigmatic lens can be considered as a combination of an axially symmetrical lens and a weak cylindrical lens. The stigmator introduces a balancing cylindrical lens field perpendicular to the unwanted one and hence cancels out its effect (Figure 1.17). (It is strictly more correct to refer to the *compensation* of astigmatism rather than its correction but the latter term is more widely used.) The stigmator can be either electromagnetic or electrostatic in nature, the essential requirement being that the magnitude and direction of the compensating field should be independently variable. An array of four electromagnets or four insulated pins of alternate polarities arranged symmetrically around the optical axis can be

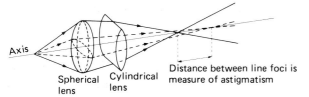

Axis

Spherical lens   Cylindrical lens   Distance between line foci is measure of astigmatism

Astigmatic lens is equivalent to combination of spherical and cylindrical lenses. Instead of a point focus, forms two line foci at right angles

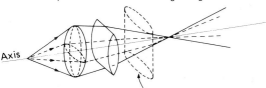

Axis

Astigmatism corrector acts as second cylindrical lens with axis at right angle to inherent defect and brings rays to common focus

Figure 1.17. Schematic diagram showing the effect of astigmatism in an electron lens, and its correction by means of a stigmator.

used (Figure 1.18), currents or potentials being adjusted to give the required amplitude of compensating field. The array is rotated about the optical axis until the lens field exactly cancels out the astigmatic field in both direction and amplitude. An equivalent effect can be obtained without the need for moving parts if a second set of four poles is added at 45° to the original one, around the optical axis. By appropriately varying the relative strengths of the two sets the effects of rotational and amplitude changes are obtained. With an eight-pole stigmator the axial astigmatism of a lens can be reduced to a negligible value. The visible effects of astigmatism in electron micrographs are illustrated later in Chapter 5, page 201.

Distortion

The final image defect we shall consider in electron lens systems is distortion. As previously seen in light-optical systems a radial change of magnification across the image field will result in pincushion or barrel distortion. In addition, a form of distortion found in electron lenses but not in glass lenses results from the variation of image rotational angle with radial distance from the optical axis. Distortions in electron lenses are usually found when the lenses are being operated at larger focal lengths than the optimum, i.e. at low magnifying powers. In a compound electron microscope it is usual to operate the objective lens at a fixed high power to minimise aberrations, and to cover a wide magnification range by varying the strengths of the projector lens or lenses. Distortions are thus liable to occur at the lower magnifications. Where several projector lenses are used in series it is possible to choose operating conditions so that barrel distortion in one lens is balanced by pincushion distortion in the other one, giving an overall low distortion. Examples of distortion are given in Figure 1.19.

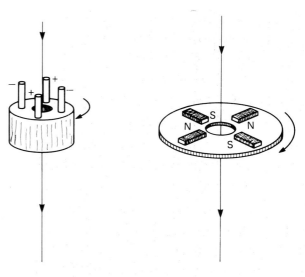

Figure 1.18. Schematic illustrations of electrostatic and electromagnetic stigmators.

Another approach is to design projector lenses with interchangeable pole-piece systems, so that the magnification range is covered by a set of, say, four different field configurations and the lens does not then have to operate under unfavourable conditions as the magnification range is covered. This practice was used on several ranges of high-grade electron microscopes but has fallen out of current use, since it is probably cheaper, easier and cleaner to put more lenses into the column than to provide

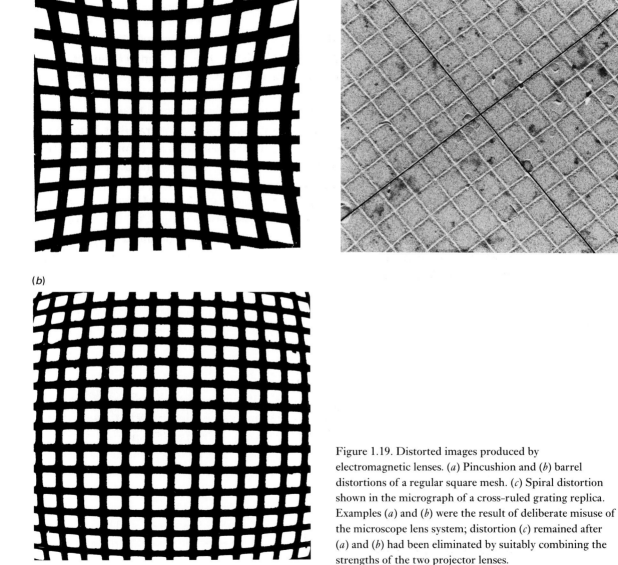

Figure 1.19. Distorted images produced by electromagnetic lenses. (*a*) Pincushion and (*b*) barrel distortions of a regular square mesh. (*c*) Spiral distortion shown in the micrograph of a cross-ruled grating replica. Examples (*a*) and (*b*) were the result of deliberate misuse of the microscope lens system; distortion (*c*) remained after (*a*) and (*b*) had been eliminated by suitably combining the strengths of the two projector lenses.

the precision engineering necessary for a turret of pole-pieces rotating in a high vacuum.

1.5.3 *Manipulating electron beams*

Lenses as described above are used for focusing beams of electrons in illuminating and imaging systems. Other devices are necessary for changing their direction of travel, i.e. shifting or tilting their axes. As with focusing elements, either magnetic or electrostatic forces may be used for deflecting electron beams. The simplest is a pair of parallel plates, one on either side of the beam axis, with a voltage difference between them. Alternatively, and the more frequent choice, magnetic fields from pairs of current-carrying wire coils, with or without magnetic cores, can be used.

A *beam tilt* control will change the beam direction in one plane; the stronger the field the greater the tilt produced. Two orthogonal pairs of tilt coils enable the beam to be tilted in any plane in space. *Beam shift* (movement to a parallel axis) is achieved by tilting one way, then an equal amount in the opposite direction. If the illumination is required to be incident on the specimen at a chosen (non-normal) angle of tilt relative to the optic axis, the double deflection is arranged so that the lower coil bends the beam back through a larger angle than the first deflection. Whereas on early electron beam instruments components were moved into alignment by physically translating or tilting them, on modern instruments the column parts are rigidly clamped together and alignments are done with deflecting coils.

Beam blanking (cutting off the beam without actually stopping the emission of electrons) is carried out by deflecting the beam away from the optic axis at some point, so that it is absorbed by a diaphragm instead of passing through it down the column.

1.5.4 *Scanned beam microscopy*

The concept of scanning microscopy, briefly touched on in connection with confocal light microscopy, gives an alternative means of imaging in which an illuminating beam is focused down to a small spot or 'probe' at the object. A 'signal', which may be transmitted or reflected light or electrons (or other radiations to be met later), is collected by an appropriate detecting device. While the illumination is stationary on the object the signal is constant. If the illumination is moved across the object in a line, or a pattern of lines, the signal will vary (modulate) as the spot interacts with object detail. If the modulated signal is applied to the brightness control of a cathode-ray tube (TV) display scanned in synchronism with the spot on the object, an image will be produced which is enlarged relative to the original object in proportion to the relative dimensions of the scans on object and display. Thus a 1 mm scan on the object, displayed on a 100 mm tube, represents a magnification of $\times 100$, without the use of magnifying lenses! Reducing the length of scan or increasing the size of the display results in a higher magnification; the useful limit is placed by the size of the smallest focused illuminating spot which can be produced in the scanning micro-scope. In practice the limiting spot size is determined by spherical and chromatic aberrations of the probe-forming lens.

1.5.5 Electron microscopes in practice

We now have a picture of the essential components of two types of electron micro-scope. A beam of electrons from an electron 'gun' is focused by a condenser lens on to a specimen. In the transmission electron microscope (TEM) the electrons trans-mitted through an entire image field are focused by objective and projector lenses into an enlarged image on a fluorescent screen. In the scanning electron microscope (SEM) a finely focused beam of electrons is scanned across the specimen and gives rise to reflected electrons and other signals which are used to form an image on a TV tube. Both types of instrument are significantly more complex than light micro-scopes because of the nature of electron sources and lenses, and the need to operate in an evacuated tube.

The instruments, their specimens and the results from them will be explained in the following chapters and appendixes.

1.6 High resolution by electron microscopy – the proof of the pudding.

The foregoing sections have shown how, given a short-focus objective lens, a stig-mator and very stable electrical supplies, it should be possible to improve on the resolving power of the best light microscope by a factor of 1000 or more. Before leaving this introductory chapter it is only right to show that this has in fact been substantiated, and some microscopes are operating which will allow single atoms to be pictured in very special specimens. A number of commercial ultra-high-res-olution microscopes are available for which their manufacturers guarantee a resolving power of below 0.5 nm on the Rayleigh criterion of separating point objects.

At this level of resolving power it becomes difficult to find suitable test objects, and the highest resolutions are demonstrated by resolving crystal lattice planes (to be met later in the book, page 49). Resolutions down to 0.1 nm and below have been demonstrated. Although there are certain reservations about this type of test, it does demonstrate that an instrument possesses excellent stability and would probably also show very good point-to-point resolution if the specimens could be found.

There is also a method of testing and explaining the imaging characteristics of electron microscopes in the sub-nm region by using image analysis and modulation transfer theory, which in skilled hands can reveal a great deal about a microscope's ultimate capability. This is beyond the scope of this book, however, and readers who want to follow it up are referred to more advanced publications, e.g. Hanszen (1971).

Micrographs showing proof of ultra-high resolution are reproduced in Figure 1.20.

1.6.1 Bonus points in electron microscopy

The driving force behind the development of electron microscopes was the desire to exploit shorter wavelength radiation in order to reveal finer structural detail in matter than could be seen by light microscopy. Attaining this goal alone would justify the effort which has been put into the development of the much more complex electron microscope. It has become evident, however, that micrographs showing size and shape are only the first stage in the complete characterisation of materials using the electron microscope. Figure 1.21 shows diagrammatically that topography, atomic

(a)

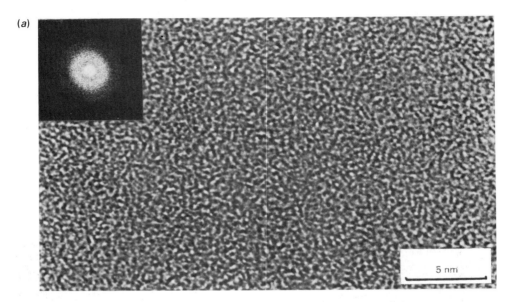

5 nm

Figure 1.20. High-resolution microscopy demonstrated. The imaging characteristics of high-performance transmission microscopes can be determined from micrographs of thin amorphous films containing very fine particulate detail with a range of very small separations. An optical diffractometer effectively analyses the particle-to-particle spacings present in the micrograph, and gives a measure of the smallest separations recorded by the microscope.

Micrograph (a) is an image of a thin evaporated carbon film recorded at 575 kV on the high-resolution electron microscope at Cambridge University (Cosslett, 1979). A diffractogram (inset) shows the imaging of detail down to below 0.18 nm.

Micrograph (b) from the same instrument operated at 500 kV shows atomic arrangements in a 14 nm decahedral particle of gold. The micrograph has reversed contrast, i.e. the columns of atoms appear white. The presence of surface steps (around the edge of the particle) and the intersections of the crystal faces are clearly visible (By L. D. Marks & D. J. Smith, 1983 with permission of *Journal of Microscopy*.)

The scanning micrograph (c) shows gold islands in an evaporated metal film images at 1 keV beam energy. Not only can narrow gaps between the islands be recognised (Rayleigh criterion) but the surface topography of the islands can be visualised. (Micrograph by courtesy of LEO Electron Microscopy Ltd.)

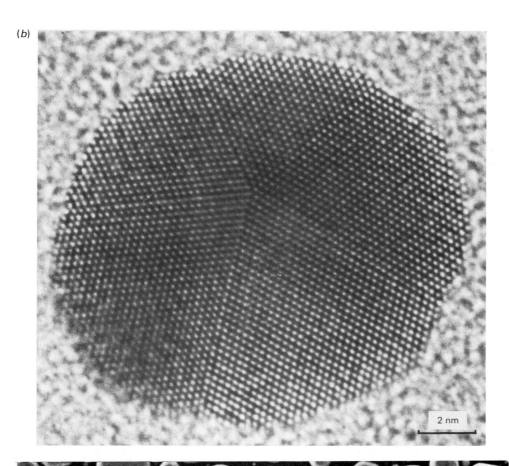

(b)

2 nm

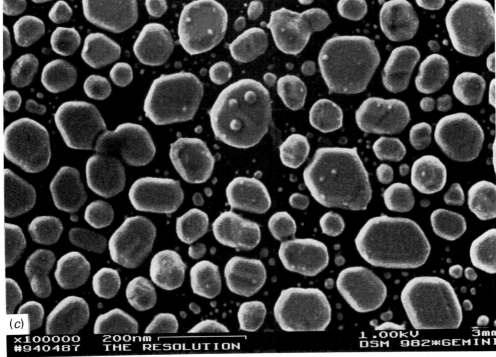

(c)

x100000   200nm   1.00kV   3mm
#940487   THE RESOLUTION   DSM 982✳GEMINI

number, elemental composition and distribution, crystallinity and surface structure can all be explored by utilising appropriate emissions from a specimen under examination in the microscope. For some investigations just one of these aspects may be of overriding importance, and the instrument will be designed around the requirements for that particular study. In most cases, however, we start with a high-performance microscope and add the appropriate attachments and accessories.

1.7 Summary

We have seen that the ultimate power, and hence the upper useful limit of magnification, of microscopes using light is limited by the wave nature of the illumination. Diffraction effects make it impossible ever to see detail finer than 100 nm with a lens-based light microscope and at the ultimate resolution the depth of field is very much less than the thickness of specimen, which confuses the final image, although this last disadvantage is overcome in the confocal microscope.

A beam of fast-moving electrons can be used as illumination in an electron microscope. The electron beam has an effective wavelength many orders of magnitude shorter than that of light, with a correspondingly smaller diffraction-limited resolv-

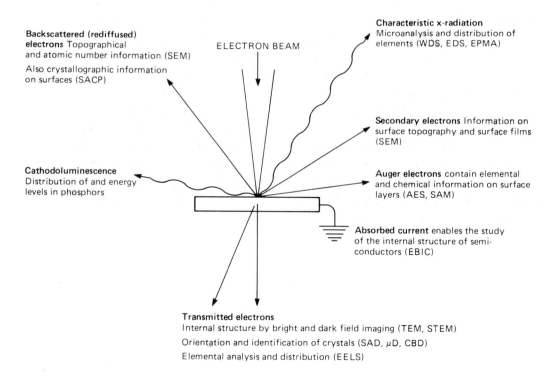

Figure 1.21. Schematic representation of the wealth of information resulting from the interaction between the electron beam and the specimen in an electron microscope. The acronyms refer to techniques which will be mentioned in later sections of the book.

ing power. Magnetic and electrostatic fields can be used as lenses to focus the electron beam and form enlarged images. The defects of electron lenses are not amenable to correction in the same way as those in glass lenses and they have therefore to be used with very small angles of illumination. Nevertheless, this still results in sufficiently good resolving power to enable specimen detail to be seen down to a near-atomic scale whilst giving a useful depth of field in the micrographs, even at the highest resolution. As in light microscopy different types of instrument are used for transmission and reflection microscopy. In electron optics, however, the reflection instrument incorporates a scanning principle and the image is built up sequentially.

A number of other phenomena occur when matter is bombarded by electrons; these can be used in electron microscopes to give additional information about the structure of the specimen.

Following chapters will describe how the special capabilities of electron microscopes are realised on practical instruments and specimens.

1.8 Suggested further reading

There is an introduction to the concepts behind electron microscopes in several of the books dealing with the applications of the instruments to specific fields of study, to be mentioned in later chapters. However, if the reader wants more details of the lenses for light-optical systems he should find Curry (1953) helpful. Slayter & Slayter (1992) describe the principles of light and electron microscope systems. Wilson (1990) gives an account of the characteristics of confocal microscopy.

Electron lenses and electron optics are dealt with in books by Hawkes (1982) and Hawkes & Kasper (1989). Straightforward introductions to the subject in general are given by Hall (1983), Wischnitzer (1981) and Meek (1976). The book by Wischnitzer contains an extensive bibliography. Authoritative accounts of TEM and SEM are given in separate books by Reimer (1985, 1993a). An account of the background and early history of the electron microscope is given by Hawkes (1985).

Details of these books and all the references in the text will be found in the general bibliography at the end of the book.

2 *Electron–specimen interactions: processes and detectors*

In light microscopy the light falling on an object is reflected, transmitted, scattered, absorbed, or re-emitted at another wavelength. The main modes of light microscopy (reflected, transmitted bright-field and dark-field, fluorescence) result from these effects.

In electron microscopy the possibilities are much wider, because the greater amount of energy carried by the illumination can cause a number of different effects to occur in the specimen. A range of microscopical and analytical modes is possible, which makes the electron microscope the basis of a very powerful analytical sytem. Before going on to deal with the microscopes themselves it is useful to consider the range of possible electron–specimen interactions, which need to be taken into account when trying to interpret the images and analyses from electron microscopes.

2.1 Beam–specimen interactions

Figure 2.1 shows schematically and very approximately the paths taken by a small number of high-energy electrons when they impinge on a specimen in an electron

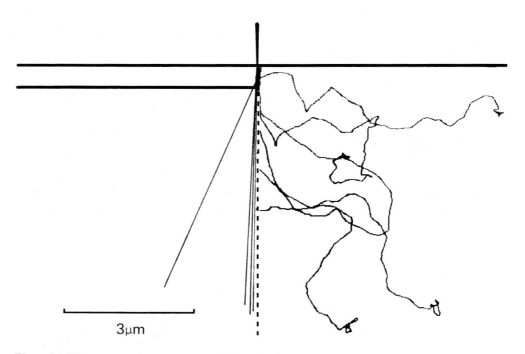

3μm

Figure 2.1. Electron–specimen encounters (schematic) for thin and bulk specimens.

microscope. Two thicknesses of specimen are considered – corresponding to a thin sample for the TEM and a bulk sample for the SEM.

Electron trajectories are deflected by collisions with specimen atoms. These may be either elastic, when the electron is deflected (even up to 180°) but no energy interchange occurs, or inelastic, when the electron interacts with the atom and supplies energy for a further process to occur, resulting in emission of an electron and/or electromagnetic radiation, but with very small deviation of track. After the collision or collisions the electrons may re-emerge into the vacuum as back(ward)-scattered or (forward) transmitted electrons, or may lose all their energy and come to rest in the specimen, contributing to specimen (absorbed) current and specimen heating.

To the primary electron a solid specimen appears to be mainly open space, populated by atoms which may be regularly spaced and uniform, in a crystalline solid, or randomly arranged in an amorphous material. The atoms themselves are open structures, consisting of a heavy nucleus surrounded by a cloud of much lighter electrons. The nucleus is made up of positively charged particles (protons) with an approximately equal number of uncharged neutrons. The number of protons corresponds with the atomic number Z of the particular atomic species, and the combined mass of protons and neutrons equals the atomic weight. The Z electrons, each only 0.05% as heavy as a proton, are held by electrostatic attraction to the nucleus and orbit around it in a precisely controlled distribution of electron 'shells' shown diagrammatically in Figure 2.2. The electron shells are at different average distances from the nucleus and hence have different attractions to it and different energy values. Moving outwards from the nucleus the shells are designated K, L, M, N etc and are also identified by numbers, $n=1, 2, 3, 4$, etc. The number of electrons in each shell is limited to a maximum of $2n^2$. The innermost, least energetic, electron shells are filled before those with higher energy. To give a few examples; helium ($Z=2$) has both of its electrons in the K shell. Aluminium ($Z=13$) will have two K electrons, 8 L electrons and 3 M electrons. Gold ($Z=79$) has two K electrons, 8 L electrons, 18 M electrons, 32 N electrons and 18 O and 1 P electron. Any of these electrons can, on the addition of sufficient energy, be excited into orbits further out provided there is a vacancy. Electrons in the innermost shells, sometimes referred to as *core electrons*, are concerned with x-ray production (as will be described) and their behaviour is influenced most by the attraction of the nucleus,

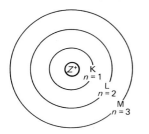

K shell 2 electrons
L shell up to 8 electrons
M shell up to 18 electrons

N, O, etc.

Figure 2.2. Simplified atomic model showing 'shells' of orbiting electrons surrounding the nucleus.

i.e. by Z. The outermost electrons are the ones involved with interactions with other atoms and with ionisation.

If an atom is bombarded by increasingly energetic primary electrons there comes a value of energy at which a K electron is given enough energy in an inelastic collision to overcome the nuclear attraction and escape from the atom. This is the *binding energy*, also called the *absorption energy* EK_{ab} of the K electron. It is a quantity which is characteristic of the atom species and increases with the atomic number of the element. The ejection of a core electron is the first stage in the production of characteristic x-radiation, and will be described later in the chapter. It also results in a characteristic loss of energy by the bombarding electron, which may traverse the specimen and emerge as a transmitted electron. Other inelastic collisions may result in removal of an outer (valence) electron, with a much smaller energy interchange. On the other hand, in an elastic collision the primary electron may pass through the atom close to the nucleus and suffer only a change in direction. Some electrons may pass right through the atom without any interaction at all.

The probability that an atom will be struck by a primary electron is conveniently visualised by a hypothetical 'cross-section'. This varies according to the energy of the electron, cross-sections being greater for low energies and vice versa. The average distance an electron can travel through the solid without a collision of some sort is the *mean free path*. This is only a few atomic diameters for low-energy (tens of eV) electrons such as knocked-off valence electrons but further for higher energy electrons.

It is possible to simulate theoretically the track of an incident (primary) electron through a solid specimen, making random elastic and inelastic collisions until it either loses all its energy and comes to rest or reaches a boundary of the specimen and leaves again. Using a modern personal computer (PC) the Monte Carlo simulation of an electron beam–solid interaction can be graphically portrayed in only a few minutes (the time depends on the total number of electron tracks to be calculated and the programme used) for a range of electron energies and materials. Figure 2.3 shows simulations for 20 keV electrons in bulk carbon (Z=6), copper (29) and gold (79), on the same scale. At 5 keV the dimensions would be reduced by an order of magnitude, and doubled at 30 keV.

It is seen that most electrons come to rest within a droplet-shaped volume which is elongated for lower Z, squashed for higher Z. Within this volume the electrons interact with sample atoms to produce electromagnetic radiations and free electrons.

The various products of the interaction are shown schematically in Figure 2.4 ; these are:

(1)   Secondary electrons
(2)   Reflected or Backscattered electrons
(3)   Transmitted electrons
(4)   X-radiation

(5) Auger electrons
(6) Cathodoluminescence radiations
(7) Absorbed (specimen) current

Of these, (3) are used for imaging in transmission electron microscopy, (1) and (2) in scanning electron microscopy, whilst (4)–(7) can provide additional analytical information about the specimen in either type of microscope.

Let us now consider the interactions individually in greater detail, so that we can

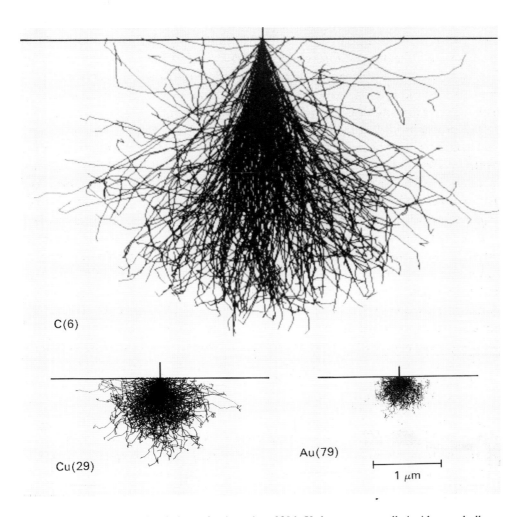

Figure 2.3. Monte Carlo simulations of trajectories of 20 keV electrons normally incident on bulk samples of low, medium and high atomic number. Elastic collisions only are considered by the programme, since inelastic collisions result in very slight, if any, change of direction of travel. Electrons are assumed to lose energy continuously at a uniform rate along their trajectories. (Programme by courtesy of D. C. Joy)

understand what information they can contribute about the specimen in the electron microscope.

2.1.1 Secondary electrons

An inelastic collision can result in outermost electrons being detached from specimen atoms, leaving behind an ionised atom with positive charge. The dislodged electrons have a low kinetic energy, less than 50 eV, and are readily captured by nearby ionised atoms. Those electrons which are created nearer to the specimen surface than the electron escape distance R, can be emitted into the vacuum and form the secondary electron (SE) emission current from the surface.

Referring to Figure 2.5(a) we see that for normally incident primary electrons only a small proportion of the secondary electrons actually formed are available for collection, and these originate from a region very close to the entry point of the primary electron beam. This means that the SE current is representative of a very small region on the specimen surface and is used in the scanning electron microscope for high-resolution surface imaging.

Referring to Figure 2.5(b) we see that if the surface is tilted relative to the incident beam a greater proportion of the secondary electrons will be able to escape and the electron emission from the surface will therefore increase. The intensity of secondary electron emission is therefore an indicator of the surface slope, and topography. Greater-than-normal emission will also occur from thin edges and spikes on surfaces which allow electrons to escape more readily from the track along which they are generated (Figure 2.5(c)).

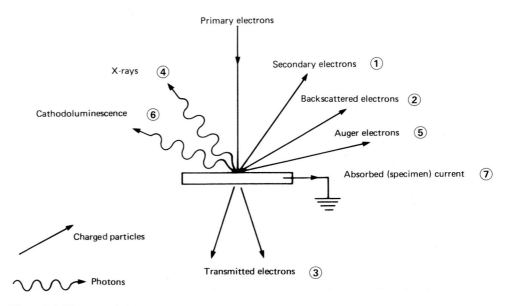

Figure 2.4. Photon and charged particle emissions from an electron-bombarded surface.

2.1.2 Reflected or backscattered electrons

These are considered to be electrons from the primary beam which have been so deflected by collisions with atoms in the specimen that they have been turned back out of the specimen again. Depending on their individual collision histories they may have energies ranging from the full primary beam energy down to the level of secondary electrons. The backscattered electrons (BSE) may be collected and used to form images in scanning microscopy. Monte Carlo simulations lead us to expect a certain degree of backscattering, and predict the variation with atomic number of the specimen. Before the theory had been refined to this degree, Bishop (1966) showed practically that the backscattering effect increases steeply with increasing atomic number (Z) of the scattering atoms (Figure 2.6). Hence local variations in Z could

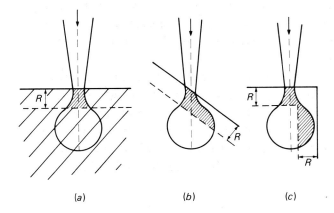

(a) (b) (c)

Figure 2.5. Secondary electron emission. (*a*) Only those created within distance R of surface can escape. (*b*),(*c*) Increase in emitted secondary electron signal from tilted surface and edges. (N.B. Diagrams not to scale; the proportions of the droplet will vary with the electron energy and the atomic number of the specimen.)

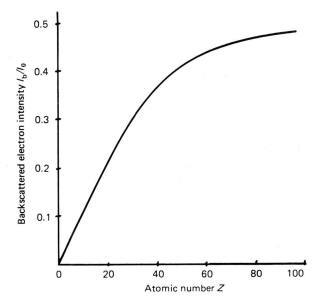

Figure 2.6. Dependence of backscatter coefficient on atomic number. (After Bishop, 1965.)

be expected to give variations in intensity of backscattered emissions. Such is the sensitivity to Z that adjacent elements in the Periodic Table can be distinguished in BS images of bulk specimens. At very low electron beam energies around 1 keV this relationship ceases to hold, and increased backscatter may not indicate higher Z material (Böngeler *et al.*, 1993).

The Monte Carlo simulation also shows that backscattering from a bulk specimen occurs from a wider area than the size of the primary spot. Backscattered electrons create secondary electrons along their whole path within the specimen, and those formed within the escape distance below the surface are emitted to form part of the secondary emission current from the surface. It is important to note, however, that they are no longer related to the surface topography under the primary electron beam and serve to dilute the informative part of the main secondary emission current.

Electron yield

The total *secondary electron yield* δ of a surface is defined as the number of electrons emitted per incident primary electron (Colby, 1969). It contains both the backscattered yield η and the true secondary electron yield Δ. If δ is plotted as a function of the incident electron energy a characteristic yield curve such as that in Figure 2.7 is obtained. The magnitude of the peak, δ_{max}, differs from material to material, being less than 2 for pure elements but as high as 10 or more from some compounds. The energy E_{max} at which the peak occurs also varies, but is generally around 1 keV.

The shape of the curve is a consequence of two opposing effects. The number of ionisations or secondary electrons created by an incident primary electron increases with the energy of the primary electron, but so also does its penetration into the specimen. At energies greater than E_{max} an increasing proportion of the secondary electrons are generated too far beneath the surface for them to escape and δ therefore decreases. At the higher voltages usually employed in scanning microscopes, e.g. 20 or 30 kV, δ is well below unity. Reduction of the voltage to a few kV, however, gives an increased SE yield and, because of reduced penetration and lateral diffusion,

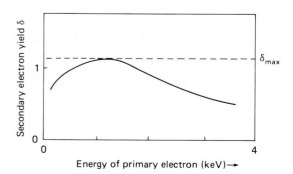

Figure 2.7. Schematic diagram of the variation of secondary electron yield δ with energy of incident (primary) electrons.

ensures that the 'backscattered' secondaries originate closer to the primary beam. In practice there are two electron beam energies at which $\delta=1$. These may be of practical use, as will be explained later.

Reflection at grazing angles of incidence

There is a special case met in the TEM when an electron beam is incident on a surface at grazing incidence and a beam of 'reflected' electrons is available to image surface effects. The reflected beam is characterised by a range of energy losses on scattering from the surface, unless the surface is smooth and crystalline so as to diffract electrons (see later section on crystalline specimens).

Detection of electrons

Electrons deposit a minute amount of charge, and in order to turn this into a useful signal it has to be amplified. The electrons used in electron microscopy are detected and amplified in several ways, differing according to the classification of the electron.

Secondary electrons have energies of 50 eV and below. This means that they are moving slowly and can be readily attracted into the detecting device, or repelled, as desired. The universally used SE detector was developed by Everhart and Thornley (Everhart & Thornley, 1960) and has been used, with only slight modifications, on all scanning microscopes ever since.

The principle is illustrated in Figure 2.8(*a*). Slow moving electrons are attracted towards and through a metal mesh at a few hundred volts positive to the specimen. Once inside the detector the electrons are accelerated by a high potential (e.g. 12 kV)

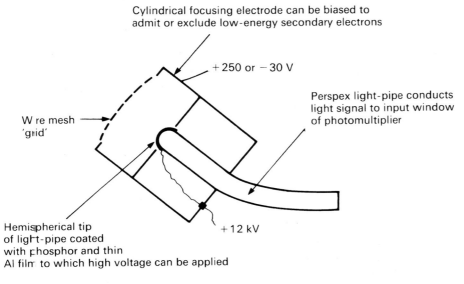

Figure 2.8. Schematic diagram of Everhart–Thornley secondary electron detector.

on to a phosphor, which emits a flash of light on impact for every electron. The electron signal is thus converted into a light signal, which is amplified by a photo-multiplier tube (PMT) into stronger electrical pulses used to create the image on the cathode-ray tube of the scanning microscope.

Backscattered electrons will also be detected by the Everhart–Thornley (E–T) detector because they cause scintillations in the same way as SE. If it is desired to exclude the SE the wire mesh grid in front of the scintillator is biased negatively, e.g. at -30 V, and secondary electrons are repelled whilst the backscattered electron current is let through. However, because of their higher energies, backscattered electrons will not be 'sucked into' the detector in the same way as low-energy secondaries; only those BSE which emerge from the specimen travelling in the direction of the scintillator will be attracted on to it. Early problems with BSE imaging using E–T detectors arose because of their inefficient collection by such a small-area detector. It was pointed out by Robinson (Figure 2.9) that if they are more efficiently collected the signal due to BSE could exceed that due to secondary electrons. Specialised detectors for BSE can operate on the scintillator/photomultiplier principle, e.g. Autrata & Hejna (1991), or use semiconductor amplification or an amplifying channel plate, the best choice depending on the geometry of the system and the energy range of the electrons to be detected. The important thing is that the detector should have a large angle of collection, e.g. by having a large area and/or being close to the specimen and facing it. The BSE image remains essentially a sub-surface image, however, except at very low beam energies.

2.1.3 X-ray photons

Photons of x-radiation with wavelength and energy characteristic of the elements in the specimen are emitted under electron bombardment in all forms of electron microscope. The spectrum of the radiation can thus be used for elemental identification, using forms of x-ray spectroscopy based on either wavelength or energy (known as Wavelength Dispersive or Energy Dispersive x-ray spectroscopy, WDS or WDX

Figure 2.9. Probable energy distribution of electrons emitted from a surface under bombardment in the scanning microscope. Curves for high and low atomic number elements are given. (After Robinson, 1973.)

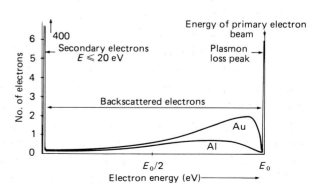

and EDS or EDX, respectively). The principles of these are outlined in Appendix 3 and the practice in Chapter 6.

X-rays are electromagnetic waves of very short wavelength, between two and five orders of magnitude shorter than visible light, which are emitted from matter being bombarded by electrons or α-particles or irradiated by even shorter wavelength radiations such as gamma rays. They are characterised by their wavelength or their energy, the two being simply related by $E=hc/\lambda$ where E is the energy, c is the velocity of light, h is Planck's constant ($6.625\ 59\times10^{-27}$erg sec), and λ is the wavelength. In practical units E (keV)$=1.24/\lambda$ (nm).

When it is bombarded by electrons an element (e.g. the molybdenum target in an x-ray tube) emits two types of x-radiation. One is a continuum or spread of wavelengths called white radiation or *Bremsstrahlung* ; the other is a series of discrete wavelengths forming the characteristic emission spectrum of that particular element. Thus, for molybdenum bombarded by 25 keV electrons, there would be a continuum stretching from 25 keV energy ($\lambda=0.050$ nm) downwards with strong emission lines at 19.6 keV (0.063 nm), 17.4 keV (0.071 nm), 2.4 keV (0.52 nm) and 2.3 keV (0.54 nm). We shall now consider, in simplifed terms, how the line spectrum and the continuum arise. To do this we must look in greater detail at the structure and operation of the atom.

We saw earlier that an inelastic collision between a K shell electron and a primary electron with energy $\geqslant EK_{ab}$ can result in ejection of the core electron from the atom. The resulting vacancy in the K shell can be filled by an electron from one of the higher energy shells, L, M, etc. The excess energy of this electron, i.e.the difference between the energy states in the two shells, is radiated during the transition as a photon of x-radiation. Thus the energy of the K x-radiation equals $EL-EK$ if the transition is between L, K shells or $EM-EK$ if it is between M, K shells. The former is designated Kα radiation; the latter Kβ. The wavelength and energy values of the x-ray emission are related by the expression $E=hc/\lambda$ given earlier.

A similar situation exists with respect to electrons in the L shell (and M for sufficiently high atomic numbers). Energy EL_{ab} will enable an L electron to leave the atom, when its place will be taken by an electron from the M, N, etc. shells, resulting in the emission of Lα, Lβ, etc. x-radiations with appropriate energy and wavelength. These processes are shown schematically in Figure 2.10.

Every atom species has, therefore, a characteristic x-ray emission spectrum which can be excited by supplying energy equal to or in excess of the appropriate binding energy, by electron bombardment or as radiation of higher energy, e.g. gamma radiation or 'harder' x-radiation. (N.B. In x-ray terminology the shorter-wavelength, higher-energy radiation is hard; conversely long-wavelength, less energetic, radiation is soft. Hard x-rays have a greater power of penetration through matter than soft x-rays, which are more readily absorbed, even by gases.)

Figure 2.11 shows a graph of the relationship originally pointed out by Moseley

Table 2.1. *Characteristic x-ray emission spectrum for silver, Ag, Z=47*

K series

	E_{ab}	β_4	β_2	β_5	β_1	β_3	α_1	α_2
E, keV	25.517	25.507	25.452	25.141	24.938	24.907	22.159	21.987
λ, nm	0.0486	0.0486	0.0487	0.0493	0.0497	0.0498	0.0559	0.0564
Rel. intensity		0.01	5	0.1	18	8	100	50

L_I series

	E_{ab}	γ_3	γ_2	β_9	β_{10}	β_3	β_4
E, keV	3.810	3.749	3.743	3.439	3.432	3.324	3.203
λ, nm	0.3255	0.3306	0.3312	0.3605	0.3612	0.3833	0.3870
Rel. intensity		2	3	0.01	0.01	11	5

L_{II} series

	E_{ab}	γ_1	β_1	η	γ_8	γ_6
E, keV	3.528	3.519	3.150	2.806		
λ, nm	0.3515	0.3523	0.3935	0.4418		
Rel. intensity		10	42	1		

L_{III} series

	E_{ab}	β_2	β_6	α_1	α_2	ℓ
E, keV	3.352	3.347	3.255	2.984	2.978	2.633
λ, nm	0.3699	0.3703	0.3808	0.4154	0.4163	0.4708
Rel. intensity		25	1	100	10	2

M series

	γ
E, keV	0.568
λ, nm	2.182
Rel. intensity	100

in 1913 of the smooth progression of K, L absorption energies with the atomic number of the element.

The picture just given of K, L, M, etc. shells of electrons is a simplification of the actual situation, in which each shell has several slightly different energy levels. The permissible transitions are controlled by selection rules; the result in practice is that there is a fine structure within the various α and β lines. Possible transitions and their nomenclature are illustrated schematically in Figure 2.12, and Table 2.1 gives more complete data for a typical element. In practice, many of the lines are weak in intensity and some are so close together that they are not detected separately by some detection systems.

The characteristic line spectrum forms the 'fingerprint' of the different atomic species, since the actual energy levels depend on the atomic number of the elements. It follows, therefore, that the elements present in an x-ray emitter can be identified by analysis of the emission spectrum (Johnson & White, 1970).

In an electron beam instrument the range of spectral lines which can be available for analytical purposes depends on the maximum kV of the electron gun. Thus, the K lines of silver will only be excited if more than 25.5 kV is available for the electron beam; K lines of cerium require more than 40.5 kV and the L lines of lead require at least 16 kV. Moreover, for the most efficient generation of x-ray spectra a voltage 3–4 times the energy of the line is desirable. In most electron beam instruments analysis will be carried out using lower energy lines (e.g. below 20 keV or longer than 0.06 nm). It is only in the TEM with gun voltages of 100 kV or higher that the K lines of heavier elements can be excited.

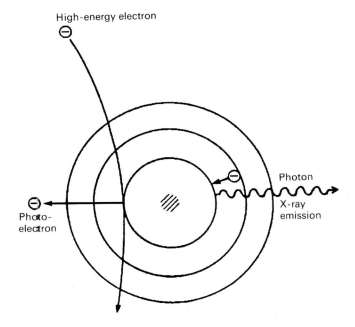

Figure 2.10. Schematic diagram of the processes leading to the emission of characteristic x-radiation under electron bombardment.

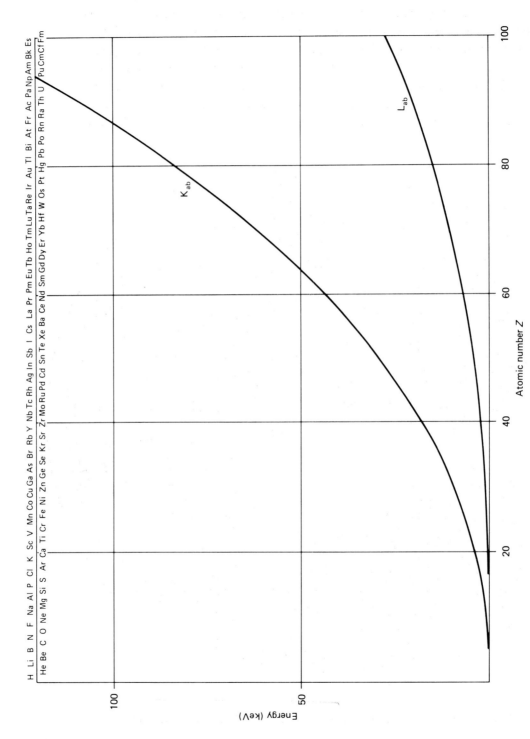

Figure 2.11. The energies of K and L absorption edges of all elements vary smoothly and progressively with the atomic number of the element, an effect first noted by Moseley.

The characteristic lines are superimposed on a background continuum whose extent is determined by the maximum energy used to excite the x-rays, and not by the atomic species itself. The latter does have some influence, however, as the intensity of the continuum is proportional to Z as well as to other factors. The explanation usually given for the continuum is that in classical electrical theory an electron which undergoes a sudden deflection in path radiates energy. Thus a fast electron which is deflected by an atomic nucleus will radiate a certain amount of energy as x-radiation. The maximum it can radiate is all it has, and hence the upper limit of

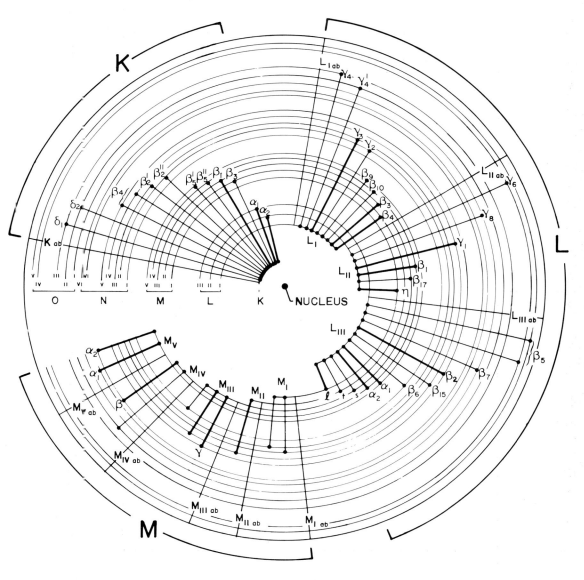

Figure 2.12. Origin of fine structure in characteristic x-ray spectra. (By courtesy of Kevex Corporation.)

Bremsstrahlung (or retarding radiation) is Ve, where V is the voltage of the electron gun, and e is the electronic charge.

Figure 2.13 illustrates an actual spectrum of characteristic x-radiation and Bremsstrahlung emitted by a specimen of platinum on alumina when bombarded by 20 keV electrons in an electron microscope. The spectrum is obtained from an energy-dispersive spectrometer (one of the two analytical systems to be described later); x-ray intensity is plotted against x-ray energy. It can be seen that the continuum or background radiation extends from a lower limit just below 1 keV, when the radiation becomes so soft that it is absorbed before reaching the detector, up to the 20 keV limit of electron energy. The overall efficiency of production and detection of radiation falls off as the upper limit is approached.

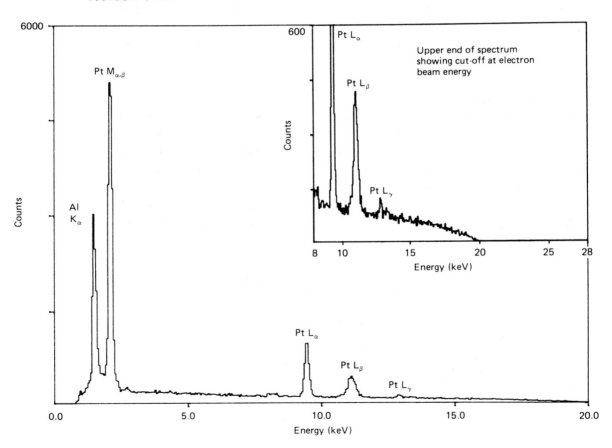

Figure 2.13. Characteristic x-ray spectrum from platinum on alumina, bombarded by 20 keV electrons. The platinum L lines are excited only weakly since the electron energy (20 keV) is only slightly in excess of the L absorption energy (13.9 keV). The platinum M and aluminium K radiations are strongly excited because of the higher 'overvoltage'.

On top of the continuum can be seen the characteristic x-ray lines of the elements present in the specimen, broadened out into Gaussian peaks by this form of spectrometer.

X-ray fluorescence (XRF)

Another way in which an atom can be ionised is by absorption of energy K_{ab} from another radiation. Thus, in Figure 2.14, absorption of x-radiation from another source can create a free electron and *fluorescent x-radiation*. Thus, absorption of characteristic radiation from one atom can result in emission of characteristic radiation of a lower energy from another atom, which may or may not be a near neighbour. It is the process of absorption of radiation and creation of fluorescence in this way which attenuates the x-ray signal emerging from a specimen being analysed in the electron microscope.

An important point to note about fluorescent x-radiation is that it is not accompanied by a Bremsstrahlung continuum, as is electron beam stimulated x-radiation. This fact can be made use of to increase the sensitivity of analysis under certain conditions.

2.1.4 Auger electrons

These are emitted together with x-ray photons as a result of a particular type of interaction between a primary electron and the inner shell of electrons around an atomic nucleus in the specimen. The electrons have energies up to 1–2 keV which are characteristic of the elements at the surface of the specimen; hence they may be used for elemental analysis of surfaces.

The Auger electron is a result of a secondary process (illustrated in Figure 2.15) following the initial ejection of an electron from an inner shell by incident electron

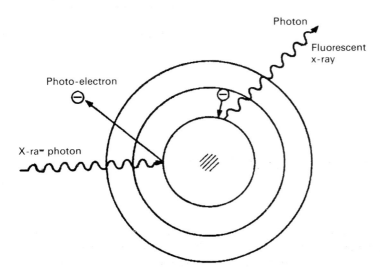

Figure 2.14. Schematic diagram of the emission of fluorescent x-radiation.

bombardment. The K vacancy is filled by an L electron but the K radiation resulting from this is absorbed in ejecting a further electron from the L shell; this is the
Auger electron and has an energy characteristic of the atom from which it came. The
two vacancies in the L shell are filled by electrons from higher bands, with emission
of L and possibly M x-radiation. The energy of the Auger electron can be measured
by means of an energy spectrometer such as a cylindrical mirror analyser; elemental
analysis down to $Z=2$ is possible from the energy spectrum.

Auger electrons are excited in the same droplet-shaped volumes as the x-radiation
we considered earlier, but only those which have not lost energy in inelastic collisions
in emerging are useful for analysis. Auger electron spectroscopy (AES) therefore
only analyses the outermost few nm of a surface, and the analysis must be carried out
in ultra-high vacuum, otherwise the outermost few nanometres are the adsorbed gas
layer. Further details are given in Chapter 6.

2.1.5 Cathodoluminescence

Many specimens emit visible and invisible light under electron bombardment, by the
process of *cathodoluminescence* (CL). This provides a means of studying the internal
energy structure of materials, since the colour of the luminescence enables energy

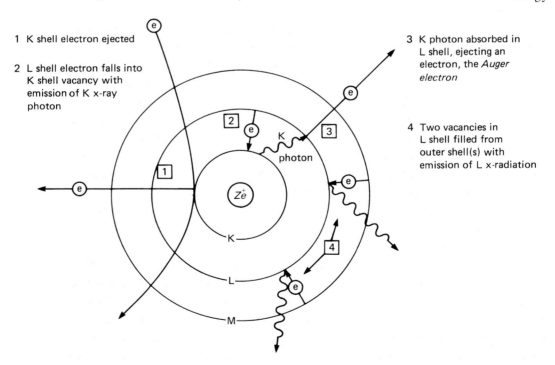

Figure 2.15. Schematic diagram illustrating the sequence of processes involved in the emission of Auger electrons.
The transition illustrated would be identified as KLL, these being the shells involved in steps 1, 2 and 3 of the
process. Other transitions are possible, and the peaks in an AES spectrum can be assigned to particular transitions.

gaps to be determined; in many materials it is very sensitively related to composition and impurities. We will see later that with appropriate detectors and displays the colour and location of differently luminescing sites can be shown with better than micrometre precision.

Without this phenomenon electron microscopy as we know it could not exist, nor would TV and computers have progressed in the way they have.

Detection of cathodoluminescence

Any detector system utilising a photomultiplier will be sensitive to visible light (the specimen chamber of an SEM must be in darkness when an E–T detector is in use, for this reason). CL signals may therefore be obtained from some E–T detectors, but the requirement for direct line of sight and high sensitivity makes them very inefficient for this purpose. An improvement is obtained by placing a curved mirror over the specimen, to collect and reflect the maximum light from the specimen towards the detector. If energy level information is desired from CL the detector must incorporate some form of colour analysis, either using a spectrophotometer or separate detectors to measure the Red, Blue, Green components of the luminescence.

2.1.6 Absorbed (specimen) current

The net electron current remaining in the specimen after the electron emission or transmission processes have added to, or diminished, the primary beam current is conducted to earth and hence back to the high tension (HT) circuit. The absorbed or *specimen current* is related to the emissive mode current (SE and BSE) and micrographs may be similar. There is a special case when the specimen is a semiconductor, and in scanning microscopy the specimen current can give information on internal structure. This technique is called Electron Beam Induced Current (EBIC).

If there is no electrical path from the specimen back to the HT circuit, i.e. if the specimen is an electrical insulator, the primary beam will charge-up the specimen negatively until the beam is suppressed. Microscopy is then no longer possible. In practice when there is inadequate conductivity through a specimen, in TEM or SEM but more frequently in the latter, the specimen is made conducting artificially, usually by applying a conductive surface layer (specimen coating, see later, page 163–4).

If the specimen is thick enough for the electron beam to be brought to rest within it the excess energy over and above that used up in ionisation and luminescence effects will be converted into heat. The visible signs of this will depend on the circumstances – type of specimen, localisation of heating and efficiency of losing heat. There may be anything from a slight localised warming of the specimen in the microscope to fusion and/or evaporation of the specimen if the rate of arrival of heat from the electron beam is considerably greater than the ability of the specimen to lose it by conduction or radiation. Examples of visible specimen deterioration due to the action of the beam are shown in Figures 4.14 and 4.20.

2.1.7 Transmitted electrons

If the specimen is thin enough some incident electrons penetrate through it and emerge as *transmitted electrons*. They may have been deflected from the line of the primary beam and may have lost energy by collisions. They carry information about the interior of the specimen, and may be used as a whole or may be separated into groups, depending on their collision history.

(a) *Elastically scattered electrons* have been deflected through wide angles (typically $1-5 \times 10^{-2}$ radian, 0.6–3 degree) by interaction with the nuclei of atoms in the specimen. They have suffered zero or very small energy loss in the scattering process. The number of electrons scattered off axis gives information on variation of atomic number Z within the specimen, since the number of electrons scattered elastically, N_e, is proportional to $Z^{4/3}$.

(b) *Inelastically scattered electrons* have lost varying amounts of energy in collisions within the specimen, e.g in exciting characteristic x-radiation. Provided the electrons have had only one inelastic collision the spectrum of energy loss is characteristic of the atoms in the material with which the electrons have interacted, so a form of elemental microanalysis, (electron energy loss spectroscopy, EELS) is possible. The number of electrons inelastically scattered, N_i, is proportional to $Z^{1/3}$.

(c) *Unscattered electrons* have suffered no energy loss or deflection within the specimen, and hence can easily be separated from the inelastic electrons by an energy analyser. The no-loss unscattered beam has intensity N_0. Thus, the total beam current $N = N_e + N_i + N_0$.

If N_i, N_e, N_0 are measured simultaneously as the specimen is probed by the electron beam different kinds of information may be obtained about the spatial distribution of Z across the specimen. All scattering material is shown up (dark-field image) by N_e/N_0. The ratio N_e/N_i is proportional to Z.

In fixed-beam microscopes transmitted electrons focused into images are detected by the light emission from a phosphor-coated screen, and by the latent image created in a photographic emulsion exposed to the beam. The fluorescent image may also be scanned by a video camera to give a remote image on a cathode-ray tube. All of these methods will be found in later chapters of the book. In a scanning microscope the transmitted beam current can be measured with a solid state electron detector, which can be shaped to collect either the scattered electrons (annular detector) or the unscattered ones (disc detector), or both.

If energy losses are to be used for elemental analysis by EELS, an energy spectrometer is placed before the electron detector. Electrons from a point or line source are spread out into an energy spectrum by passage through a magnetic field. Electrons of a particular value of energy loss are measured by putting the detector at the appropriate part of the spectrum. The energy loss spectrum is obtained by deflecting the whole spectrum in steps across the detector, or by placing a linear array of identical detectors along the spectrum (Parallel EELS, or PEELS).

Plasmons

An additional beam–specimen interaction which is detectable only by its effect on the EELS spectrum is the collective excitation by high-energy electrons of the free electrons in metals and alloys. The excitation or oscillation forms a *plasmon* and extracts a characteristic small amount of energy, e.g. 14.9 eV for aluminium, from the beam. There is a mean free path for plasmon production which is shorter than that for inner shell ionisation, and which increases as the electron beam energy increases; multiple peaks may be found for thicker specimens. The characteristics of EELS spectra are illustrated in Chapter 6.

2.2 Crystalline specimens

So far we have considered only the effects which result from the general interaction between an electron beam and a specimen, whereby its surface morphology and elemental composition may be deduced from electron emissions and analysis of x-radiation, Auger electrons or transmitted electron energy loss spectrum. The electron beam interacts with the specimen in a very specific way when the specimen is crystalline.

2.2.1 Crystalline structures

In these the constituent atoms are geometrically arranged in a regular pattern, based on a unit cell which is repeated in all three dimensions throughout the crystal. The simplest unit cell is a cube, and the simplest crystal form based on this is the primitive or simple cubic structure. Figure 2.16(*a*) is a schematic representation of the eight atoms in a unit cell, and a clarified diagram of the same arrangement with the atom centres represented by points. In a simple cubic crystal a number of these elementary building blocks are stacked together as in the larger shape of Figure 2.16(*b*).

If we study the crystal in Figure 2.16(*b*) we can find a number of planes in it which contain a high density of atoms. The cube faces and various diagonals are readily seen, and when the crystal becomes more extensive a number of oblique planes can be found. In our simple cube similar sets of planes exist based on each of the three orthogonal axes x, y, z. Thus, in an extended crystal a system of sets of parallel planes is obtained, with readily calculable distances separating the planes in a particular orientation. One set of planes is shown shaded, the spacing between the planes being $1/\sqrt{2}$ times the length of the side of the cube. The three-dimensional array of atoms in a crystal is known as a *crystal lattice*, and the atom-rich planes are referred to as crystal planes or *lattice planes*. The perpendicular distance d between parallel planes in a particular series is called the '*d-spacing*' for that set of planes.

The lattice planes are identified by a group of three numbers in the form (hkl), called *Miller Indices*, where h, k, l are integers or zero. These numbers can be regarded as reciprocals of the intercepts which the plane makes on the x, y, z axes of the unit cell. Thus planes with $h=1$ pass through the corner atom, $h=2$ intersect the x-axis midway to the corner atom, and $h=0$ intersect at infinity, i.e. they are parallel to the x-axis (Figure 2.16(c)). The shaded planes in Figure 2.16(b) are therefore (110) planes, and $d_{(110)}=a\sqrt{2}$, where a is the length of the side of the unit cube.

Just under one-half of the solid elements in the Periodic Table form crystals based on a cubic unit cell; either the simple cubic cell just described or the variants face-centred cubic (f.c.c.) or body-centred cubic (b.c.c.) with extra atoms at the centres of the faces or the centre of each cube, respectively. The other elements crystallise in one of the other 11 forms, hexagonal, monoclinic, rhombohedral, etc., which will not be described here. For fuller details of these and of Miller Indices for them the reader should refer to books on crystallography, x-ray- or electron diffraction, or metallurgy, e.g. Hammond (1992) or Jackson (1991).

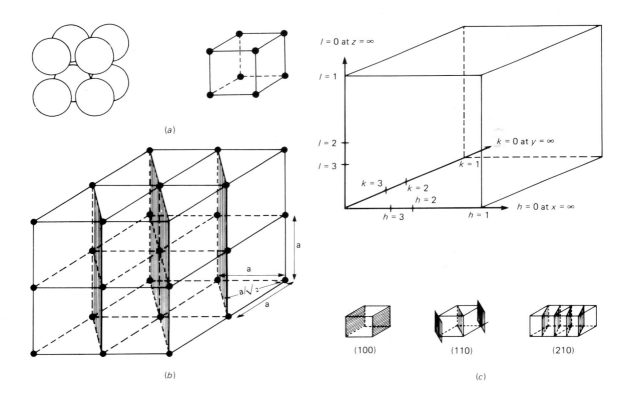

Figure 2.16. (a) Representation of atomic arrangement in the unit cell of a simple cubic crystal. (b) Eight unit cells in a larger crystal of cubic habit. Separation between adjacent shaded planes is $a/\sqrt{2}$. (c) Miller Indices: to obtain plane (h,k,l) join the points $(1/h, 1/k, 1/l)$ on the x, y, z axes, e.g. (100), (110), and (210) planes.

2.2.2 Diffraction by crystal planes

In the present context the importance of the regular arrays of atoms in crystal planes lies in their ability to act as selective reflectors of incident radiations such as x-rays or electrons. It was shown by Bragg that there is a reinforcement of reflections from successive parallel planes when the angles of incidence and reflection θ satisfy the relationship $2d \sin\theta = n\lambda$, where d is the lattice spacing between atomic planes, λ is the wavelength of the radiation, and n is an integer. At other angles of incidence the reflected waves from successive planes cancel each other out.

This relationship is illustrated diagrammatically in Figure 2.17. It should be noted that the directions of the incident and reflected (or diffracted) beams make an angle of 2θ with one another. Bragg's Law was originally propounded for x-rays but was later shown to apply also to electron waves. Since the wavelengths of fast electrons are some 100 times shorter than x-rays the angles θ are correspondingly smaller. For example, when 100 keV electrons ($\lambda = 3.7$ pm) are diffracted from the faces of a simple cubic crystal of thallium chloride with cube side 0.3843 nm the angle θ is $0°16.5'$, and the angular deflection of the electron beam will be $0°33'$ or 9.6×10^{-3} radian.

We shall see later (Chapter 5) that the fulfilment or otherwise of the Bragg condition for diffraction has important consequences on the interpretation of electron micrographs from some specimens in transmission microscopy.

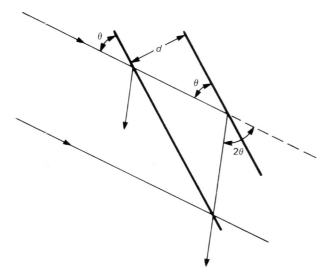

Bragg's Law: $2d \sin \theta = n\lambda$
Incident beam deflected through 2θ

Figure 2.17. Bragg's Law of diffraction of x-rays or electrons from crystal planes.

2.2.3 The electron diffraction camera

Suppose we have a parallel beam of electrons falling on a crystal at the correct angle of incidence for diffraction to occur from a set of lattice planes ('at the Bragg angle' in diffraction terminology). The diffracted beam is deviated through an angle 2θ from the undeviated beam. If a fluorescent screen or photographic film is placed at a distance L from the crystal (Figure 2.18) the diffracted beam will be displaced from the undeviated beam by a distance R such that $R=L\tan2\theta=2\theta L$ (since for very small angles θ in radians$=\sin\theta=\tan\theta$). Combining this relationship with Bragg's Law we see that $R/L=n\lambda/d$ or $d=\lambda L/R$ for first order ($n=1$) diffraction. So if λ, L and R are known or measured, d may be calculated.

If the single crystal is now replaced by a cluster of randomly orientated crystals, diffraction of electrons will occur from those crystals satisfying the Bragg relationship with the incident electron beam, and a hollow cone of diffracted rays will be formed which intersects the screen as a ring of radius R. The breadth of the ring is determined by the size of the diffracting crystals; at its narrowest it is the width of the electron beam falling on the crystal, and it becomes broader and more diffuse as the crystals become smaller. Nanometre-sized crystals would still give recognisable rings but it would be unwise to estimate their radii or width without making a quan-

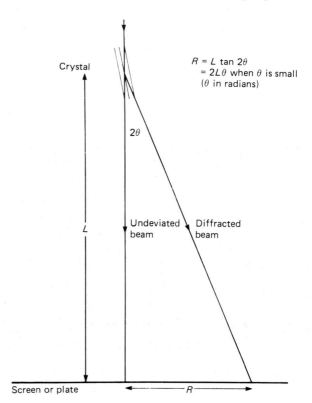

Figure 2.18. Geometry of the diffraction camera.

titative measurement, e.g. by using a microdensitometer on the photographic record. Materials are often described as crystalline when their diffraction rings are sharply defined or amorphous when the rings are broad and diffuse.

The arrangement of electron gun, crystal holder and screen or film, all within an evacuated enclosure, constitutes an *electron diffraction camera*. The distance L from crystal to screen is the *camera length* and λL the *camera constant*. When λL is known the *d*-spacings of crystal planes responsible for particular diffraction spots or rings can be calculated once R is measured. Every crystalline solid contains a number of crystallographic directions giving rise to diffraction at particular angles. A *diffraction pattern* consists of an arrangement of spots or a concentric series of rings, each R being related to a characteristic value of d in the crystal. The crystal *habit* (cubic, hexagonal, etc.) and dimensions of the unit cell may be deduced from measurements on a diffraction pattern, and the identity of the crystalline material can be ascertained, since each element and compound has its own characteristic set of *d*-spacings. Lattice spacings and unit cell data for thousands of elements and compounds are contained in the Powder Diffraction File (JCPDS, 1994) issued by the International Center for Diffraction Data. This information is available on individual data cards, collected in books, or on various systems, including CD-ROM, for rapid retrieval, and is regularly updated. Although compiled primarily for x-ray identification the data on lattice spacings also applies in electron diffraction. Reflections which are weak in the x-ray spectrum will be stronger for electrons, and some additional reflections will be found in electron diffraction. Figure 2.19 shows ring patterns from polycrystalline cubic specimens, with a table of corresponding *d*-spacings.

Crystalline specimens must obviously be thin enough in the direction of travel of the electrons for the electrons to penetrate them without significant energy loss (and

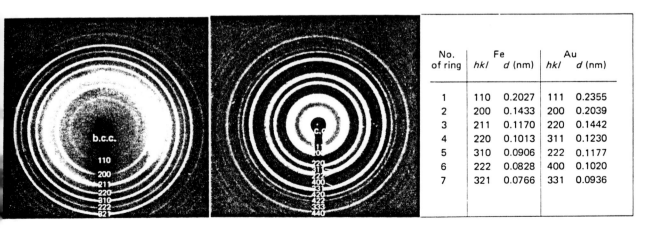

| No. | Fe | | Au | |
of ring	hkl	d (nm)	hkl	d (nm)
1	110	0.2027	111	0.2355
2	200	0.1433	200	0.2039
3	211	0.1170	220	0.1442
4	220	0.1013	311	0.1230
5	310	0.0906	222	0.1177
6	222	0.0828	400	0.1020
7	321	0.0766	331	0.0936

Figure 2.19. Electron diffraction patterns from polycrystalline specimens of body-centred cubic (Fe) and face-centred cubic (Au) habit, with the Miller Indices of each ring identified.

hence wavelength change). Thus, the specimen requirements are similar to those for normal microscope specimens. However, it is also possible to diffract electrons from the smooth face of thicker specimens inclined at the appropriate small angle to the electron beam as in Reflection Electron Microscopy in the TEM. In this case (reflection electron diffraction) only one-half of a diffraction pattern is obtained, but its geometry is still governed by Bragg's Law (see Figure 7.12).

2.2.4 Convergent beam electron diffraction

The previous section has referred ideally to the result of parallel illumination falling on a crystalline region of specimen in the TEM. Sharp diffraction spots are formed, whose geometrical arrangement and spacing convey information about the crystal structure.

If the illumination is strongly convergent, falling as a cone on the crystal, the diffraction 'spots' become enlarged discs containing a great deal of fine structure. Under the correct instrumental conditions such convergent beam diffraction patterns can yield a full crystallographic analysis of the specimen, and are increasingly used by materials scientists who need this form of identification. See, for example, Spence & Zuo (1992).

2.2.5 Electron channelling patterns

There is also an effect of preferred reflection of electrons from bulk crystalline specimens. Backscattered electrons are channelled along preferred directions related to the crystal lattice of the surface atoms. In the scanning electron microscope channelling patterns can be recorded and analysed from broad areas or selected smaller regions, 'selected-area channelling patterns' (see later, section 6.1.2).

2.3 Effect of variation of electron energy

On all but the simplest electron microscopes there is a range of operating voltages provided. These enable the conditions of examination to be suited to the type of specimen and the information to be obtained from it. As a very broad generalisation, the higher the voltage the better will be the performance of the microscope, and the lower the voltage the greater will be the interaction between beam and specimen.

2.3.1 Microscope performance

The operating voltage affects the illumination and the lenses. An electron source designed for a given voltage, e.g. 30 kV for SEM or 100 kV for TEM, will be less bright at lower voltages, and this will impair operation at higher magnifications in both TEM and SEM. In the former, a much enlarged image will have too few electrons in it to be visible, and focusing and selection of field of view will be impossible.

In the SEM the scanning electron probe cannot be made as small as at higher voltages, because there will be insufficient electrons in it to form an image. Also, the wavelength of the electron beam will be increased at lower energies, resulting in poorer detail resolution in a TEM. We saw earlier (Chapter 1) that the resolution varies as $C_s^{1/4}\lambda^{3/4}$ and hence for a given spherical aberration constant the relationship between resolving power and electron gun voltage will be as shown in Figure 2.20. For better resolution, higher kVs are desirable, hence the exploration of 1 MV and 3 MV High Voltage TEMs.

In a practical electron microscope the electron beam will be influenced by magnetic fields from extraneous sources (e.g. fluorescent lamp chokes in the microscope room) as well as the lenses. The degradation of performance because of extraneous effects is worse for slower, low-energy, electrons than at higher energies. This fact is used in the design of scanning microscopes for use at low voltages, when the beam is kept at a higher energy most of its way down the electron-optical column, and is only slowed down just before the specimen.

2.3.2 Beam–specimen interactions

The probability of electron–specimen interaction occurring can be thought of in terms of a hypothetical cross-section which varies inversely with voltage, so that there is a greater chance of interaction at lower voltages. Thus in the TEM the contrast of an image is increased by changing to a lower operating voltage, and vice versa (Delong *et al.*, 1994).

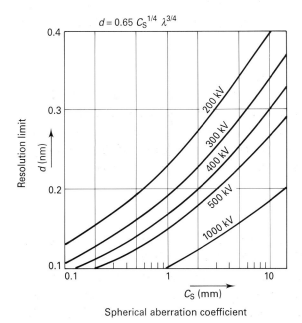

Figure 2.20. Relationship between resolving power, voltage, and C_s. (By courtesy of JEOL Ltd.)

Decreasing the beam–specimen interaction results in a longer mean free path in the specimen and a greater penetrating ability for the electron beam without losing so much energy that the image sharpness deteriorates due to chromatic aberration in the lenses. The variation of penetrating power with electron energy is shown in Figure 2.21.

The increased mean free path at higher voltages means that analytical techniques such as electron energy loss spectroscopy, EELS, can be carried out on thicker specimens in higher voltage TEMs. A factor of 3 increase between 100 kV and 1 MV has been found. In addition, the efficiency of collection of inelastic electrons is increased because of the reduced angular scattering at higher energies.

Although penetration is still rising above 1 MV the greatest rate of gain is in the range 100–400 kV; hence Intermediate Voltage TEMs working in this range have become the preferred instruments for performance at the highest level.

There is also a serious effect found in MV instruments of complete atoms being displaced by inelastic collisions with the electron beam, a phenomenon which is much reduced at Intermediate Voltages.

In the SEM reducing the voltage reduces the electron penetration into, and lateral diffusion in, the specimen (as the Monte Carlo calculations indicate) and increases the amount of surface information in the image (Müllerová & Frank, 1993). Also, for bulk specimens it reduces the specimen heating.

2.4 Electron interference effects
2.4.1 Fresnel fringes
In transmission electron microscopy, a discontinuity in a specimen can act as a secondary source of electrons. Hence, in any plane above or below the dis-

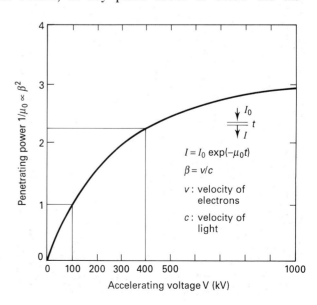

Figure 2.21. Penetration–voltage relationship for transmission microscopy (By courtesy of JEOL Ltd.)

continuity electrons will be arriving both directly and from the scattering source. At periodic distances across this plane the two wave motions will interfere with the production of bright and dark regions. If the discontinuity is a circular hole, for example, the interior of the hole will contain a set of circular light and dark rings, whose separation will vary with the separation between the plane of the hole and the plane under consideration. If the objective of the microscope is focused in any plane other than that of the hole, therefore, it will see the hole with a ring pattern inside and outside, the separations increasing as the focused plane becomes further from the plane of the hole. These patterns are known as Fresnel fringes, and provide a very convenient guide to focusing in the TEM. They also have a profound effect on the detailed appearance of images at high magnification, which will be described at greater length in the section on focusing effects in Chapter 5.

The number and visibility of Fresnel fringes is increased if the electron illumination is coherent, as from a pointed filament or field emitter. An example of multiple fringes from a pointed filament is shown in Figure 2.22.

2.5 Suggested further reading

Books on the theory of scanning electron microscopy contain more or less rigorous treatments of beam–specimen interactions, e.g. Goldstein *et al.* (1992) and Lee

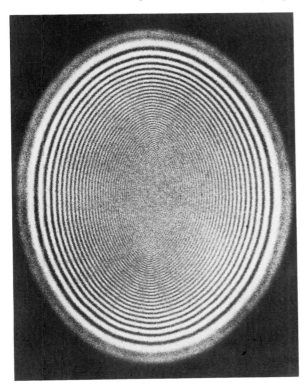

Figure 2.22. Transmission micrograph of a hole in a carbon film, showing many orders of Fresnel fringes, from a microscope using a pointed thermionic filament. (By courtesy of Siemens Ltd.)

(1993). There are also several relevant chapters in Hawkes & Valdrè (1990). It can be instructive, as well as fascinating, to spend some time in front of a computer with a disc of Monte Carlo programmes and a manual, e.g. Joy (1995).

The elements of crystallography are explained by Vainshtein (1964), Hammond (1992) and Jackson (1991).

More information can be obtained from suitable specimens by the use of *electron holography*. This has not yet reached the stage of general applicability and will not be dealt with here. The technique is very much associated with the work of Tonomura; see, e.g. Tonomura (1994).

The electron microscope family

3.1 Introduction

Just as 'light microscope' is a generic term covering a range of instruments for producing magnified images using glass lenses and visible or ultra violet light, so the name 'electron microscope' does not refer to a specific design of instrument but to a family of instruments which produce magnified images by the use of electrostatic or electromagnetic lenses with fast-moving electrons as illumination. They all share the ability to give images of high or very high resolution over a very useful depth of field.

Within the electron microscope family there are two well-defined ranges of microscope, corresponding to transmission and reflection (metallurgical) light microscopes, which look directly at the internal structure of translucent specimens and the outside features of bulk material, respectively.

The transmission electron microscope (TEM) is a direct derivative of the compound light microscope, making use of the shorter wavelength electron illumination. In its simplest form its magnification is achieved by exactly the same lens arrangement as in its light counterpart. It has also been developed further to make fuller use of the special properties of electron illumination; principally the higher resolution, but also the ability to carry out various forms of elemental and crystallographic microanalysis. Its variants include higher voltage and higher resolution instruments (HVEM and HREM) and, with the addition of probe-forming and scanning facilities, the very versatile analytical electron microscope (AEM).

The electronic equivalent of the metallurgical or reflected light microscope, used to study the outside of specimens rather than their internal arrangement, is the scanning electron microscope (SEM). This finds applications throughout the magnification range between a hand lens and the TEM. This instrument differs from traditional microscopes in forming its image sequentially, i.e. not all at one time. Its most advanced forms owe much to developments in electron sources and short-focus electron lenses. Whereas the TEM has developed by going up in voltage, the SEM has done the reverse in order to approach true surface imaging. Another direction of development is environmental, or low-vacuum, scanning microscopy (ESEM), in which hitherto 'awkward' moist specimens can be examined closer to their normal state.

The highly refined field emission scanning transmission electron microscope (FESTEM) is a specialised SEM competing with the high-resolution TEM at studying the internal structure and localised composition of specimens.

Although cross-bred light and electron microscopes have appeared, with interesting features and capabilities, these have not caught on, and will not be dealt with in detail.

Microprocessors and computers have had a great impact on commercial electron microscopes of all types, both in their design and operation and in viewing, recording and analysing images.

The various members of this large family will now be described and illustrated, to give the reader an idea of what the instruments are like and how they appear in commercial form. Since electron microscopes seldom wear out and can continue to operate for tens of years, readers may encounter instruments of a range of ages and designs. A description of the operating principles will appear more directly applicable to older instruments than to modern computerised microscopes in which the operation is embedded in software rather than control knobs.

3.2 The transmission electron microscope (TEM)
3.2.1 The two-lens TEM

The simplest electron microscope has two image-forming lenses and is an exact analogy of the compound light microscope, as shown diagrammatically in Figure 3.1. The illumination coming from an electron gun is concentrated on the specimen by a condenser lens. After passing through the specimen the electrons are focused by the objective lens into a magnified *intermediate image*. This image is further enlarged by a projector lens and the final image is formed on a fluorescent screen or a photographic film.

Electron lenses can conveniently be made to magnify 50–200 times, and the simple two-stage TEM would have a magnification of about 10 000 times. The objective lens would typically resolve detail down to 5 nm, which would require an overall enlargement of 50 000 to be readily seen by the eye. The additional enlargement is obtained by using a simple magnifying glass on the fluorescent image in the microscope, or a photographic enlargement of the image recorded on the photographic film.

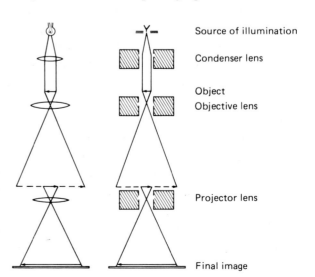

Source of illumination

Condenser lens

Object
Objective lens

Projector lens

Final image

Figure 3.1. Light microscope and equivalent two-stage electron microscope.

The practical realisation of an electron microscope is larger and more complex than the diagram suggests, however, for several reasons:

- The electron beam illumination must be generated and used in a high vacuum. The body of the microscope is an airtight tube continously evacuated to a pressure of 10^{-4} mbar or below. The electron path length within the tube may be more than one metre.
- Electronic lenses are more bulky than glass lenses, and must be powered from highly stabilised electrical supplies.

Thus, as well as the *electron-optical column* of lenses with associated vacuum pumps, there will also be an electric power unit to supply the lenses with current and provide a high voltage of 100, 200 or even 1000 kV to accelerate the electrons. Consequently, although a light microscope can be accommodated on a square metre of laboratory bench all but the smallest electron microscopes bring their own bench with them and require a separate room or even their own building! Early commercial two-stage electron microscopes are illustrated in Figure 3.2.

3.2.2 Higher-resolution microscopes

When the design and construction of electron lenses is sufficiently good to reveal details beyond the 5 nm resolution level it is desirable to employ more than two stages of magnification in order to see and record the finest detail. Three imaging lenses enable magnifications of up to about 500 000 or so to be achieved, and resolving powers of fractions of a nanometre utilised. Three-lens microscopes would typically cover magnification ranges from ×1000 to ×200 000 or ×2500 to ×500 000. Additional flexibility of operation can be obtained if a fourth imaging lens is added; several present-day microscopes employ five or six imaging lenses, and cover ranges from ×50 to over one million times without change of specimen position, with zero- or controlled-image rotation and distortion, and permitting selected-area electron diffraction (see later, Chapter 6) to be carried out over a wide range of magnifications. There is a further advantage to be gained by sharing the magnification between a larger number of lenses, insofar as the overall length of column can be reduced and the construction can be made more rigid, an essential feature when very high magnifications are involved.

'High-resolution electron microscope' is a term rather loosely applying to all modern TEMs which have a potential resolving power of one nanometre and smaller. The *resolution* actually obtained on all but the thinnest specimens is specimen dependent and probably poorer than 1 nm: where details such as crystal lattices or atomic placements are being studied, resolving powers of two- or three-tenths of a nanometre are being used, and the microscopes are then described as having *ultra-high resolution*.

3.2.3 The high-resolution TEM in practice
All high-resolution microscopes have a basic structural make-up which may be modified or added to in individual designs. This is illustrated schematically in Figure 3.3. The essential components, working from one end of the column to the other, are as follows:

(1) Electron gun: In its simplest form this consists of a heated tungsten filament and electrostatic lens which provide a well-defined beam of electrons apparently coming from a source (the crossover) 50–100 µm across. Accelerating voltages are typically

(a)

(b)

Figure 3.2. Stepping stones in the development of electron microscopes. (*a*) RCA Type B two-stage electron microscope, 1940. 30–60 kV, 500–16 000 diameters magnification, 10 nm resolving power guaranteed. (Hillier & Vance, 1941; by courtesy of Radio Corporation of America.) (*b*) Philips EM 75 two-stage microscope, 1955. 10–75 kV, 1200–12 000 diameters magnification, resolving power better than 5 nm. (van Dorsten & le Poole, 1955; by courtesy of Philips Electron Optics.)

variable by 20 kV steps in the range 40–125 kV, sometimes extending downwards to 20 or even 10 kV and upwards to 200, 300 or even 400 kV. The total current emitted by the electron gun may be up to 100 μA but only a fraction of this reaches the final image, the remainder being absorbed by diaphragms within the column.

When a microscope is required to operate at the higher end of the magnification range a brighter electron gun will be desirable. Lanthanum hexaboride gives a ten times improvement with little additional complication; the ultimate, the field emission gun, offers three orders of magnitude increase in brightness but demands the same degree of improvement in operating vacuum. Electron guns are described in greater detail in Appendix 4.

(2) Condenser lens: This provides a means of varying the strength of illumination in the specimen plane to suit the type of specimen and the magnification of the final image. If the same final image brightness is required over a range of enlargements between ×500 and ×250 000 then the intensity of illumination at the specimen must be varied by a factor of 500^2 or 2.5×10^5. The maximum intensity occurs when the electron source or crossover is focused in the specimen plane. Lower intensities are obtained by focusing the source above or below the specimen plane, (usually above), as shown in Figure 3.4.

The illumination from a single condenser lens usually falls on a much larger area of specimen than appears in the final image. Thus, for example, the focused image of the electron crossover will be at least as large as the crossover itself, i.e 50–100 μm across, whereas the field of view actually appearing on a micrograph taken at

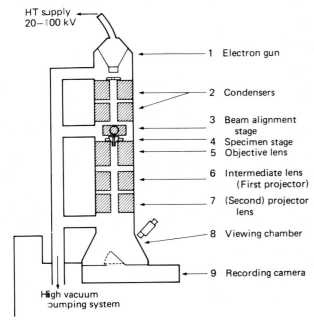

1 Electron gun
2 Condensers
3 Beam alignment stage
4 Specimen stage
5 Objective lens
6 Intermediate lens (First projector)
7 (Second) projector lens
8 Viewing chamber
9 Recording camera

Figure 3.3. Schematic diagram of high-resolution transmission microscope. Numbers refer to description in the text.

×100000 will only be about 1 μm across. The electron illumination outside this field is therefore wasted and, more importantly, may actually be damaging the specimen. Most high-resolution microscopes, therefore, use a double condenser lens system which enables the focused patch to be reduced to a diameter of a micrometre or less. The additional condenser lens, C1, is strongly excited and forms a diminished image of the source which is then projected onto the specimen by the second condenser lens, C2. This is shown diagrammatically in Figure 3.5. The brightness of illumination is not increased by double condenser operation since only a small proportion of

Figure 3.4. The most intense illumination at the specimen occurs (*left*) when the condenser forms an image of the crossover of the electron gun in the plane of the specimen. For all other strengths of the condenser lens (or C2 of a double condenser system) the illumination is spread over a larger area at the specimen plane, i.e. its intensity is reduced. The diagram (*right*) shows how reduced brightness is obtained by a stronger condenser lens (overfocus) or a weaker one (underfocus, dotted).

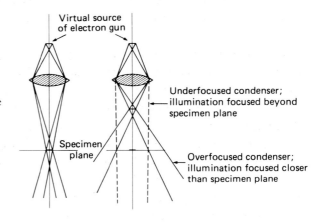

Figure 3.5. Control of the size of the illuminating spot by double condenser system. In the upper diagram C1 is strongly excited, forming a reduced image of the crossover. This is projected as a small spot in the specimen plane. In the lower diagram C1 is weakened and images the crossover only slightly diminished, and closer to C2. The illumination projected into the specimen plane is now a larger disc. Its brightness is no less than in the first case, however, since less of the electron beam from the crossover is wasted betwen C1 and C2. The dotted lines in the upper diagram show the smaller solid angle of illumination which is actually used in illuminating the specimen.

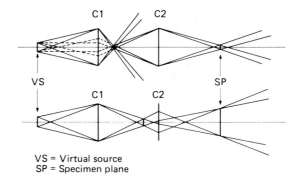

the electrons in the demagnified image of the source are actually projected by the second lens. Unlike the compound light microscope the illumination aperture in the electron microscope does not have to be matched to the aperture of the objective lens; in the electron microscope it is usually much smaller. The maximum aperture of illumination occurs when it is focused on the specimen and is determined by the size of the diaphragm (condenser aperture) in the second (or only) condenser lens. An increasing number of microscopes now use three or more condenser lenses, under computer control, whereby the brightness and size of the illuminating patch at the specimen can be varied independently. Constant image brightness with change of magnification may also be provided automatically. More will be said about multi-lens condenser systems in a later section on the Analytical TEM.

(3) Beam alignment stage: The microscope will only give its best performance when the illumination axis coincides with the axes of the objective and following lenses. Misalignments may be either a lateral displacement between two parallel axes or an angular difference (tilt) between the two axes. The first is corrected by translating the illumination sideways until the axes coincide; the second is corrected by tilting the illumination into parallelism and then translating it into coincidence.

On early microscopes misalignments were corrected by physically moving or tilting the parts of the microscope. Nowadays, in most cases, the lenses of the optical column are aligned ('pre-aligned') and rigidly clamped together during manufacture. Any slight re-alignments which may be necessary during use are performed by electrical tilt- or shift coils built into the column (Figure 3.6) and already described in Chapter 1.

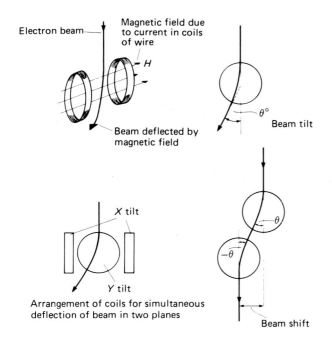

Figure 3.6. Principle of electromagnetic beam-tilt and beam-shift stages.

A beam alignment stage as described, with X, Y tilts and shifts, will enable any slight change in illumination axis with alteration of operating voltage to be readily compensated.

(4) Specimen stage: The specimen in a transmission microscope is usually supported on a thin circular metallic mesh or specimen grid 3 mm in diameter, to be illustrated more fully later. In the microscope this must be positioned at right angles to the optic axis, near the focus of the objective lens. In order that as much as possible of the specimen be available for examination it must be capable of at least ± 1 mm movement from its centred position on the microscope axis. It is usual to provide orthogonal x and y movements in the plane of the specimen grid. The function of the specimen stage is reception and translation of the specimen grid. The latter is mounted in a holder outside the microscope and introduced into the specimen stage through an airlock, so that the high vacuum in the column is not affected by insertion of the specimen.

There are two quite distinct types of specimen stage in common use, illustrated schematically in Figure 3.7. The specimen holder in a *top-entry* stage enters the bore of the objective lens from above; in a *side-entry* stage the holder, in the form of a flattened rod, is introduced into the pole-piece gap from the side. Each type of stage has certain advantages and disadvantages. Obviously a top-entry cartridge can be made to position the specimen at any desired level in the lens bore, whereas the side-entry arrangement is only suitable when the object plane lies between the pole-pieces. Much therefore depends on the design of the objective lens. Top-entry cartridges for a normal 3 mm grid cannot be less than 4–5 mm in diameter, so the bore of the upper pole-piece must be at least 6 mm to permit the necessary ± 1 mm of specimen stage movement. Again, top-entry stages require a sophisticated airlock mechanism, especially if several cartridges are to be loaded at the same time, as in some microscopes. Side-entry stages have the advantage of easier loading into the column and it is easy to provide for multiple specimens on one rod. There is very limited space between

Figure 3.7. Schematic diagrams of top- and side-entry specimen stages, showing also the principles of a multiple specimen rod for side-entry and a cartridge for top-entry stages. Because of the present use of very short focus objective lenses the space for specimen manipulation has become very limited, and some of the special stage movements have to be constructed with watch-making precision.

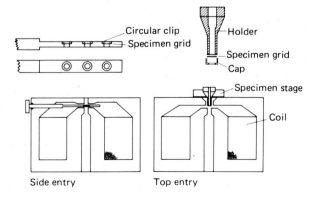

the pole-pieces of a very short focus lens, however, and operational techniques requiring the specimen plane to be tilted about the axis of the rod may be restricted by the narrowness of the gap. *Goniometer* stages are available with side- or top entry, permitting the specimen plane to be tilted up to $\pm 60°$, combined with rotation of $360°$ about the lens axis, or alternatively, simultaneous and independent tilts about two axes at right angles.

Sometimes the stage movements are motorised with hand- and foot controls so that specimens can be traversed, tilted and rotated simultaneously to obtain particular effects. In a *eucentric goniometer* stage the axis of tilting is arranged to pass through the field of view, so that the field of view and image focus remain unchanged when the specimen plane is tilted.

Special specimen stages and holders are available for many microscopes which enable a range of specimen manipulations to be carried out under observation in the microscope. Thus, there are stages and cartridges which allow specimens to be heated, cooled, stretched, compressed or magnetised under observation. In all cases it is of the utmost importance that the stages and mounted specimens be mechanically and thermally very stable, and the stage movements free from backlash, wobble and drift. The edge will be taken off a high-resolution micrograph if the specimen moves as little as 1 nm during a typical exposure duration of 4 seconds, which is a drift rate of less than 1 cm per year!

(5) Objective lens: This is the most important lens in the column. It has a focal length in the range 1–5 mm; in general the shorter the focal length the smaller the aberrations and the better the resolving power. Several different designs of objective lens were illustrated earlier (Figure 1.14). The objective lens is operated at or near to maximum strength for all but the lowest magnifications, when it may be switched off or run at very low power. In or near to the back focal plane of the objective lens is placed the *objective aperture* (Figure 3.8), an opaque metal disc with a central circular hole centred on the lens axis. This diaphragm intercepts electrons scattered by the specimen which would otherwise reduce the contrast of the final image because of the combination of chromatic and spherical aberrations. The angular aperture 2α of the objective lens is given by the expression $2\alpha = d/f$, where d is the diameter of the objective aperture and f is the focal length of the lens. Inserting the values $f = 4.0$ mm, and $\alpha_{opt} = 5 \times 10^{-3}$ radians for the objective lens we met in Chapter 1 (page 19) we find that the optimum objective aperture diameter for that lens is 40 µm. For most practical specimens, however, the objective aperture size is chosen for its effect on image contrast (smaller aperture, higher contrast and vice versa) rather than its effect on image resolution. For consistently working with low-contrast specimens, microscopy will be easier in a microscope with a longer focus objective lens.

The *stigmator* or astigmatism corrector (Chapter 1) is fitted between the objective aperture and the next lens in the column. With this the effects of imperfections in the objective lens (including the objective aperture) may be compensated to give a final image truly symmetrical about the optical axis of the microscope.

(6) Intermediate or first projector lens: In a straightforward three-lens column the intermediate lens generally operates over a wider range of focal lengths than the other lenses and may play either a reducing or a magnifying role. In a four-lens column an additional weak lens, usually called the diffraction lens, is fitted between the objective and the intermediate lens. This enables top magnifications of ×500 000 upwards to be achieved more conveniently, and introduces greater flexibility into the magnification range as a whole. Its main purpose, in the technique of selected-area electron diffraction, will be described in Chapter 6.

(7) Second projector lens: The overall magnification of the microscope may be varied in predetermined steps between hundreds and several hundred thousand diameters (Figure 3.9). All but the lowest values of magnification are obtained by altering the strengths of one or both projector lenses. The magnification is divided between the two projectors in such a way as to balance the distortions of the two lenses against each other and obtain distortion-free images at all magnifications.

With fixed object position and final image plane the achievement of predetermined fixed steps of magnification involves the pre-selection of projector lens current values. There may also be a slight re-adjustment of objective lens strength needed through the magnification range. On many recent microscopes all lens cur-

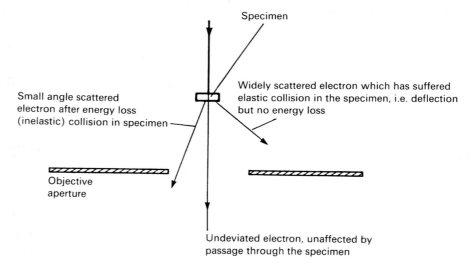

Figure 3.8. The objective or contrast aperture, situated after the specimen, removes from the image all electrons which have been scattered through more than a certain angle in passing through the specimen. It therefore modifies the degree of light and shade, or contrast, of the final image. It is positioned symmetrically about the beam axis in normal bright-field imaging.

rents, including the objective, are pre-programmed into a microprocessor so that the image is made to remain essentially in focus throughout the entire magnification range. Manufacturers describe this completely programmed operation as 'zoom focusing', by analogy with light-optical variable focus systems in which magnification is varied without the need to refocus. The electron-optical situation is made more complex by the rotation of the electron image about the lens axis as the lens powers are varied. Once the numerous calculations have been made for five and six magnifying lens columns, computer control of all lenses can be made to balance out rotations over a wide magnification range, or even combined to give controlled image rotation at will.

(8) Viewing chamber and fluorescent screen: The electron image becomes visible when it falls on a screen coated with a phosphor which luminesces green or yellow-green under electron bombardment. The operator selects and focuses the field to be recorded whilst observing this fluorescent image. The particle size of the phosphor may be up to 100 μm and the maximum magnification of the microscope should be such that the smallest resolvable distance is magnified to at least this dimension. The fluorescent screen or a suitably tilted auxiliary screen can be viewed through binoculars at an additional magnification of 8–10 times so that the fine detail may be seen

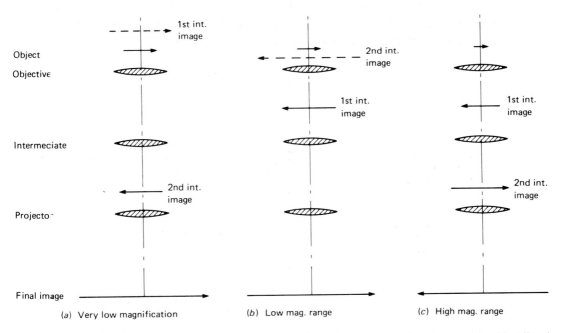

Figure 3.9. Magnification ranges of three-stage microscopes. (*a*) Objective lens weakened so that the object lies closer than the focus and the first intermediate image is virtual. (*b*) Objective lens operated at full strength, intermediate lens weakened so that second intermediate image is virtual. (*c*) All lenses form real, magnified images. Note that the image is reversed on switching from low- to high-magnification range.

and focused by the operator without undue eye strain. The viewing chamber usually has sufficient window area to permit several observers to see the fluorescent image at the same time, although not all through the binoculars. The microscope room is darkened so that image detail can be seen at full contrast without distraction. The windows of the viewing chamber will be made of lead glass to absorb x-radiation generated within the chamber by the high-energy electron beam.

A recent development, which not only relieves the TEM operator of the need to peer through binoculars at the fluorescent image in the viewing chamber but allows a higher level of lighting to be used in the microscope room, uses closed-circuit TV (CCTV) to transfer the image from the fluorescent screen in the viewing chamber to an external TV monitor. 'Real-time' imaging can be used for selection of field and for the operations of focusing and astigmatism correction. The TV camera used will frequently use a solid-state Charge-Coupled Device (CCD) as photocathode. The video-signal may be recorded on video-tape or a single frame can be 'grabbed' and recorded on magnetic or optical disc or printed on a video-printer, resulting in a near-instant, low-cost, hard copy. The simple equipment which is adequate for setting-up the microscope is not, however, capable of providing a high-quality recorded image. The various options for this are described after the traditional methods in the next section.

(9) Photographic recording camera: The fluorescent screen in the viewing chamber is hinged at its front edge; when it is tilted out of the electron beam the electron image falls upon a photographic plate or film situated below, to record a permanent image (Figure 3.10). The *camera* of a transmission electron microscope is merely a means of transferring a piece of electron-sensitive material from a light-tight box into the position where the electron image falls on it, and back into the same or another

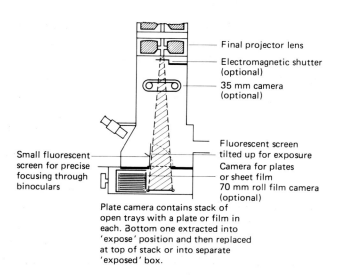

Figure 3.10. Image recording on the TEM (not to scale).

storage box whence it can be retrieved later for processing to a permanent silver image. Cameras with a capacity for up to 36 glass plates or 60 sheets of cut film are commonly used, as well as 70, 60 or 35 mm roll films. The 35 mm camera may be positioned higher up the column, where it records the same field of view at a lower magnification than the fluorescent image. As an added facility, when the screen is raised the electron beam can be intercepted by an electrically operated shutter with a range of pre-set opening times. Additionally, the electron current to the image plane can be monitored, and the shutter opened and closed automatically to give the correct electron exposure to the photographic material.

Any silver halide photographic emulsion may be used for recording electron images, but those generally employed for electron micrography are slow, fine-grained, blue-sensitive materials (Valentine, 1966, Farnell & Flint, 1975). Exposure to fast electrons results in a latent image which is developed to metallic silver in the normal way. Since the photographic material is actually introduced into the high vacuum of the microscope it is necessary to remove as much as possible of the water vapour and other volatile matter from the film before putting it into the camera. This is usually done in a subsidiary vacuum chamber, which may be situated for convenience adjacent to the column of the microscope. A method for putting the photographic material outside the vacuum is described later (page 77).

Accessories for electronic imaging or electron energy analysis may be fitted at this end of the column. The depth of focus of the final lens is very great and the positioning of the recording medium is not critical, except from the point of view of reproducibility of magnification.

Although photographic images are the traditional output from transmission electron microscopes, and are capable of producing the highest quality of result with the simplest of equipment, there are other competing systems for image recording, which will be described below.

The solid-state Electron Image Plate is effectively a replacement for photographic film (Mori et al., 1990). It is loaded and exposed in the same way as photographic film but, instead of being developed chemically, the image is 'read-off' the surface of the plate by a scanning light beam in a special processor, giving a digital image which may be stored in a memory or printed out. The Image Plate may be 'wiped clean' and re-used. Image Plates are more sensitive than a photographic emulsion, and can record a wider brightness range (Shindo et al., 1990). The special processing equipment is expensive, and is probably capable of further development (Burmester et al., 1994).

The more common alternative is to use electronic (digital) imaging. For the highest-quality digital image the electron image formed by the microscope falls upon a combination of scintillator, fibre-optic interface plate and high-resolution, slow-scan, cooled-CCD camera (Krivanek, 1992), which produces a high-quality digital image suitable for storage, processing and analysis, but doesn't require chemicals, water or a darkroom – only considerable investment in a technology which is still developing. This subject will be discussed further in Section 3.6 and in Chapter 5.

Power supplies

The components which have been described above together constitute the electron-optical column, which is normally in a single unit with the console and control panels. In some cases the electrical supplies to the lenses and the electron gun are generated in a separate cubicle situated several metres away from the column and console; recent instruments with fully solid-state electronic circuitry have all the power supplies and the vacuum pumps integral with the console and require electrical mains and cooling water supplies to be provided directly to this. The weight of a complete high-resolution microscope with its power supplies and vacuum pumps may be between 650 and 1500 kg.

The environment in which it is sited can have a fundamental influence on the quality of microscopy obtainable from an electron microscope. The performance of all microscopes is adversely affected by more than a minimum level of vibration and magnetic field strength, and a room with a solid floor and controlled temperature is necessary if the instrument is to give of its best.

3.2.4 The commercial TEM

There are currently more than a dozen different models of high-resolution transmission electron microscope available commercially throughout the world, coming from five major manufacturers. These are the latest in lines of development which have produced fifty different models over as many years.

Compact, simple to use microscopes well suited for examining and recording micrographs of thin biological sections will have the basic column described above, with double condenser illumination, and three or four imaging lenses. The operating voltage will be 100 kV, possibly with lower voltages available. Image recording will be available on sheets of cut photographic film, but the option of a TV camera and video imaging may also be offered. Figure 3.11 shows a cross-section and photograph of a modern 100 kV microscope.

Above this basic level there is the possibility of more and higher operating voltages, more condenser and/or imaging lenses providing a greater flexibility of operation, specimen stages with a range of tilts and/or rotations, and accessories for a wider range of operating modes, including electron diffraction, x-ray analysis, and specialised scanning and dark-field operation.

The Intermediate Voltage TEM, with electron energies of 200, 300 or 400 keV, provides the materials scientist with greater penetrating power for denser specimens and, since the electron wavelength reduces as the square root of voltage increase, there is the possibility of higher resolution operation.

The ultimate in voltage is the High Voltage TEM (HVEM) to be described later, in which one million volts are used for accelerating the electron beam. A comparatively small number of these microscopes are in use around the world, since they are very expensive and large.

All but the simplest TEMs are offered with brighter electron sources, necessary for working at the highest magnifications and resolution (Figure 3.12).

Features which naturally differ between manufacturers are the styling of the instruments and the layout of controls. In overall layout, however, most microscopes

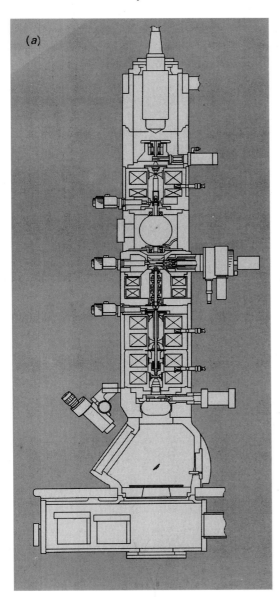

(a)

Figure 3.11. High-resolution TEM: (*a*) Cross-section, (*b*) (*overpage*) Photograph of the same microscope, JEOL JEM 1010, 40–100 kV, ×600–×500 000, 0.45 nm point resolution. (By courtesy of JEOL Ltd.)

are made to a similar pattern; a flat-topped desk with a knee-hole, an upright instrument panel at the back and the microscope column rising vertically from the desk, usually the middle. The operator looks diagonally downwards through a window at the fluorescent screen. Either a very large front window or several side windows are provided so that several people can watch the image simultaneously. If the microscope has computer control, or uses closed-circuit TV to display the image, there will be one or more imaging displays and a keyboard, and the traditional TEM operating procedure will be modified accordingly.

Operational aids

(1) Multiple specimens: Of the features designed to make the operator's life less onerous the possibility of loading several specimens at one time has already been mentioned. This is possible with both top- and side-entry specimen stages. However easy it may be to reload specimen holders and insert them in the microscope it is an

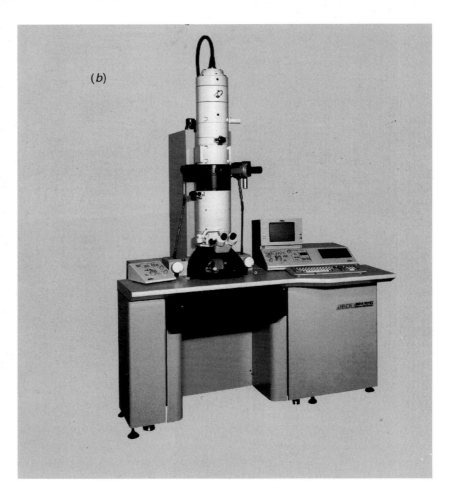

Figure 3.11 (*cont.*)

attraction to be able to compare two specimens without having to reload the same holder several times.

(2) Wobbler focusing: Zoom focusing has also been referred to; once the specimen has been focused at one magnification the image stays close to exact focus at all other magnifications. This system is now widely used, as also is the 'beam wobbler' focus-

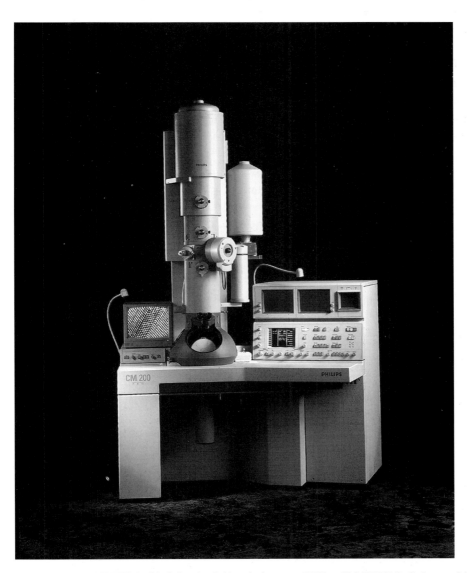

Figure 3.12. 200 kV TEM with Schottky field emission gun, Philips CM 200FEG. Column with five illumination and five imaging lenses under computer control permits analytical electron microscopy on a nanometre scale. The basic specification provides 20–200 kV, ×25–×750 000, 0.27 nm point resolution. (By courtesy of Philips Electron Optics.)

ing aid originally devised by le Poole for Philips microscopes (van Dorsten *et al.*, 1950). This latter device rocks the illuminating beam alternately between two tilted positions on opposite sides of the optical axis at a frequency of 10 or 20 hertz. The pivot point of the beam tilt is in the specimen plane. If the objective lens is focused exactly on the specimen plane the wobbler has no effect on the image. If, however, the objective lens is focused on a plane above or below that of the specimen (*under-* or *overfocused*, respectively) the final image oscillates between two positions equally displaced on either side of the normal 'rest' position. The closer the objective lens is to exact focus the smaller is the double-image effect, and the operation of focusing is simply one of switching on the wobbler and adjusting the objective focus control until the image is reduced to a single one (Figure 3.13). The wobbler, with an abrupt (square-wave) oscillation of illumination angle, is effective over the whole magnification range, although its amplitude should be reduced at higher magnifications. Paradoxically, most micrographs are not taken with the objective exactly focused at the specimen, but are slightly 'underfocused'. However, the wobbler gives a good starting point and the operator can underfocus or not according to his experience and the requirements of the micrograph. A microprocessor may make this focus adjustment automatically. The wobbler image shift can be made the basis for automatically focusing a TEM.

(3) Vacuum systems: As has already been mentioned the production and maintenance of a high vacuum in the electron microscope gun and column is essential, and details of just what is involved are given elsewhere in the book (Appendix 1). On earlier microscopes the operator was responsible for operating pumps and valves in the correct order, but most modern instruments have automatic control of these operations so that all the operator has to do is turn the cooling water on and press the *start* button.

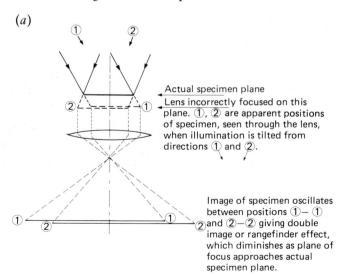

(*a*)

Actual specimen plane
Lens incorrectly focused on this plane. ①, ② are apparent positions of specimen, seen through the lens, when illumination is tilted from directions ① and ②.

Image of specimen oscillates between positions ①– ① and ②–② giving double image or rangefinder effect, which diminishes as plane of focus approaches actual specimen plane.

Figure 3.13. Beam wobbler focusing aid. (*a*) Schematic diagram of principle of operation. (*b*) The wobbler in practical use.

There is an increasing emphasis on the quality of vacuum inside the microscope column. Although routine high-resolution transmission microscopy can be carried out with a simple vacuum system, a lower pressure and a cleaner environment are essential for work in ultra-high-resolution and analytical instruments, particularly if brighter electron guns are used. In these instruments manufacturers ensure that the specimen and electron gun are maintained at up to a thousand times lower pressure than the 'dirty' end of the column. One manufacturer sought to eliminate a major source of contamination from the column by placing the photographic material outside the vacuum and relaying the fluorescent image to it via a plate composed of very fine optical fibres (Gütter & Menzel, 1978)(Figure 3.14).[N.B. This trans–fibre-optic photography (TFP) camera enables 'instant' transmission micrographs to be made using Polaroid-type materials]. The trend towards TV displays and digital

(*b*)

Out of focus

In focus

Wobbler off   Wobbler on

imaging has the same effect of taking photographic materials outside the vacuum, and also enables 'instant' micrographs to be obtained from a video printer.

(4) Other operational aids: Other ways in which manufacturers have eased the actual 'driving' of microscopes are by replacing function switches by push-buttons, and by automating the recording of micrographs. The operator has only to select the field of view and magnification, trim up the focus and astigmatism and press the *expose* button. The instrument will then advance an unexposed film or plate, correctly expose it to the electron image (printing on it at the same time the exposure number, operating voltage and magnification) and will return the exposed film to a magazine and restore the image to the viewing screen. All manufacturers will now supply a 'low dose unit' or 'minimum dose focusing' accessory designed to help the microscopist whose specimen would be damaged or even destroyed by the electron bombardment received during setting up and recording a micrograph. The accessory generally provides a controlled beam deflection so that focusing and astigmatism correction are carried out on an adjacent area of the specimen, the beam only being switched on to the selected area for the minimum duration necessary for an exposure to be made. In the extreme case the area recorded on the micrograph may never even have been inspected by the operator.

Finally, some manufacturers are helping the microscopist by providing an indication at desk level of the specimen stage position at any time, so that he can see readily how his present field of view is related to the specimen grid as a whole. The information may be presented as an illuminated screen with a roving spot to represent the

(*a*)

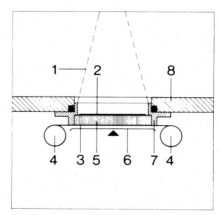

Figure 3.14. Trans-fibre-optic photography (TFP). (*a*) Schematic diagram and (*b*) example of micrograph recorded externally through a fibre-optic plate. (Diagram and micrograph by courtesy of LEO Electron Microscopy Ltd. The micrograph is of an ultra-thin section of the cardiac muscle of a rat.)

1 Electron beam
2 Transparent fluorescent screen
3 Roll film
4 Film spools
5 Fibre optics plate
6 Film contact pressure plate
7 Retaining flange with vacuum seal
8 Viewing chamber

(*b*)

beam position on the specimen, or there may be a digital read-out of the x, y, coordinates of the stage position. In a logical extension of the latter system the coordinates of noteworthy features found on an initial survey of the grid may be stored in memory and recalled for more detailed study later. If the stage movements are motorised, with stepping motors, the whole recall operation can be automated.

It is of course possible to take first-class micrographs on an instrument without any of the aids and conveniences mentioned above, provided the instrument has adequate resolving power and stability for the job in hand. It requires experience, however, and the manufacturers' aim in providing these conveniences is to make it easier for the non-specialist microscopist to take good micrographs for himself, with minimum risk of spoiling either the picture or the instrument in the process.

Computer-aided microscopy

The banks of knobs and switches on early electron microscopes have gradually been reduced in number by replacement by push-buttons and multi-function controls, e.g. the mono-knob focusing control in which a single rotary control can be switched between coarse, medium and fine sensitivity of operation.

The use of localised microprocessors to control functions of the microscope leads naturally to the software control of microscope operation, even to the extent of reproducing pages from the operating manual on the small visual display unit (v.d.u.). Once programmed into the software, complex operations like the control of multi-lens columns and the setting up of conditions for specialised operation (e.g. tilted-beam dark-field, or conical scan) are just a matter of selecting a function from a menu and pressing a button. The great advantage of this situation is that the more specialised techniques are more likely to be used, with their potential benefits to microscopy.

An interesting concept made possible by computerisation is that operation can be customised for individual operators of the same instrument, according to status or operational prowess. For example, changing the operating voltage or beam current, or correcting astigmatism may not be offered to a beginner, and the complete range of adjustments may only be accessible to an authorised service engineer.

Finally, changing the practices of half a century, viewing the final image of a TEM on a video monitor instead of through binoculars on a small fluorescent screeen permits a higher level of lighting to be used in the microscope room. For a discussion of non-photographic imaging from both SEM and TEM see Section 3.6. Saxton & Chang (1988) provide a short introduction to the concept of automated setting up and focusing of the TEM via the TV image.

The energy-filtering microscope

Two commercially available transmission microscopes which incorporate electron energy spectrometers in their columns are made by LEO (formerly Zeiss). The earlier one, containing a Henry–Castaing energy filter (Castaing & Henry, 1962), is shown in Figure 3.15. The beam is bent through 90° by a magnetic sector and reflected by a mirror electrode back through the same sector. The electron beam is dispersed according to the electron energy, and electrons which have lost a chosen amount of energy in the specimen can be allowed through a diaphragm to form the final image. This design of energy analyser is limited to 80 kV.

The second design, called an Omega (Ω) filter after the shape of the electron path through it, performs a similar function but can be fitted to higher energy columns. The version fitted to a 120 kV column is shown in Figure 3.16. These filters can be used either as spectrometers or filters. With suitably thin specimens the techniques of EELS or PEELS can be used for chemical analysis of the specimen (see Chapter 6 for more details). Alternatively, they can be used to counteract the effects of chromatic aberration on images of specimens which are on the thick side. In such cases the clarity of transmission images and diffraction patterns are greatly improved.

An Ω-filter may be fitted in the column of an HVEM and used as spectrometer or

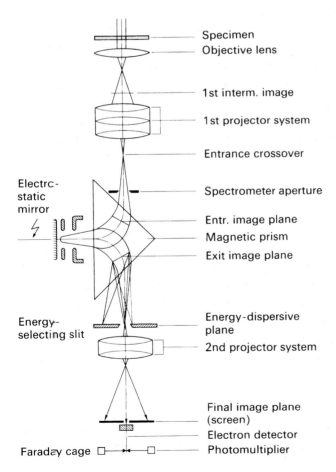

Figure 3.15. Essential components of energy-filtering TEM (LEO 902) using Castaing magnetic prism to form an electron energy spectrum. Electrons for final image may be selected by energy-selecting slit. Alternatively, by withdrawing the slit the energy spectrum can be projected on to the viewing screen for rapid assessment of peaks. (By courtesy of LEO Electron Microscopy Ltd.)

(*a*)

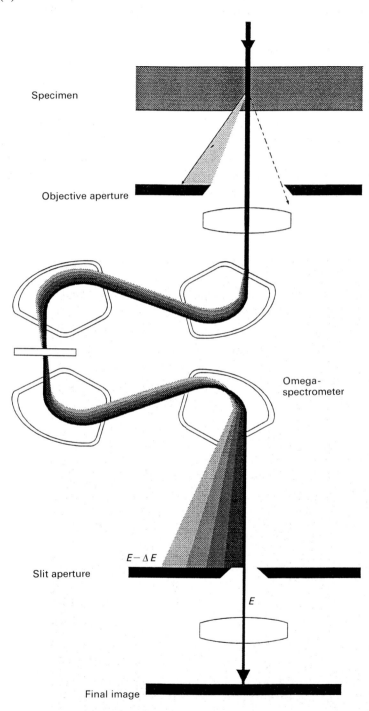

Specimen

Objective aperture

Omega-
spectrometer

$E - \Delta E$

Slit aperture

E

Figure 3.16(*a*). Omega energy
filter (schematic) in LEO 912
column; (*b*) the filter is in the
rectangular box on the left-hand
side of the column. (By courtesy
of LEO Electron Microscopy
Ltd.)

Final image

energy filter to take advantage of the improved visibility of absorption edges in electron energy loss spectroscopy at 1 MeV (Zanchi *et al.*, 1983).

3.2.5 The analytical TEM

Transmission electron microscopes in which the capabilities for crystallographic and elemental analysis have been highly developed are sometimes classified as analytical electron microscopes (AEM).

The first such instrument was designed by Duncumb (Cooke & Duncumb, 1969) and produced commercially by the AEI as EMMA-4 (Figure 3.17). This was a 100 kV high-resolution TEM to which two vertical x-ray spectrometers were fitted so that high-resolution elemental analysis could be carried out on normal TEM specimens. A third condenser lens was incorporated into the illuminating system,

(*b*)

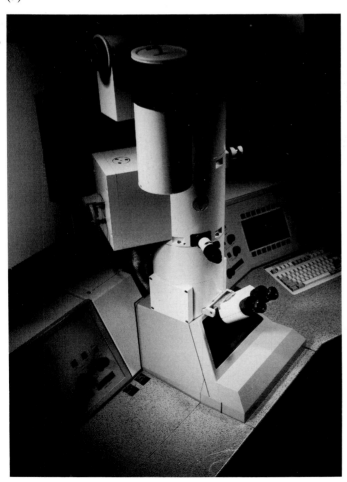

enabling the beam to be focused down to a 0.2 μm spot for selecting features for microanalysis.

More recently, it has been found much simpler and cheaper to mount a solid-state Si(Li) x-ray energy spectrometer on standard high-resolution TEMs with scanning capabilities. The Si(Li) detector may be mounted very close to the specimen to maximise the collection efficiency of x-radiation from the specimen (Figure 3.18). At the same time the smallest area which can be analysed has been reduced to a nanometre in the TEM and even smaller in the field emission STEM. The analytical TEM is now able to combine transmission and surface microscopy with crystallographic and elemental microanalysis down to the nanometre scale. A more detailed description

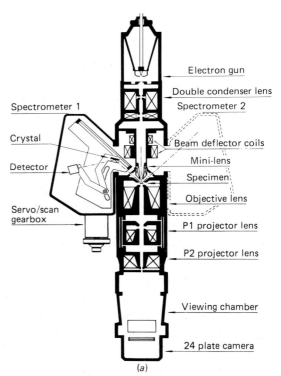

(a)

Figure 3.17. Diagrammatic cross-section of the AEI EMMA-4 combined transmission microscope and microanalyser. (By courtesy of Kratos Ltd.)

Figure 3.18. Solid-state x-ray detector with high acceptance angle in analytical TEM. The spacing between upper and lower pole-pieces is only 9 mm. Also fitted into the gap but not illustrated are a backscattered electron detector and the objective aperture. (By courtesy of Philips Electron Optics.)

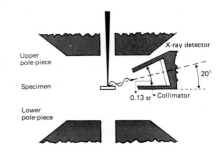

will be given later in this chapter, after the concept of SEM and STEM attachments has been introduced.

3.2.6 The high-voltage electron microscope (HVEM)

The 100 or 125 kV high-resolution transmission microscope gives its best performance with specimens which are very thin by light microscopy standards; a few nm thick for ultra-high (0.5 nm or less) resolution work and not thicker than tens of nm for good, sharp, fine detail. Thicker specimens than this, e.g. from 0.1 μm upwards, start to cause difficulties because of two effects. Firstly, the specimens scatter electrons so strongly that not enough are available to form a bright image even at modest magnifications. Secondly, those electrons which do come through have undergone multiple inelastic collisions within the specimen, so that the sharpness of image is degraded because of the chromatic aberration of the lenses. Apart from the obvious remedy of preparing thinner specimens, the microscopist wishing to examine thick specimens at high resolution can either remove inelastically scattered electrons from the beam with an electron energy filter or use a higher voltage microscope, which will simultaneously provide brighter illumination and more penetrating electrons.

The latter option was the initial stimulus for the development of megavolt electron microscopes. While the norm for TEM operation was 100 kV, a number of 1 MV TEMs were built in the UK and Japan, and in France 3 MV was used to evaluate the benefits of going above 1 MV. Altogether, about 60 microscopes operating at 500 kV and above were built worldwide. The resulting experience showed that there was indeed a worthwhile improvement in going to 1 MV. Experimental results in the range 1–3 MV show that there is little further gain in penetrating power through heavier materials, although lighter elements may be penetrated up to 10 μm or more at the highest energies.

On the debit side, thicker specimens may be damaged by heating or ionisation during examination; at high enough energies there can be serious damage to specimens when the electrons knock complete atoms out of position. A further disadvantage is that the size and cost of the 1–3 MV instruments make them suitable only for large research organisations or central, shared, facilities.

Consequently, although it is still possible to obtain 1 MV instruments, there is greater commercial potential in the Intermediate Voltage TEM operating at 200, 300 or 400 kV.

The electron-optical column of a high-voltage microscope follows the same arrangement as a four-imaging-lens TEM except that above about 150 kV the normal electron gun has to be replaced by a more elaborate electron accelerator. The accelerator and high-voltage generator increase in scale with the voltage and in a 1 MV design are enclosed in separate tanks of inert gas. The accelerating voltage is usually variable in steps of 100 kV between 200 kV and 1500 kV so that the optimum voltage may be chosen for each specimen to be examined. A resolution of 0.34 nm or less is obtained at 1500 kV.

The size of the electron-optical column in a megavolt microscope is larger than that of a 100 kV instrument for several reasons. One is that stronger lenses are required to focus the faster electrons; another results from the dangerous x-radiation emitted when 1 MeV electrons strike any solid surface in the microscope column. The stronger magnetic fields in the lenses are obtained by using higher currents, more turns of wire, or both, and so the external dimensions of the lenses are at least doubled. The need to screen the operator from x-rays generated in the column means that the lenses must be encased in lead. X-radiation from the final viewing screen is absorbed by a 25 cm thick lead glass window. Radiation monitors sited at strategic positions on the column keep a continuous check on the x-ray level and will turn off the electron beam if a safe level is exceeded during operation.

The great size of the 1 MV microscope means that a number of operations normally carried out directly by the operator on a 100 kV instrument, e.g. adjustment of diaphragms across the optical axis and specimen movement and tilt, can often be performed more conveniently by servomotors controlled remotely from the operator's desk. Adjustments to the electron gun during operation must of course be carried out by remote control, since it is at high tension (HT) negative potential; meters are read by closed-circuit television. A photograph of the column and desk of the EM 7 is shown in Figure 3.19. The column of the 3.5 MV microscope at Toulouse (Dupouy, 1973) is shown in Figure 3.20.

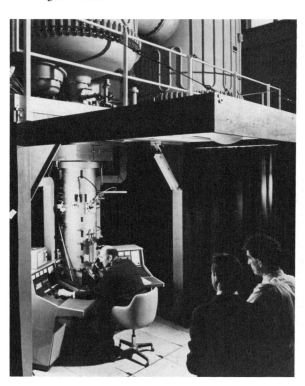

Figure 3.19. Operating the EM 7 megavolt microscope. Parts of the tanks housing the high-voltage generator and electron accelerator are seen above the top of the lens column. (By courtesy of Kratos Ltd.)

The large size and weight of a megavolt microscope with associated power supply makes it difficult to accommodate in a normal sized laboratory; when possible a special building should be provided for it. Nevertheless, in spite of its formidable appearance and size, the actual operation of the instrument is very similar to that of a conventional microscope.

3.2.7 The atomic resolution microscope (ARM)

If the main objective of transmission microscopy is to resolve the finest possible detail in specially prepared specimens it is advantageous to use very short wavelength illumination, an objective lens with very low spherical aberration, and a column and supplies with extremely high mechanical and electrical stabilities. For regular working with atomically resolved specimen detail, versions of commercial 200–400 kV microscopes with specially low C_s values are available (Figure 3.21).

Experience has shown that to attain the highest resolutions extreme care has to be taken to isolate the microscope from mechanical vibrations. Even the acoustic vibrations of the column due to speech can have a degrading effect on performance.

Figure 3.20. Column and desk of the Toulouse 3.5 MV electron microscope. The column has double condensers and four imaging lenses. (By courtesy of B. Jouffrey.)

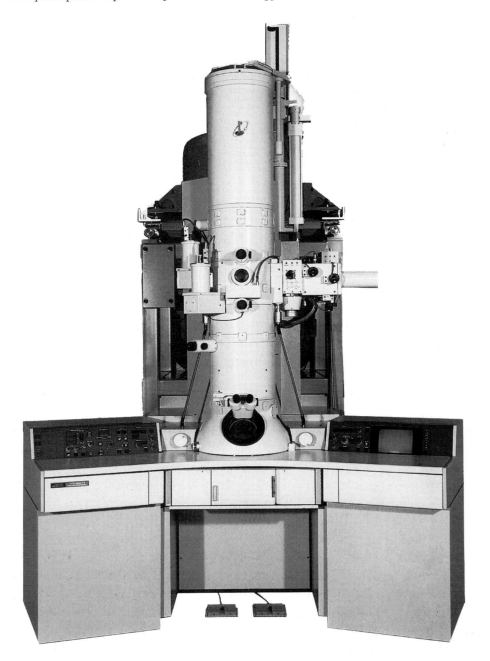

Figure 3.21. The JEOL 4000 EXII Atomic Resolution Microscope. 200–400 kV, guaranteed resolution 0.165 nm (point), 0.10 nm (lattice), C_s = 0.9 mm, C_c = 1.6 mm. (By courtesy of JEOL Ltd.)

Valuable pioneering work was carried out on a 600 kV instrument constructed at Cambridge University, based on components from the EM 7 1 MV microscope (Cosslett *et al.*, 1979).

A small number of atomic resolution microscopes, aiming at a point-to-point resolving power of 0.1 nm, have been installed in Japan and Germany (Phillipp *et al.*, 1994). These can all operate at up to 1250 kV, in order to gain the advantage of the shortest-wavelength illumination.

3.3 The scanning electron microscope (SEM)

Historically the basic principle of the scanning electron microscope is practically as old as the transmission instrument (Knoll, 1935) and various experimental scanning microscopes were constructed in the late 1930s and early 1940s (McMullan, 1995). In 1942 a scanning microscope with resolving power of 50 nm was operational in the RCA Laboratory at Camden, NJ, USA (Zworykin *et al.*, 1942) but it suffered from deficiencies in the collection and amplification of the signal from the specimen. It was some years later, mainly as a result of improvements in signal collection and amplification systems in the Engineering Laboratory at Cambridge University (Oatley *et al.*, 1965; Oatley, 1972), that the scanning microscope as we know it today was made possible. These improvements were embodied in an almost foolproof commercial instrument by Cambridge Instrument Company (subsequently Leica Cambridge Ltd and now LEO Electron Microscopy Ltd) and introduced as the Stereoscan Mk 1 in 1964 (Stewart & Snelling, 1964). Since then the potentialities of the scanning principle in electron microscopy and analysis have been appreciated by manufacturers and users throughout the world, and today the potential purchaser of an SEM has a range of over 40 different models from which to choose. Major changes in operating procedures have come with developments in computer systems and digital image storage, whilst field emission guns and short-focus lenses have made possible resolving powers at the nanometre level.

3.3.1 Basic principles

The basic principle of the SEM was introduced in Chapter 1. A very fine 'probe' of electrons with energies from a few hundred eV to tens of keV is focused at the surface of the specimen and scanned across it in a 'raster' or pattern of parallel lines (Figure 3.22). As described in Chapter 2 a number of phenomena occur at the surface under electron impact, and a 'signal' due to electrons or radiations can be collected for every position of the incident electron probe. This signal is amplified and used to vary the brightness of the trace on a cathode-ray tube being scanned in synchronism with the probe. There is thus a direct positional correspondence between the electron beam scanning across the specimen and the fluorescent image on the cathode-ray tube.

The magnification produced by a scanning microscope is the ratio between the dimensions of the final image display and the field scanned on the specimen; e.g. if the electron probe scans an area 1 mm square on the specimen and the output is displayed on a screen 100 mm square, then the linear magnification is 100 times. Changes in magnification are brought about by altering the extent of scan on the specimen whilst keeping the size of the display constant. In practice the magnification control on the microscope actually varies the angle through which the beam is deflected and the linear extent of the scan depends on the *working distance* of the specimen from the final lens. In a typical microscope the magnification range may be ×20 to ×100000 at a working distance of 11 mm, but using working distances between zero and 35 mm the actual magnification range will be extended to ×10 to ×200000. (N.B. The reader may like to bear in mind that the minimum necessary magnification to utilise a resolving power of 5 nm would be about 40000 times).

Resolving power of the SEM

In the SEM object detail is sampled point by point and the resolving power of the instrument is directly determined by the smallest area which it can sample, i.e. by the diameter of the electron probe. Minimising the probe diameter calls for a system

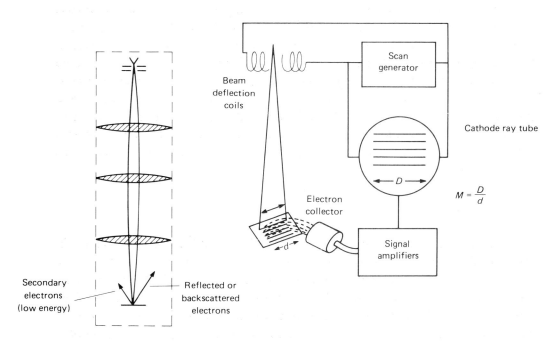

Figure 3.22. The principle of the scanning electron microscope. A succession of lenses produces a finely focused electron spot on the specimen. A current of emitted electrons is collected, and amplified and used to modulate the brightness of a cathode-ray tube. The electron spot and the c.r.t spot are scanned in synchronism across the specimen and the tube face, respectively. A microscope results whose magnification is the ratio of scanned dimensions on specimen and c.r.t.

of lenses with small aberrations and high electrical stability. In practice the smallest probe which can be formed with enough electrons in it to be useful depends on the focal length or working distance of the final probe-forming lens, the brightness of the electron gun and the operating voltage. For a conventional tungsten–filament gun and a working distance of about 10 mm the smallest usable probe size is about 3–4 nm, and the secondary electron resolving power between 3.5 and 7 nm. Better resolving power is only obtainable by working with a stronger lens or brighter electron source. Both of these are done in scanning microscopes, as will be seen later. With conventional lenses and a field emission electron source, resolving powers of 1.5 nm at 30 kV are attained, and with novel electron optical configurations including 'in lens' specimens this reduces to 1 nm or below. It is perhaps ironic that scanning microscopes with the best resolving powers use electron-optical designs originally developed for scanning attachments on transmission microscopes.

As with the TEM we must make a distinction between the resolving power of the instrument and the resolution of fine detail of a practical specimen. In the SEM the practical resolution in secondary and backscattered electron images will generally be poorer than the probe diameter would suggest, because the electron beam penetrates and diffuses sideways in the specimen, releasing electrons from a wider area than is actually illuminated by the focused probe. Practical resolution of detail in bulk specimens is dependent on the material of the specimen, its orientation in the microscope, the working distance and the signal used for the final image. It is not a quantity which is amenable to the conventional Airy disc/Rayleigh criterion for microscope systems and there is as yet no objective standard for measuring or comparing the resolving powers of scanning microscopes. The minimum width of gap visible between adjacent features, such as oxide particles in magnetic tape or islands in an evaporated gold film, is frequently used to demonstrate a particular degree of edge resolution, as was illustrated in Figure 1.20(*c*).

3.3.2 The SEM in practice

Electron-optical column

The column of the SEM is essentially a double condenser illuminating system as in the TEM followed by a third lens which projects the fine illuminating spot on to the specimen, over a working distance between a few millimetres and a few centimetres. There are no further lenses, only electron or radiation collectors (detectors). The essential parts of the SEM column are shown schematically in Figure 3.23, together with a cross section of an actual SEM. The components are numbered as follows:

- *(1) Electron gun:* This has traditionally been a triode gun with a tungsten thermionic filament but may alternatively be a brighter emitter such as lanthanum or cerium hexaboride or a Schottky field emission gun (Oatley, 1975). (See Appendix 4 for a detailed comparison). It will deliver a total electron current of up to 250 μA at energies adjustable between less than 1 keV and 30 keV (sometimes up to 40 or 50 keV). The ability to vary kV is a very useful feature, as will be seen later. However, a gun optimised to work at 30 or 40 kV will not be as effective at lower voltages, either in terms of brightness or probe size. In addition, at low voltages the beam will be more sensitive to the effects of stray fields in the column. The combined effects on a typical instrument are such that a guaranteed resolution of 1.5 nm at 30 kV may be reduced to 7 nm at 1 kV.

(2) Double condenser lens system: A 10^4–10^5 times demagnified image of the 'crossover' of the electron source is formed close after the second lens.

(3) Final or third condenser lens (usually called the objective lens): This lens projects the diminished image of the electron crossover as a spot focused on the surface of the specimen (Figure 3.24). The solid angle of the focused beam is defined by a diaphragm immediately following the lens. The size of this diaphragm and the

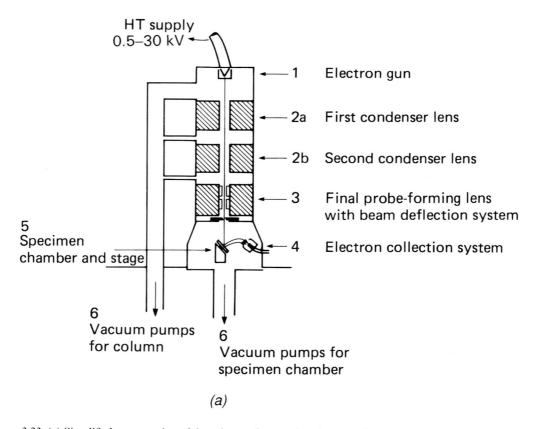

(a)

Figure 3.23. (*a*) Simplified cross-section of the column of a scanning electron microscope; numerals refer to descriptions in the text. (*b*) A commercial realisation of (*a*) (JEOL JSM 6300; by courtesy of JEOL Ltd).

working distance together determine the depth of focus of the probe beam (Figure 3.25 *a,b,c*).

The bore of the final lens also contains the stigmator, necessary to ensure that the scanning spot is truly circular, and the deflection system responsible for scanning the electron probe. Pairs of beam deflection coils are supplied with current from the *scan generator*, an oscillator which proves separate frequencies for the *line scan* (*x*-deflection) and the *frame scan* (*y*-deflection). The simplest scanning systems use single sets of coils for both *x* and *y* deflections (single-deflection scanning). Most SEMs, however, use the more refined double-deflection system in which the beam pivots in the plane of the final aperture, as illustrated in Figure 3.26.

The outside shape of the objective lens frequently is a truncated cone, in order to permit the examination of highly tilted large specimens such as semiconductor wafers and to give improved access for x-ray detectors.

The aberrations of the objective lens, and hence the smallness of electron probe which can be focused, decrease as the lens power increases. For the highest resolutions a modified lens and electron collector are used, which will be described in Section 3.3.4.

(b)

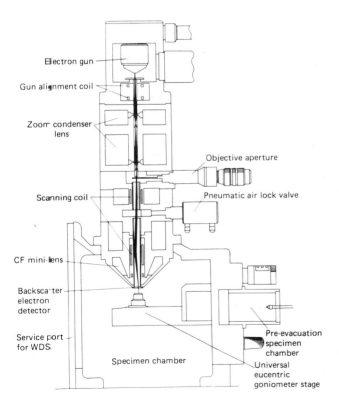

Figure 3.24. Probe-forming system of a scanning
microscope (diagrammatic). In (*a*) the demagnifying
lenses C1 and C2 are only weakly excited; the final
probe has a large diameter and the current in it is also
high. In (*b*) the illumination is strongly demagnified
into a small spot on the specimen, but it can be seen
that only a small proportion of the available electrons
actually end up in the spot. The resolving power of
(*b*) is better than (*a*) but the signal from the specimen
is reduced, giving a lower signal-to-noise ratio in the
image.

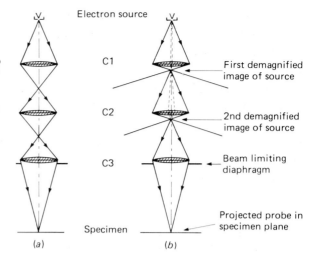

Figure 3.25. Effect of final aperture and working
distance on depth of field of scanning microscope.
The depth of focus in a probe system is the distance
D along the optical axis between planes in which the
probe diameter reaches an arbitrarily chosen value
(e.g. twice the minimum spot size, as in the diagram).
D can be increased by reducing the angle of the
probe, e.g. by using a smaller final aperture, as in (*a*)
and (*b*). Even with a smaller aperture the depth of
focus becomes shorter again if the working distance is
reduced (*c*) and the probe attains the same angle as
previously with the larger diaphragm.

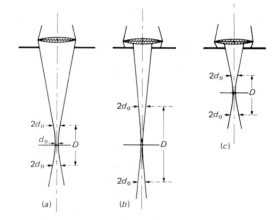

   The distance *D* described here as the depth of focus
of the final lens is in fact the depth of field of the
scanning microscope regarded as a magnifying system.
Readers who use cameras will be familiar with the effect
of lens aperture on depth of field, which is the same
effect as we have here. In considering the minimum size
of the electron probe it must be remembered that
because of the spherical aberration of the final lens the
probe is focused visually to a circle of least confusion
larger than the ideal demagnified spot. Spherical
aberration becomes smaller as the lens becomes
stronger and the working distance less. In some SEMs
(e.g. those with in-lens specimen stages) the working
distance is fixed at a low value to minimise spherical
aberration. In others, a wide range of working distances
is available and the resolving power consequently varies
according to the working conditions.

(4) Electron collection system: The secondary electron (SE) detector common to all SEM systems is based on the design of Everhart and Thornley (Everhart & Thornley, 1960), already described in Chapter 2 and Figure 2.8. As fitted in the original 'Stereoscan' this was mounted on the removable specimen stage (see next section, Figure 3.27). It consisted of a Perspex (Lucite) cylindrical light-guide with a hemi-spherical end coated with a phosphor and a thin aluminium film maintained at + 12 kV to attract electrons whose impact causes the phosphor to scintillate. The Perspex rod guides the light impulses to the cathode of a photomultiplier, which re-converts them into an amplified electrical signal. A metallic mesh in front of the scintillator can be biased either + 250 or −30 eV so that secondary electrons may be attracted or excluded in addition to the current of high-energy backscattered electrons. The efficiency of collection of backscattered electrons was very poor. In many SEMs the bias is continuously variable between the two extremes so that the proportions of SE and BSE can be varied. The detailed construction of the scintillator and 'light-pipe' have been modified by SEM manufacturers, and they are usually mounted permanently inside the specimen chamber, but the basic principles are unchanged. Electron collection in a high-resolution SEM is described in Section 3.3.4.

Dedicated BSE detectors, which may be of the Robinson or Autrata scintillator types, solid-state silicon annuli or channel-plate electron multipliers, are positioned

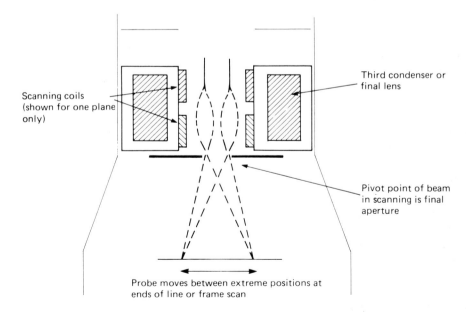

Scanning coils (shown for one plane only)

Third condenser or final lens

Pivot point of beam in scanning is final aperture

Probe moves between extreme positions at ends of line or frame scan

Figure 3.26. Principle of double-deflection scanning system. A set of coils tilts the electron probe between two extreme angles on either side of the axis of the final lens. The coils are arranged so that the tilting action takes place about a point in the plane of the final aperture. Two sets of coils are provided, to give orthogonal frame and line scan deflections.

between the specimen and the final lens, where they collect a much higher proportion of BS electrons emitted from untilted specimens than the E–T detector could. Two- or four-quadrant BSE detectors can be used to obtain topographical as well as compositional information (illustrated in Figure 5.24(c), page 223).

(5) Specimen chamber and stage: This chamber accommodates the specimen holder and mechanisms for manipulating it, as well as detectors for the various emissions (electrons, light, x-radiation) which will form the scanning microscope image in its different operating modes. The specimen stage may be fitted into the chamber in several different ways. It may be retractable from the chamber on rails, like a drawer, fixed permanently inside the chamber and loaded through an airlock or, as is increasingly the practice, it may be attached to one face of the specimen chamber block and hinged to open like a door. It can also be incorporated into the bottom part of the chamber, and be removable with it. Whatever the design, the highest mechanical rigidity is necessary, or the high- resolution performance of the microscope may be adversely affected. Two examples are illustrated in Figure 3.27.

A five-axis goniometer specimen stage is able to move in three mutually perpendicular directions, *X, Y, Z*, the last being parallel to the axis of the column, with specimen rotate and tilt. This allows every point on the specimen surface (within the size limitation of the particular stage) to be brought under the electron beam and examined. A scanning micrograph shows the specimen as it would appear to an observer situated in the bore of the final lens. The specimen tilt axis is such that the tilted specimen faces the secondary electron collector. The action of tilting introduces perspective into the micrograph and increases the electron signal to the E–T detector, improving the quality of the image.

In a simple goniometer specimen stage the shifts, tilts and rotation are unrelated to the optic axis of the column, and any movement of the specimen is liable to move it out of focus and out of the field of view. A *eucentric goniometer stage* is constructed so that once the specimen height has been adjusted to the eucentric point, all successive tilts and rotations take place about the centre of the field of view. This is illustrated schematically in Figure 3.28.

Special accessory specimen stages are available which allow the specimen to be heated, cooled or mechanically stressed whilst under observation.

The specimen chamber has to be a darkened light-tight box (since the E–T detector is sensitive to visible light). A TV camera and infrared light source may be fitted inside the specimen chamber to show the positions of the specimen in relation to the final lens and any other detectors (Figure 3.29).

(6) Vacuum pumps: The electron-optical column and specimen chamber are evacuated to a high vacuum of below 1×10^{-4} mbar by a rotary pump/vapour diffusion pump combination or by a turbomolecular pump. (For details of these see Appendix

1.) The electron-optical column and electron gun can usually be isolated from the specimen chamber by a valve, so that the chamber can be vented to atmospheric pressure and opened for a specimen change without disturbing the high vacuum in the remainder of the instrument. After re-evacuation of the chamber the isolation valve is reopened. Pumping and valving sequences are normally carried out automatically.

Supplies and controls of the SEM

The electron-optical column with its associated vacuum pumps form only a part of the essential components of an SEM. Electrical supplies are needed to power the

(a)

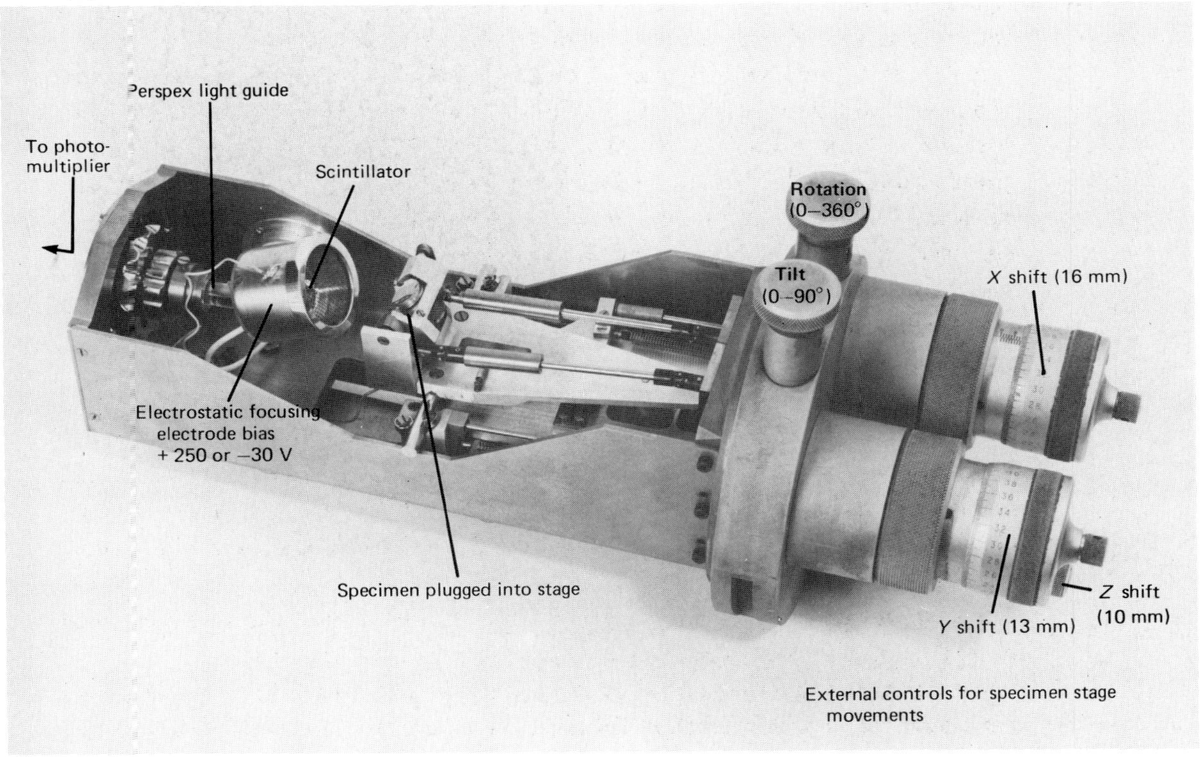

Figure 3.27. Examples of specimen stage. (a) Simple drawer-type stage of the Stereoscan 2a, with specimen manipulation controls identified. One 13 mm diameter specimen was accommodated. The Everhart–Thornley detector was mounted on the stage and withdrawn with it. It would nowadays be mounted permanently in the wall of the specimen chamber. (b) (*overpage*) Large capacity door-mounted specimen stage with multiple specimen turret (carousel). The stage is eucentric (see next figure), and tilts towards a fixed E–T detector (not shown). (Example (b) by courtesy of Gresham–CamScan Ltd.)

electron gun, lenses and vacuum system. Also controls for the various detectors and signal processing circuits, and the one or more image tubes on which the micrograph is eventually displayed. Early microscopes had separate power supply cubicles which were set apart from the column and control console. Miniaturisation and the wide-spread use of solid-state circuitry has resulted in the modern SEM being completely self-contained in one unit in most cases. In the latest instruments there will also be a computer and hard disc drive, with a keyboard and probably a 'mouse'. The actual form of modern commercial SEMs will be illustrated in Section 3.3.5.

We shall describe next what functions are controlled by the operator, illustrated

(b)

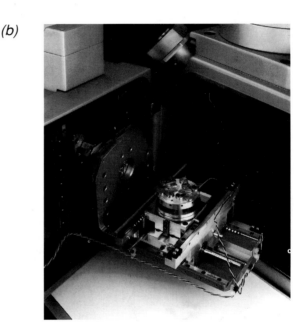

Figure 3.27 (*cont.*)

(*Opposite*)

Figure 3.28. Eucentric goniometer stage. Three types of motion are desirable in a specimen stage for TEM and SEM, i.e. translation in X, Y directions, tilt about an axis perpendicular to the optical axis of the microscope, and (primarily in the SEM) rotation about the optical axis. These are illustrated as separate movements in the left-hand diagram. The easiest way of combining these motions, shown in (*a*), is to mount the tilting and rotary mechanisms on a stage which translates in X and Y directions. However, any departure of the stage from its centred position results in the tilt and rotate axes moving away from the field of view so that rotation and/or tilt results in a loss of focus and a change in the field of view. This disadvantage is overcome in the eucentric goniometer (*b*) in which the optical axis (i.e. the centre of the field of view) lies permanently on the tilt axis and the centre of rotation of the stage; the X and Y traverse mechanisms are mounted on the tilting and rotating framework, so that whatever movement is carried out the image remains in focus and merely tilts or rotates as desired. It is necessary to have a Z (height) adjustment also, so that the specimen plane can be made coincident with the tilt axis. In practice the construction of a fully eucentric specimen stage is complex, and it is usual to provide eucentricity with respect to tilt but not rotation. An example of such an SEM stage is shown. (By courtesy of LEO Electron Microscopy Ltd.)

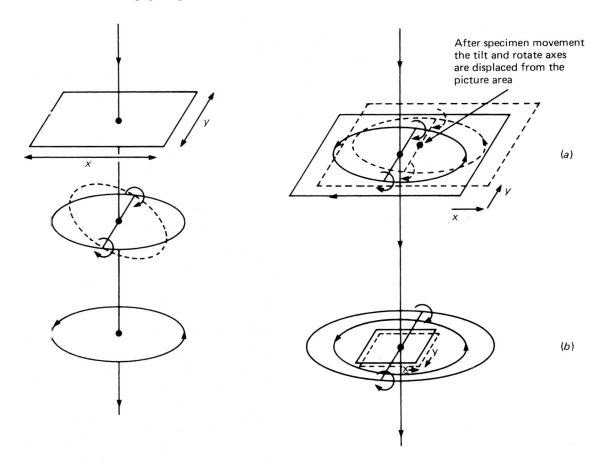

After specimen movement the tilt and rotate axes are displaced from the picture area

(a)

(b)

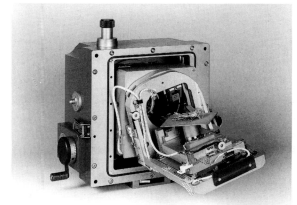

(Figure 3.30) with a photograph of the Cambridge Stereoscan 2a (1968–70) on which the essential control switches and displays are labelled. The operating procedures are more readily followed on such an instrument than on later models in which they have been 'streamlined' or even totally embedded in computer software.

3.3.3 Signal generation, display and image recording
Line- and frame-scan options
Every image from a scanning microscope results from a controlled interaction between an electron beam and a specimen. In the scanning process the small focused electron spot is scanned at a uniform rate in a straight line across the specimen. At the end of the line the beam is rapidly returned to its starting point and moved sideways by a small amount. A second line-scan parallel to the first is followed by another flyback and sideways deflection, and so on until the specimen has been scanned by one frame of N lines. If the line-scan duration is t, the frame-scan occupies a time $F=N\times t$.

On early SEMs the operator had independent control of the line-scan time, e.g. 1–300 ms, and either the frame duration (which would determine the number of

Figure 3.29. Interior of specimen chamber showing relative positions of multiple specimen holder and objective lens pole-piece, seen by a Chamberscope infrared TV system mounted in chamber wall. (By courtesy of K. E. Developments Ltd.)

lines in a frame, F/t) or the number of lines in a frame, which would control the frame-scan time $N \times t$. For selecting a field of view and focusing the image a comparatively coarse image with 100–200 lines is quite adequate, whereas for photographing the image the number of lines will be ten times as many per frame. The more slowly each line is scanned the less granular or 'noisy' the image will appear, since there will be more electrons per line and the statistical fluctuations along the line will be relatively smaller. However, although slower scan rates of one or two frames per second yield 'smoother' images they can be tiring to watch and too slow to enable specimen translation, tilting or rotation and focusing of the image to be followed as they are undertaken. It is customary, therefore, to focus using a reduced raster, say a quarter of a picture each way, or one-sixteenth of a frame, in which the same frame- and line–

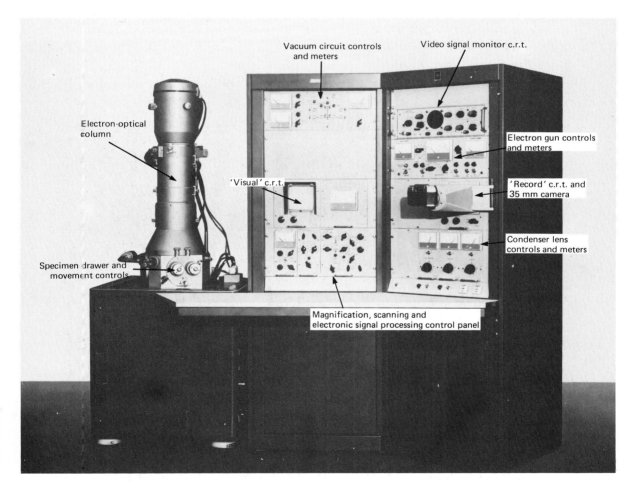

Figure 3.30. 'Classic' SEM showing separate units for controlling the operation of an SEM. These (and more) are still necessary in modern SEMs but have been integrated into the operating procedures. (Stereoscan 2a; by courtesy of LEO Electron Microscopy Ltd.)

scan speeds create a more continuous display. It is now universal, also, to find TV-rate scanning provided in addition to the slower rates. This most commonly uses 525 or 625 lines/frame, 25 or 30 Hz frame scan so as to be compatible with closed-circuit TV equipment and video recorders. TV scanning cannot be used in all situations, particularly at high magnification when the image is excessively grainy, but this objection can be overcome if the signals from successive frames are added together and averaged in a frame store. To facilitate this sort of manipulation the signal needs to be digitised. This is done either by using a digital scan generator, in which the beam deflecting fields are increased in discrete steps instead of being continuously ramped, or the 'analogue' signal from a continuously varying scan can be chopped into small picture elements (pixels) by an analogue to digital converter.

The magnification control reduces the extent of the scan on the specimen in fixed steps, to give magnifications of, e.g. $\times 20$, $\times 50$, $\times 100$, increasing by factors of 10 up to several hundred thousand or more. As the focusing of the electron probe on the specimen is not altered by change of magnification it used to be advisable to focus at one or two steps higher magnification than the micrograph was to be recorded at, and then return down again, since the effects of errors in focus and astigmatism could be judged more readily at the higher magnification. On present-day micro-scopes the same effect is obtained by having the display image on a video monitor several times larger than the recorded image dimension. The operator can also select the type of detector being used (secondary electron, backscattered electron, etc.) as well as the photomultiplier and amplifier characteristics to give higher, lower or reversed contrast images. He or she generally has a monitor screen on which can be displayed the signal amplitude along each line as it is scanned. Any distortion of signal (due to overloading the amplifier, for example) can be detected on this display.

When a digital image store is provided with the microscope there are several vari-ations possible in its manner of use. If the signal from each line is placed into memory as it is scanned, after the first frame the displayed image will be steady and uniformly bright, even with a frame-scan duration of a second or more. If faster scanning is used and the image would be grainy or noisy, it can be arranged for a predetermined number of scans, e.g. 6 or 12, to be put into the memory and 'averaged'. If n frames are being averaged, the $(n+1)$th scan is arranged to replace the data from the 1st scan, so that the image is being continuously updated. Digital storage of images on the SEM is possibly the most valuable technical advance since the replacement of thermionic valves in the scan circuitry.

Signal processing and display

(a) Gain, gamma and black level: The electrical signal in the SEM has to be ampli-fied and electronically processed before being displayed. Signals from a photo-multiplier can be modified by gain, gamma and black level controls. The SEM operator will use these appropriately to give a uniform level and brightness range of

image signal to suit the photographic or other recording medium, from specimens which might have widely different characteristics.

A variable *gain* or *PM voltage* control is provided to alter the amplification of signals by the photomultiplier. A small signal variation on top of a high average level from a bright specimen is made more visible by subtracting a proportion of the steady input with a *black level* control and amplifying only the informative, 'contrasty', part of the input. On the other hand, if the specimen has too high an inherent contrast it is necessary to amplify the signal but to compress the overall brightness range to lie between the permitted limits. A *gamma* control, which modifies the response of the video amplifier, achieves this. If the normal response is linear with the input voltage V, the gamma control may modify it to be proportional to $V^{1/2}$, $V^{1/3}$, $V^{1/4}$ in successive stages. With these three controls the operator can 'tailor' the signal processing to give him uniform micrographs, observing the effects on the signal level monitor and 'visual' display tubes. The effects of these controls are illustrated diagrammatically in Figure 3.31. On a digital image these effects can be achieved by image-processing operations using a range of 'look-up' tables.

In addition to the normal brightness- or intensity-modulated images the SEM may have other display modes which are appropriate for use on specialised subjects. Those to be described are intended to emphasis particular features of the SEM image which might be overlooked in a normal brightness-modulated display. This often results from the presence of both fine and coarse detail on the same specimen, which is difficult to show together adequately under the same display conditions.

(b) In *y-modulated* images the video signal is used to give a proportional vertical deflection to the c.r.t. spot instead of varying its brightness. The number of scan lines in the frame is reduced so that individual line traces can be seen clearly. In this way relatively small changes of brightness, which might have escaped notice on an intensity-modulated display, are made more obvious. The technique is therefore used to good effect on images of low contrast. A further advantage of *y*-modulation is that quantitative comparisons of signal strength may be made. Figure 3.32 shows the characteristic differences between brightness- and *y*-modulated displays of the same signal.

(c) Derivative images provide another way of drawing attention to sudden small changes of signal intensity. Such changes may be superimposed upon a slowly changing background, and can best be separated from such a background by emphasising the regions where the rate of change of video signal is greatest. This procedure is known electronically as differential- or derivative-processing, and it provides a brightening (or darkening if more appropriate) of the c.r.t. trace whenever a sudden change of signal occurs. Thus an outlining effect is produced which highlights edges, scratches, cracks, etc. on more expansive areas of specimen detail, whatever their

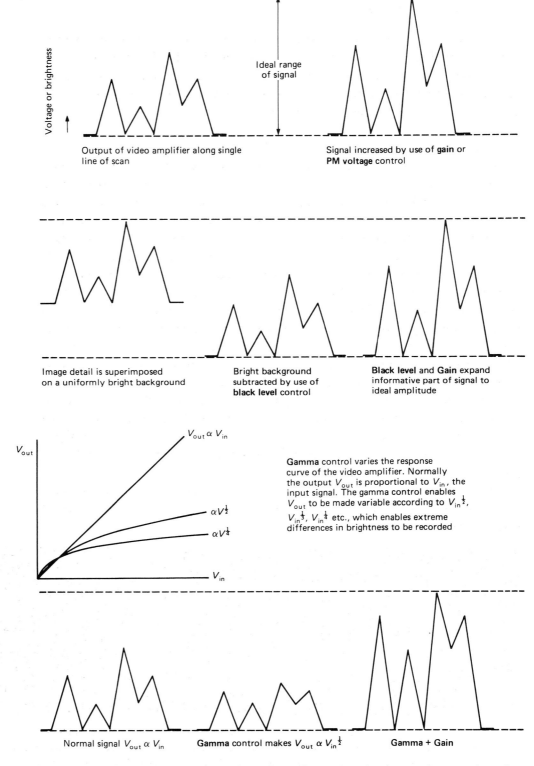

Voltage or brightness

Ideal range of signal

Output of video amplifier along single line of scan

Signal increased by use of **gain** or **PM voltage** control

Image detail is superimposed on a uniformly bright background

Bright background subtracted by use of **black level** control

Black level and **Gain** expand informative part of signal to ideal amplitude

V_{out}

$V_{out} \propto V_{in}$

$\alpha V^{\frac{1}{2}}$

$\alpha V^{\frac{1}{4}}$

V_{in}

Gamma control varies the response curve of the video amplifier. Normally the output V_{out} is proportional to V_{in}, the input signal. The gamma control enables V_{out} to be made variable according to $V_{in}^{\frac{1}{2}}$, $V_{in}^{\frac{1}{3}}$, $V_{in}^{\frac{1}{4}}$ etc., which enables extreme differences in brightness to be recorded

Normal signal $V_{out} \propto V_{in}$

Gamma control makes $V_{out} \propto V_{in}^{\frac{1}{2}}$

Gamma + Gain

Figure 3.31. Diagrammatic illustration of the effects of the various signal-processing controls on the SEM.

tone. A derivative-processing unit usually provides a facility for adding together the outlining signal and a chosen fraction of the unmodified signal. Figure 3.33 illustrates a derivative image and compares it with the original unprocessed signal. Similar results may be obtained by applying special filters to a recorded digital image, given appropriate image-processing software.

Second channel, split screen and window presentations

Developments in computers and image-processing software have made many 'tricks' possible in the initial scanned raster and in image displays. The *split screen* display, usually in two halves but occasionally in four quadrants, enables a side-by-side presentation of signals resulting from two different, interlaced scans, or images from different detectors stimulated by the same scan. The former type includes the use of zoom magnification in which a chosen portion of a main field is shown at a higher magnification, often continuously variable up to ten times that of the main field. The area for more detailed study is selected by positioning a 'window', either a bright outline or a brighter rectangle within the main field (illustrated in Figure 3.34). When the 'dual magnification' button is pressed every alternate scan across the specimen will be shortened in the zoom ratio (1/2, 1/5, etc.) and will only scan the region within the window. The images from the full and the abbreviated scans will be displayed separately on the split screen. The second type of split screen presentation is possible when the microscope has separate detectors for, for example, secondary or backscattered electrons, or x-rays, and enables two different responses of the same area of specimen to be compared side-by-side. The effect of signal processing can be studied in this way, the 'raw' signal being compared with *y*-modulation or derivative-processed images to see whether the processing is showing up new information.

Splitting the screen involves sacrificing part of the field of view, although the side-by-side images are convenient to record. A less restricting alternative is to fit a second display channel and view the full-sized images.

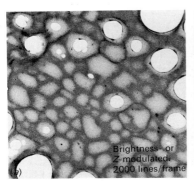 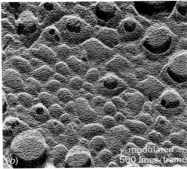 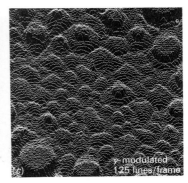

Figure 3.32. Brightness- and *y*-modulated displays of the same information. The subject is a scanning transmission image of a 'holey carbon film' containing thin patches and perforations (*a*), which acquire a pseudo-three-dimensional appearance in the *y*-modulated images. The latter are shown with (*b*) 500 and (*c*) 125 scan lines to the frame.

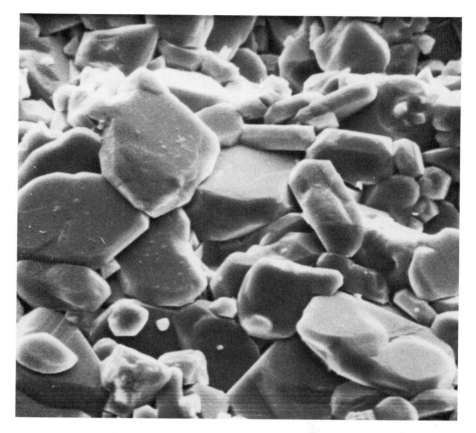

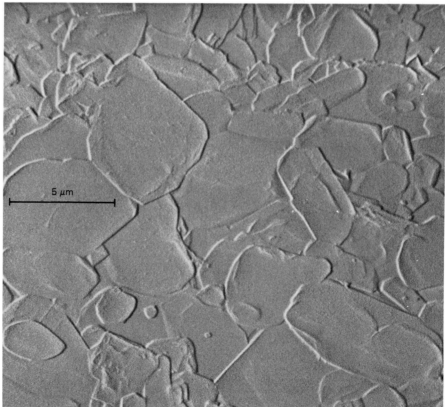

Alphanumerics and display aids

It is normal for printed information to be automatically added to images, in the form of a 'micron-marker' or line showing the scale of the micrograph, and other data such as micrograph number and kV used. The microscope may have an additional keyboard with which lines of additional descriptive data may be added anywhere on the micrograph (Figure 3.35).

Modes of operation of the SEM

As will be evident from Chapter 2 a range of different currents and radiations is available in the SEM. All can be used for imaging provided the microscope is equipped with the appropriate detector. The principal operating modes are:

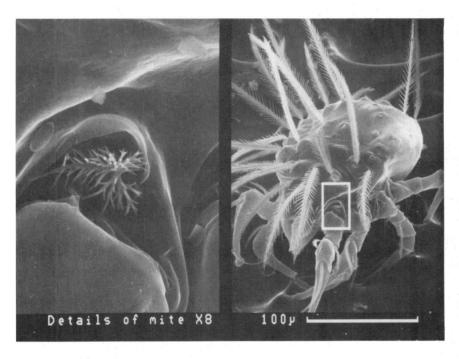

Figure 3.34. Dual magnification split screen presentation. The left-hand image is eight times enlarged from the field in the rectangle on the right. (By courtesy of Gresham–CamScan Ltd.) N.B. The electron probe size will be the same for both magnifications, which may result in poorer resolution of detail at the higher magnification than could have been obtained in a separately set-up high-magnification micrograph.

(Left)
Figure 3.33. Emissive mode micrograph of a tilted alumina surface and the same signal after derivative processing. This has typically removed all 'modelling' or three-dimensional effect and left only the major outlines of the grains. The same procedure could have been applied more strongly to show up finer structure on individual grains. Some roundness can be introduced by adding a proportion of the original signal to the differentiated signal.

(i) *Emissive*: The most commonly used mode shows surface topography when secondary electrons and/or backscattered electrons are detected by an E–T detector. If backscattered electrons only are detected, by E–T, Robinson, solid-state (silicon), yttrium-stabilised aluminium garnet (YAG) (Autrata) or multichannel plate detectors, the image shows compositional detail.

(ii) *X-ray*: Very common and useful mode for elemental analysis of specimens. Characteristic x-radiation analysed by either wavelength dispersive (WDX) or energy dispersive (EDX) spectrometers (see also Chapter 6).

(iii) *Absorbed current*: For most specimens this is the inverse of the emissive mode, but semiconductors give information related to internal structure, crystal defects, etc. (EBIC).

(iv) *Cathodoluminescence*: Can give information on spatial and spectral distribution of cathodoluminescence radiation, distribution of impurities, energy levels, etc. An E–T detector may be used, but a dedicated mirror/spectrometer system is highly desirable.

(v) *Transmitted electron*: Not a frequent requirement, but can examine TEM specimens by scanning transmission electron microscopy. Further information is given later in this chapter.

Examples of these modes are given in Chapters 5 and 6.

Figure 3.35. Lines of descriptive data superimposed on a micrograph. (By courtesy of Gresham–CamScan Ltd.)

Image recording on the SEM

Unlike the TEM, in which the whole image field is recorded at the same time, the SEM image is recorded sequentially, generally by photographing the face of a c.r.t. in a time exposure as the image is drawn on it line by line for one complete frame. It is usual to record from a tube with a screen coated with a blue phosphor (P11) with a short persistence, having a higher resolution than the yellow-green (P7) or colour phosphors of the 'visual' tube. The 'record' tube has its own brightness and contrast controls which can be set to suit the characteristics of the photographic material being used for recording. The image is scanned with more lines per picture than the screen can resolve, so that no scanning lines are seen in the photographic record. The 'record' scan is made more slowly than the visual scan used for selection and focusing, in order to give a smoother, grain-free micrograph. The frame time in recording may be between 30 and 200 seconds, depending on how 'noisy' the image is. High-magnification images made with finely focused probes will require the longer times since the signal-to-noise ratio will be low. Under these conditions, too, the image may be too grainy for TV rate scanning to produce a recognisable image without signal averaging, and slow scanning will be required at all stages. If the microscope has a digital frame store the averaging can be done beforehand, and a super-slow record scan may be unnecessary.

The photographic record may be made on fine-grained black-and-white negative film of 35 mm or larger format; alternatively instant prints may be made on Polaroid-type material. It is increasingly common for essential data such as exposure number, magnification, kV and signal mode (SE, BSE, CL, etc.) to be recorded on the film at the same time. Early scanning microscopes used a square picture format on a picture tube 100 mm square, but nowadays a rectangular format is used which is compatible either with the 1:1.25 aspect ratio of 4″×5″ photographic material or with the TV format.

If the image has been digitised and stored in memory, there is a wider choice of recording. Hard copy can be produced very economically and quickly on a monochrome video printer, although the image may not have archival permanence. The image may also be recorded on an optical or magnetic floppy disc for later retrieval and preparation of hard copy. Images can be transferred in appropriate format (e.g. as TIF files) into a computer network and used for report preparation.

For a discussion on digital image recording see Section 3.6.

The manner of photographic recording leads to one important difference between micrographs made in the TEM and SEM. It is generally possible to enlarge a portion of a well-focused transmission micrograph to examine a particular feature in greater detail, since the electron image contains all the resolution of which the microscope is capable, and the photographic emulsion on which it falls is very fine-grained. A scanning micrograph, however, will only contain image detail which is comparable with the grain size of the phosphor of the recording tube; finer structure than this must be magnified in the microscope at the time of taking, and cannot be obtained by photographic enlargement afterwards. In other words, it is normally pointless to use a magnifying glass on a scanning micrograph already enlarged to, say, twice the size of the record screen. This argument does not apply to digital images, for which the pixel resolution of an image will determine the enlargement a micrograph will stand.

Operational aids

Several scanning modifications, now usually fitted to microscopes as standard, can be very helpful when the appropriate requirement arises, and can be kept switched off at other times. These are described below.

(a) Automatic focusing and astigmatism correction: Detail in a line-scan across an 'in-focus' image is more sharply defined than on either side of focus (Figure 3.36(*a,b*)). An electronic circuit which will assess sharpness whilst varying the focusing current can be made to find the sharpest setting automatically. This is the basis of the self-focusing facility available on many scanning microscopes. As with the wobbler focusing aid on the TEM the device will enable the novice operator to obtain sharply focused pictures, and will give a good starting point for manual fine adjustment. An extension to the programme will automatically detect and correct asymmetries in the image caused by astigmatism (Figure 3.36(*c*)).

(b) Dynamic focusing: If the electron probe is focused at the centre of the field of view of a plane specimen which is tilted, the probe goes out of focus at the top and bottom of the picture. The effect is greater at higher specimen tilts, short working distances and with beams of larger angular aperture. Dynamic focusing is an electronic technique whereby the power of the final lens is progressively altered from line to line of the scan in order to maintain the probe in focus over the entire surface of the tilted specimen. In the usual arrangement of the SEM, shown diagrammatically in Figure 3.37, at the top of the picture the lens is weakened and at the bottom the lens is stronger than it would be without dynamic focusing. The advantage of this technique lies in the fact that the effective depth of field can be increased without degrading the image quality by using a smaller final aperture or increasing the working distance. The degree of modulation of the final lens current depends on tilt angle and extent of scan, or magnification. A variable control is usually provided and is increased until both top and bottom of the picture are sharp, after first focusing the microscope in the centre of the field with the dynamic focusing control set to zero. Naturally this technique is ineffective if the specimen has a rough, irregular surface; depth of field in this case can only be increased by the traditional method of using a small final aperture and long working distance.

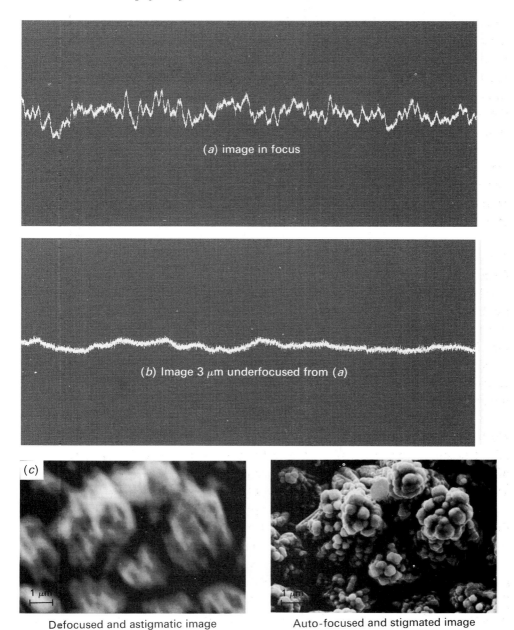

(a) image in focus

(b) Image 3 μm underfocused from (a)

(c)

1 μm

1 μm

Defocused and astigmatic image   Auto-focused and stigmated image

Figure 3.36. Profile of the video signal along single scanned lines in an image which is (a) focused and (b) 3 μm underfocused, illustrating the contrast changes which can be used by a microprocessor to automatically assess and correct the focus of a scanning microscope. The microprocessor varies the strength of the final lens in steps from one end of the focusing range, assessing the sharpness of signal at each step until it reaches a peak and stops, diplaying the final focused image on the screen. (c) The programme can be extended to include automatic astigmatism correction. ((c) by courtesy of Amray Inc.)

(c) Scan rotation: As the electron beam travels down the column the orientation of the scanned frame rotates. Therefore the relationship between the orientation of the scanned frame and the X and Y axes of the specimen stage will vary with working distance. For some purposes, e.g. stereoscopic microscopy (see later, Chapter 5) it is necessary to align the axes of the displayed image precisely with the X, Y shifts and the tilt axis of the specimen stage. A scan rotation unit rotates the scanned frame in its own plane around the optic axis of the microscope. The rotation control may be calibrated in working distances, or the effect can be corrected visually by rotating the scan until operating the X, Y shifts produces image movement parallel to the appropriate edges of the screen. This type of adjustment is made very much more easily on a live image at TV scanning rates.

(d) Tilt correction: The effect of perspective is to make a square appear to be foreshortened into a rectangle when tilted about one edge, the degree of foreshortening increasing with the angle of tilt. The magnification on a scanning micrograph of a tilted specimen is therefore different in all directions on the image. Tilt correction is an artifice to produce constant magnification at all orientations on the micrograph, by foreshortening the scanned frame on the specimen in proportion to the angle of tilt. The control is calibrated in terms of tilt angle. It must be borne in mind that: (i) tilt correction should only be applied to a perfectly plane specimen; (ii) the specimen tilt axis must be accurately parallel to the scanned frame on the specimen (see *scan rotation*, above); and (iii) the 'correction' creates its own artefact since square-on proportions are combined with an inclined viewpoint. Illustrations of correct and incorrect applications of tilt correction are shown in Figure 5.22.

(e) Dynamic stereo-microscopy: This fascinating and useful manipulation is only available on a limited range of microscopes. It will be described in Section 5.4 (page 253).

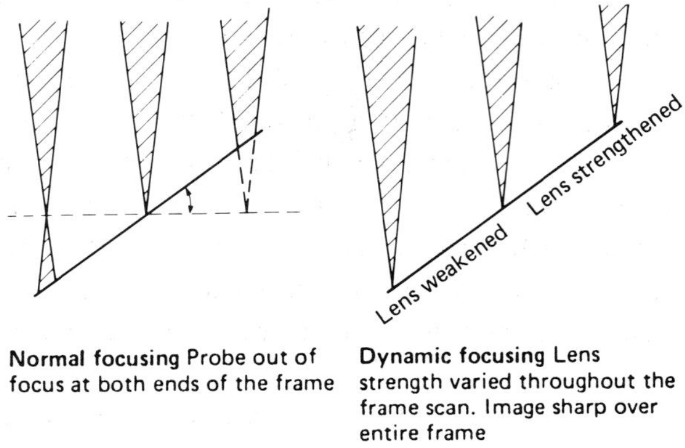

Figure 3.37. Dynamic focusing (schematic). The diagram shows a cross-section through the specimen normal to the direction of line scan. The lowest part of the specimen in the microscope forms the top of the micrograph in the SEM. Dynamic focus is illustrated in Figure 5.18(*b*).

Normal focusing Probe out of focus at both ends of the frame

Dynamic focusing Lens strength varied throughout the frame scan. Image sharp over entire frame

Lens weakened   Lens strengthened

(f) Electron-optical alignment aids: No electron microscope will give its best performance unless its column is correctly aligned, i.e. the optical axes of the lenses are coincident and aligned with the electron beam from the gun. The beam-limiting diaphragms in the column must also be accurately centred on the optic axis. The components of the column are mechanically aligned during assembly in the factory and clamped together; no mechanical re-alignment is possible nowadays during operation but electromagnetic coils are placed in appropriate parts of the column so that the effects of physical movements of the components can be simulated by purely electrical means. Any slight changes in alignment which develop during use of the microscope can thus be corrected without taking it to pieces. Manufacturers have worked out routines for periodically checking the alignment and correcting it as required. Sometimes checking routines are built into the instrument, which automatically set up suitable conditions for observing and correcting particular maladjustments or misalignments. For example, if the diaphragm in the final lens is not accurately centred about the beam axis, changing the focus of the lens results in an image shift perpendicular to the axis. The operator can manually vary the strength of the final lens through focus and back again, moving the diaphragm shift knobs until there is no image movement as the focus is changed. The manufacturer can make the procedure easier to carry out (and therefore more likely to be used!) by fitting a 'wobbler' circuit which superimposes an alternating current on the final lens current, causing the image to go through focus repeatedly; any image shift can quickly be eliminated by repositioning the aperture using both hands. Alignment and saturation of the electron gun, which are of great importance if the best resolution and longest filament lifetime are desired, can also be aided by procedures built into the microscope.

3.3.4 High-resolution SEMs

The scanning microscopes so far described have used the illumination provided by a tungsten or LaB_6 thermionic cathode to examine specimen detail down to the 3 or 4 nm level. For higher resolutions and, especially, high contrast resolution of surface detail, the concept of a fine, bright electron probe is combined with operation at low beam energies, e.g. 5 keV and below.

Several electron-optical solutions have been provided to the requirement of producing a focused nanometre-sized electron probe at normal and low voltages. All are based on using the small, bright source of electrons provided by a field emission gun (see Appendix 4 for details), from either a cold emitter or a Schottky thermally assisted field emission source. Minimising the size of the focused probe demands the use of a short-focus objective lens with low spherical aberration, and can place restrictions on the size and form of the specimen and the ease of collecting the signal from it.

A conventional objective lens at a working distance of 8 mm from the specimen can be used with specimens at least 10 cm across and tilts of up to 60°, with resolutions of 1.5 nm at 30 kV and 7 nm at 1 kV. The narrow bore of the objective lens (pinhole lens) screens the specimen from the lens field, and secondary electrons can be collected by an E–T detector placed at one side of the optic axis.

Alternatively if the lower lens bore is opened up, the lens magnetic field spills out on to the specimen, and magnetic specimens can no longer be examined. However, the brightness of the probe allows small probes to be formed at lower voltages; 1.5 nm resolution at 15 kV and 4.5 nm at 1 kV. Because of the magnetic field at the specimen, secondary electrons cannot be extracted to the side of the specimen, but are forced to spiral up through the lens bore, to be collected by an E–T detector above the objective lens. This modification to the objective lens is shown schematically in Figure 3.38.

Another approach (Figure 3.39) which still allows full-sized specimens and useful

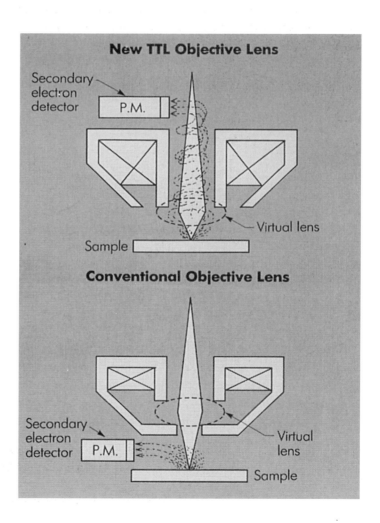

Figure 3.38. Modification of objective lens design to assist high-resolution, low-voltage, operation. (By courtesy of Hitachi Ltd.)

tilts is to combine a pinhole magnetic lens with an electrostatic lens. Full specimen facilities, eucentric movements and tilts to 50° are retained, down to a working distance of 6 mm, when resolutions of 1.2 nm at 20 kV and 4 nm at 1 kV are attained. Some secondary electrons are collected within the objective lens – others from a normally offset E–T detector. The two signals may be added together in variable proportions.

The final approach is the completely 'in-lens' specimen stage, using a condenser/objective lens as in high-resolution TEM and placing a side-entry specimen stage in the centre of the lens field. The size of specimen is, by SEM standards, restricted, but the strong magnetic field of the lens enables a resolution of 0.9 nm to be obtained at 30 kV with a Schottky field emission gun, and even 2.5 nm with a tungsten filament gun at 40 kV. Secondary electrons spiral up the lens bore, and are collected outside the lens. The Topcon SM-700 Series SEMs (Figure 3.40) offering this 'in-lens' stage also have a more normal stage for large specimens at the bottom of a five-lens column, with 'poorer' resolutions of 4 nm at 40 kV (tungsten) or 1.5 nm at 30 kV (FEG).

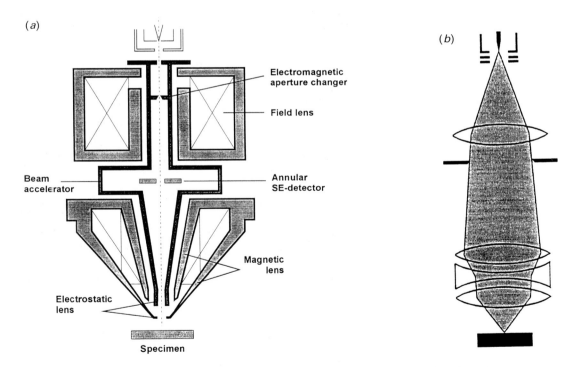

Figure 3.39. LEO 982 Gemini Field Emission SEM has a novel combination of electromagnetic and electrostatic lenses to provide low aberrations at low voltages, and 4 nm resolution at 1 kV. The beam travels with high energy until it is decelerated to the desired voltage in the objective lens. (*a*) is a schematic cross-section of the column, whilst (*b*) shows the optical effect of the lens combination. (By courtesy of LEO Electron Microscopy Ltd.)

There are now more than a dozen commercial models of SEM offering 2 nm resolution or better, by the use of field emission sources.

Use of a field emitter in the electron gun imposes a very severe requirement on the pressure and cleanliness within the gun. All high-resolution SEMs must be evacuated to 10^{-8} mbar or below (10^{-9} for cold field emission), i.e. about three orders of magnitude lower pressure than that necessary for a conventional thermionic tungsten emitter. This usually involves a separate sputter-ion pump on the gun chamber itself, in addition to high-grade pumping of the rest of the column.

3.3.5 Commercial SEMs

In the 30-plus years since the first commercial SEM was introduced the growth rate of this side of the electron microscope family has been rapid. There are more companies making scanning microscopes than transmission instruments, with Hitachi, JEOL, LEO, Philips, Tesla and Topcon having a foot in both camps. Currently there are over 40 different models of SEM, of which more than a dozen are field emission instruments.

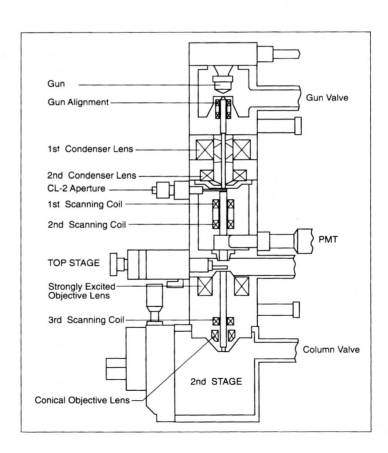

Figure 3.40. Cross-section of the Topcon SM-701 scanning microscope. The top stage accepts specimens up to 10 mm diameter by 5 mm thick, and has access ports through which detectors for backscattered electrons or x-rays can be introduced. (By courtesy of Topcon (Europe) BV.)

In this section some of the variations on the basic concept of column and controls will be mentioned. Because it has fewer and simpler lenses, and much of the rest is microelectronic circuitry and display, the basic SEM is appreciably cheaper than the TEM. A research-quality SEM could be purchased for the cost of a routine TEM. There are, of course, many add-on accessories, detectors and analysers which would increase the versatility of any instrument, as well as its cost. If the x-ray analysis is more important than the observation of very fine detail or very large specimens the x-ray accessory could well cost more than the microscope to take it!

The majority of commercial SEMs are still laid out along the same pattern as the original Stereoscan, with the column and specimen stage on the left and the desk with electron-optical controls on the right, but there is no right or wrong about this layout and one manufacturer has always arranged his microscopes the opposite way round. The size of the control console of commercial microscopes has reduced with the elimination of many of the control knobs and switches, and the streamlining of control electronics systems by the use of integrated circuitry.

A new approach, first introduced with the Philips XL Desktop SEM in 1990 and since followed by other manufacturers, has been to replace all physical knobs and switches by icons, buttons and sliders in Windows™ menus on a computer screen, selected and actuated by a 'mouse'. The electron-optical column and pumping system are still there, but the various boxes supporting a large flat desk contain the power supplies and the computer. In a completely integrated system the same computer may be responsible for controlling the electron microscope, for performing x-ray analysis and for digital image recording and analysis. What goes on inside the column, though, is much the same as in the very first SEM. The mouse-controlled equivalent of the Stereoscan 2a shown earlier (Figure 3.30) is illustrated in Figure 3.41.

The guaranteed resolving power of conventional SEMs has stabilised at the 3.5–5 nm level for high-voltage (20–30 kV) operation.

As in the TEM, quality of vacuum in SEMs is at least an order of magnitude better than twenty years ago, partly because better pumps and fluids have been introduced and partly because lenses and coils have been taken out of the vacuum by the use of column liner tubes. There is the incentive that a cleaner vacuum brings improved operation with less trouble from contamination of specimens and diaphragms and an extended lifetime of tungsten filaments in the electron gun. Vacua must be improved at least in the gun area if brighter electron emitters are to be used, and additional pumping is provided for lanthanum hexaboride and field emission sources.

There is often a choice between turbo-molecular pumps or vapour diffusion pumps to evacuate the microscope. The increased cost of the turbo pumps is counterbalanced by saving on automatic high-vacuum valving systems, since the single pump will operate all the way from rough to high vacuum (see Appendix 1 for vacuum pumping routines). There is undoubtedly a great attraction in having a microscope which does not require cooling water for vacuum pumps, although some types of turbo-molecular pump require running water to cool the bearings. Other commercial developments have resulted in the Environmental or Low-Vacuum microscopes (see the next section), in which specimens may be examined in specimen chamber pressures so high that even fresh biological samples can be studied in a near-normal state without any specimen preparation.

The smallest expenditure nowadays provides a three-lens column, power supplies and vacuum pumps, and a far from Spartan specification. A choice of five or more voltages up to 25 kV will be provided, and push-button or computer operation will supply a restricted range of scanning line and frame speeds, although TV-rate scanning is universal. The same cathode-ray tube may have to be used for displaying the image and recording it. Limited signal processing will be available; dynamic focus is now a universal feature. The lowest end of the magnification range (i.e. down to 5–10 times) will be missing. Even the simplest microscope will handle much larger specimens than hitherto, and stage movements will allow most, if not all, of their surface to be examined, even with eucentric movements in some cases. A newcomer, the RJ Lee PSM-75 (Figure 3.42) has an unusually broad specification, with choice of five operating voltages from 2 to 20 kV, and computer control via a mouse. The screen

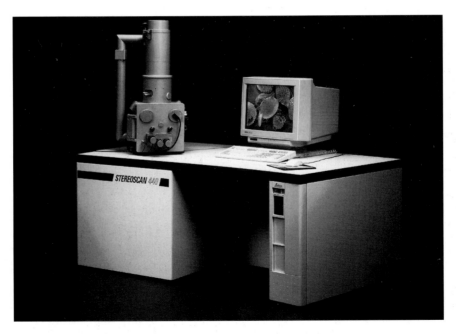

Figure 3.41. The LEO 440 SEM, the 1994 successor to the instrument shown in Figure 3.30. 0.3–30 kV; ×5–×300 000; resolution 3.5 nm (tungsten); range of specimen chambers and stages available, including variable-pressure model; fully software controlled *via* a 'mouse'. (By courtesy of LEO Electron Microscopy Ltd.)

display contains simulated knobs to control filament, magnification, scan rotation and spot size, and these 'rotate' as the settings are changed with the mouse.

More money will buy an increased range of operating voltages, higher resolving power, larger and/or motorised specimen stages, downward extension of the magnification range, more complete facilities (e.g. scan rotation, foreshortening correction, derivative processing) and separate visual and record tubes. In a digital SEM some of these facilities are added by purchasing the relevant software programmes. Naturally, different manufacturers have their own priorities on improvements, and the customer's requirements will also dictate whether his money is better spent on more comprehensive examination possibilities (detectors, signal processing and display modes) or larger or more specialised (e.g. cryogenic) specimen stages. Such is the wide applicability of scanning microscopy nowadays that no two persons' requirements will be identical, and manufacturers' lists of optional facilities run into tens of items.

Figure 3.43 shows examples of commercial SEMs with varied specifications.

3.3.6 Specialised SEMs

The low-vacuum and environmental SEM

Anticipating Chapter 4 on specimen requirements and preparation it must be admitted that the need to have the specimen in a high vacuum provides problems for anyone wanting to examine moist or volatile samples, especially biological matter

Figure 3.42. Simple 'mouse'-driven SEM with a number of novel features including removable specimen chamber, and change of column liner with filament exchange. (By courtesy of RJ Lee Instruments Ltd.)

such as plant or mammalian tissues. Such problem material can normally only be examined and/or analysed either frozen or dried: preparatory procedures are the subject of a whole technology, to be met in another chapter.

A recently developed possibility for surface examination of bulk moist specimens is to use one of the environmental or low-vacuum scanning electron microscopes now available from all manufacturers. (Low-Vacuum, Variable Pressure, WET-SEM, ECO-SEM and EnVac are the terms used for these systems by JEOL, Hitachi

(*a*)

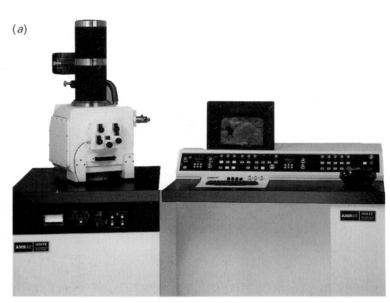

Figure 3.43. Examples of currently available commercial SEMs.

(*a*) Amray 1850 FE. One of a family of SEMs with a range of specimen chambers and electron guns; also ECO-SEM low-vacuum model. 0.5–30 kV; Schottky thermal field emission gun; 1.5 nm resolution at 30 kV; $\times 10$–$\times 500\,000$.

(*b*) Gresham–CamScan CS 44. 0.5–40 kV; W or LaB$_6$ gun; $\times 3$–$\times 400\,000$; 4 nm at 30 kV (W), 3 nm (LaB$_6$). Thermal field emission and low-vacuum models available.

(*c*) Hitachi S-3200N Variable Pressure SEM, chamber pressure 1.3×10^{-2}–2.7 mbar in high-pressure mode; 0.3–30 kV; $\times 20$–$\times 300\,000$; tungsten filament gun; 3.5 nm resolution in high vacuum, 5.5 nm at 2.7 mbar specimen chamber pressure; range of specimen chambers and stages available.

(*d*) Philips XL 40 FEG. 0.2–30 kV; Schottky thermal field emission gun; $\times 10$–$\times 300\,000$; 1.0 nm resolution at 30 kV. (Photographs by courtesy of Amray Inc., Gresham–CamScan Ltd, Hitachi Ltd, Philips Electron Optics, respectively.)

(*b*)

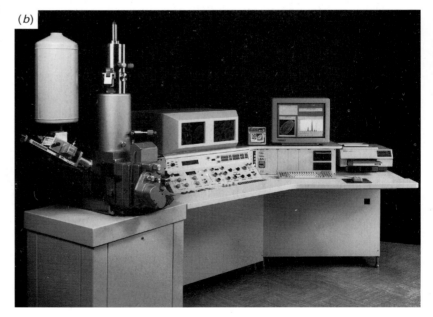

(c)

(d)

and LEO, Topcon, Amray and Gresham–CamScan, respectively.) These differ from the high-vacuum SEM in allowing operation with a gas pressure in the specimen chamber higher than the usual 10^{-4} to 10^{-5} mbar; up to about 3 mbar for the low-vacuum microscopes and 50 mbar for a dedicated instrument such as the ElectroScan ESEM™ 2020 (Figure 3.44). (N.B. The saturated vapour pressure of water at 20°C is 23.4 mbar.)

These microscopes differ from the normal high-vacuum instrument in two main respects: (1) Vacuum pumping, and (2) Signal detection.

(1) Vacuum pumping: In the environmental SEM the electron-optical column at 10^{-5} mbar pressure or below and the specimen chamber must be evacuated separately, with the pressure differential of up to seven orders of magnitude maintained across a small aperture linking the two regions. This is a severe requirement, which can be satisfied more easily if intermediate pressure zones can be placed between column and chamber.

In the ElectroScan the differential between 10^{-7} mbar in the electron gun and up to 50 mbar in the specimen chamber is maintained through two diffusion-pumped regions at 10^{-6} and 10^{-4} mbar respectively, and a rotary-pumped region at 10^{-1} mbar (shown schematically in Appendix 1, Figure A1.11). Chamber pressures are adjusted automat-

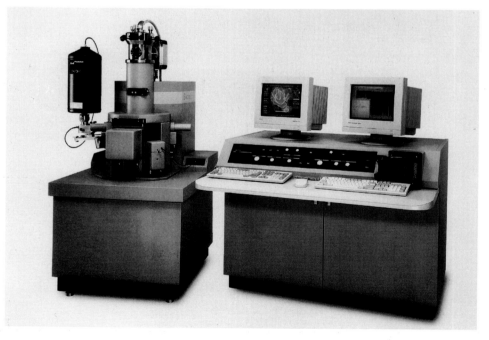

Figure 3.44. ElectroScan ESEM™ 2020, capable of operating with specimen chamber pressures between 10^{-5} and 50 mbar. Resolution 5 nm at 30 kV (4 nm with LaB_6 gun) with 10 mbar chamber pressure. 0.1–30 kV; ×30–×600 000; range of specimen stages available. (By courtesy of ElectroScan Corporation.)

ically by bleeding in gases or vapours to a pre-set pressure. To monitor and maintain the necessary pressure differentials is technically demanding and costly; hence, apart from the *ElectroScans* all other commercial low-vacuum SEMs have one differential pumping diaphragm only, and offer a more limited range of high pressures.

(2) Signal detection: The normal E–T SE detector has a voltage of more than 10 kV applied to its scintillator, and it is unsafe to use this at high chamber pressures because of the risk of gaseous ionisation and voltage breakdown. Most low-vacuum microscopes therefore use only backscattered electrons for imaging. The combination of BSE imaging and gaseous ionisation gives images free from specimen charging effects, but also low in topographical contrast unless a split or off-centre detector is used.

The ElectroScan uses a positively biased solid-state detector mounted directly under the objective lens, coaxial with the primary electron beam, which collects backscattered and secondary electrons which will have been amplified by ionising collisions with the gas in the specimen chamber (Figure 3.45).

The mechanisms of signal amplification and collection within a really-low-vacuum microscope are still the subject of debate; Shah and Danilatos are two prominent workers in this field (e.g. Durkin & Shah, 1993; Danilatos, 1991).

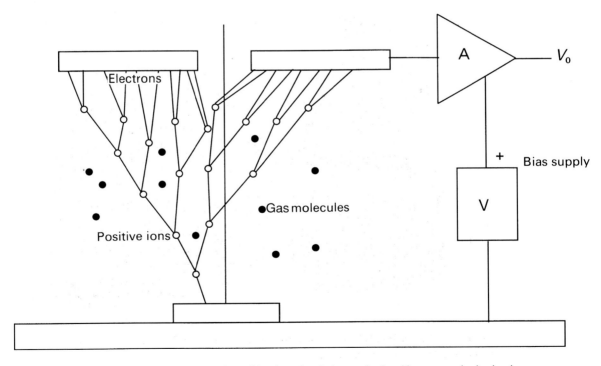

Figure 3.45. Schematic diagram of process of amplification of emissive mode signal by gaseous ionisation in Environmental SEM. (By courtesy of ElectroScan Corporation.)

The Mobile SEM

Although the concept of taking the microscope to the job is difficult enough in light microscopy, which does not have to worry about maintaining the specimen in a vacuum, a small but robust scanning electron microscope is available which can be used on any specimen to which its vacuum interface can be sealed. The Tesla Mini-SEM, shown in Figure 3.46, operates at 15 kV and has a magnification range of ×10 to ×50 000. Detail resolution is stated to be 16 nm at a working distance of 8 mm. It operates at TV scanning rate, and a commercial TV set may therefore be used as a large display monitor. Vacuum pumping is by turbo-molecular pump bolted to the specimen shamber; no cooling water is required.

3.4 Scanning transmission electron microscopy (STEM and FESTEM)

The acronym STEM can be taken to mean: (a) Transmission microscopy on a dedicated SEM; (b) Scanning microscopy on a dedicated TEM or analytical TEM; (c) Ultra-high-resolution transmission microscopy on the field emission (gun) scanning transmission electron microscope, sometimes FESTEM or FEGSTEM.

3.4.1 *Transmission microscopy on a dedicated SEM*

We have already considered this as one of the operational modes of the SEM. A scanning microscope owner who has occasional need of micrographs of submicron particles, shadowcast surface replicas, or thin sections of biological materials might be able to satisfy his needs by fitting a transmission stage and detector to his SEM. This

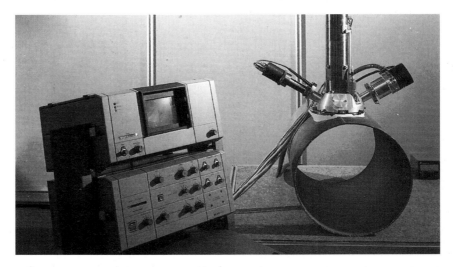

Figure 3.46. Tesla BS 343 Mini-SEM Mobile SEM. This features several forms of specimen chamber, including a 'suck-in' flange permitting the microscope to be mounted on a specimen such as the 0.45 m diameter welded pipe shown in the photograph. (By courtesy of Tesla Elmi a.s..)

would consist of a holder for standard 3 mm TEM grids and an electron detector, either a disc of silicon or a scintillator/'light-pipe'/ photomultiplier sequence. Very good results can be obtained by this approach, provided that the microscope has at least 30 kV available and is able to form a very small scanning probe of electrons. Since there are no imaging lenses after the specimen, the transmitted image is free from chromatic effects and it may be possible to examine materials which would require a higher voltage (e.g. at least 60 kV) in a dedicated TEM. A marked improvement in the spatial resolution for x-ray analysis will also be obtained on transmission specimens. Because of the almost universal availability of scan rotation on SEMs the STEM worker can orientate his image to suit a rectangular recording medium in a way in which relatively few TEM workers can. The high quality of result possible can be seen in Case Study 1, Chapter 8.

3.4.2 Scanning microscopy on a dedicated TEM

Probably the most common usage of STEM covers the scanning facility added to a dedicated TEM, which enables a finely focused illuminating probe to be scanned across the specimen, with as many operating modes of scanning microscopy to be used as there are detectors fitted. These are usually for secondary, backscattered and transmitted electrons and x-rays; the combined instrument forms a very powerful analytical electron microscope.

The column design used by a number of manufacturers for a very adaptable analytical TEM is based on the symmetrical condenser/objective lens with low aberrations developed by Riecke and Ruska (Kunath *et al.*, 1966) and shown earlier in Figure 1.14. In this lens the specimen is positioned at the centre of the pole-piece gap. The Riecke–Ruska lens acts as a double lens – a condenser lens before the specimen (*pre-field*) and a magnifying objective lens following it. The pre-field lens will operate as an additional demagnifying lens for probe formation for scanning microscopy or, if illumination is focused in its front focal plane, will give optimum parallel illumination at the specimen for transmission microscopy. Another condenser lens (C3) may be used as an active interface between the usual two-condenser illuminating system of the TEM and the objective pre-field, forming either the very small focused probe or parallel illumination at the specimen, according to its mode of operation.

The Riecke–Ruska lens allows highly convergent probes down to about 1.5 nm in diameter to be formed from tungsten filament guns, and smaller probes from brighter sources (Figure 3.47).

Compared to the dedicated SEM, and even to the upper specimen stage of an in-lens high-resolution SEM, the space available around the specimen in a high-resolution TEM is very limited; however room is found for a solid-state backscattered electron detector, an energy-dispersive x-ray detector, and a scintillator secondary electron detector, all within a centimetre or two of the specimen, giving very good collection efficiency for the various emissions (Figure 3.48). The secondary electron image is formed in the same way as the in-lens SEM, by electrons spiralling up the lens bore, and is therefore formed mainly by true secondaries, with very little backscattered component. It may therefore be a strictly surface image, without the depth contribution provided by the higher energy electrons. Electrons and x-rays undergo very little lateral scattering within a transmission specimen, and emissive signals of electrons and x-rays emerge from a region not much larger than the

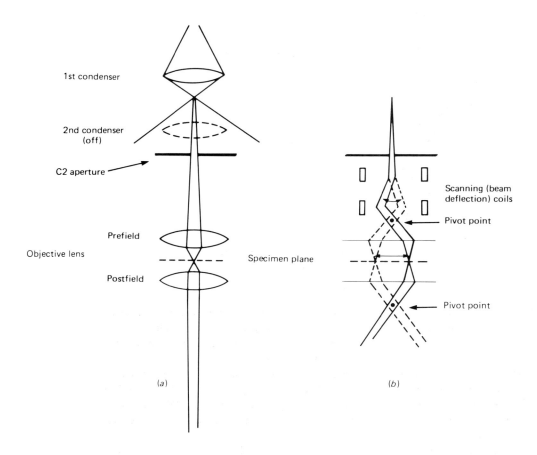

Figure 3.47. (*a*) Formation of a small electron spot for scanning microscopy in an analytical TEM with a symmetrical Riecke–Ruska condenser/objective lens. With the second condenser switched off the demagnified image of the electron source formed by C1 is further demagnified in the specimen plane by the pre-specimen field of the strongly excited objective lens. The effect of spherical aberration on the spot size is minimised by use of a small aperture in C2. (*b*) Double-deflection scanning of the beam in the STEM mode uses the beam deflection coils of the illuminating system to tilt the beam between the extremes shown, so that the beam pivots in the front focal plane of the objective lens. The electron beam is then travelling parallel to the optical axis of the column in the specimen plane, and scans a line across the specimen between the extremes shown. After passing through the specimen the beam pivots in the rear focal plane of the objective lens. A second set of deflection coils may be located here to 'de-scan' the beam, should it be deired to form a static transmitted beam for analysis by electron energy loss spectroscopy (EELS, see Chapter 6). There will be a second set of deflection coils at right angles to those shown, to provide the frame-scanning deflection.

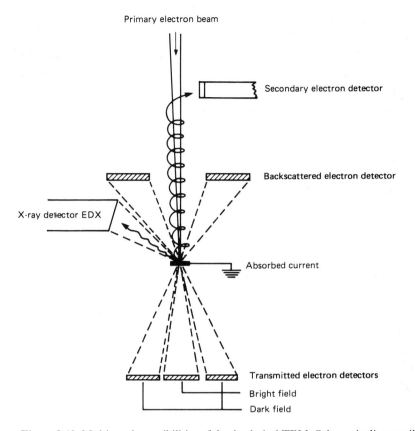

Figure 3.48. Multi-mode possibilities of the Analytical TEM. Schematic diagram illustrates a typical disposition of detectors in and around the specimen stage of an analytical TEM. Access to the specimen to collect radiations and electrons is very restricted, especially in top-entry stages. It may not be possible to fit all the detection systems illustrated on a particular microscope.

Backscattered electron and x-ray detectors require closest approach to the specimen, and maximum collection angle at the specimen. The solid-state BSE detector is ideally mounted on the underside of the upper lens polepiece, probably less than 0.5 cm above the specimen plane. The EDX detector is preferably mounted above the specimen plane so that x-radiation can be collected with the specimen in its normal, untilted, position.

SE and STEM detectors may be mounted further from the specimen plane, since the electron paths are constrained by the lens fields. Secondary electrons, having very low energy, are forced to spiral around the optical axis and are pulled out of the upper lens bore to a positively charged scintillator (of Everhart–Thornley type). The STEM signal is detected at any convenient point on the axis below the specimen, e.g. below the final projector lens or beneath the fluorescent screen in the viewing chamber. It may be a single disc or a concentrically mounted disc and annulus for simultaneous and separate detection of bright- and dark-field images. If the specimen holder is suitably isolated electrically from the stage the net electron current to it can be amplified and used for specimen-current imaging.

diameter of the incident electron beam. A qualitative elemental analysis of 10^{-17} g of material is readily possible under these circumstances.

STEM modes can be added as accessories to suitable transmission microscopes, with the additional supplies for scan coils and detectors housed in subsidiary modules. On purpose-designed AEMs the coils and circuits will be integrated with the controls for the other operating modes. Extra visual and record cathode-ray tubes, together with the EDX display and computer keyboard make up the visible signs of the most important development in transmission microscopy for many years.

3.4.3 The field emission STEM (FESTEM)

This differs from the previous two techniques in not being an attachment or an additional mode of operation but an instrument dedicated specifically to obtaining near-atomic resolution from a detailed analysis of transmitted electrons, as outlined in Chapter 2 (page 48).

The concept and first realisation of the FESTEM was the work of A. V. Crewe (Crewe, 1971) and collaborators in the USA. It was the first electron microscope designed around the high brightness and small source size of the cold field emission electron gun. It differs from the field emission SEM described earlier in being designed specifically for very thin transmission specimens and in analysing the direction and energy of the transmitted electrons. It is shown schematically in Figure 3.49. Separate solid-state detectors are fitted for on-axis electrons (bright-field) and those scattered off-axis (annular dark-field). In addition, an electron energy analyser is fitted before the bright-field detector.

The electron beam transmitted by the specimen has three components.

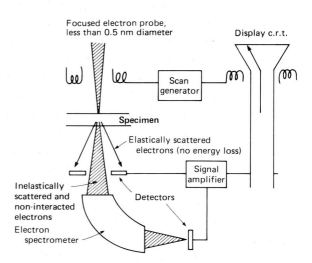

Figure 3.49. Schematic outline of the Field Emission Scanning Transmission Electron Microscope FESTEM.

N_e is the current of elastically scattered electrons collected by the annular detector.

N_0 is the electron current which has not interacted with the specimen, collected by the on-axis detector when the energy filter is set to transmit electrons with the full beam energy, i.e. with zero energy loss.

N_i is the current of inelastically scattered electrons, which have lost energy to the specimen, and whose trajectories have been deflected through small angles in the process. They are collected by the on-axis detector when the energy filter is tuned to energy losses greater than zero. Plotting N_i against energy loss gives a spectrum usable for elemental analysis by electron energy loss spectroscopy (EELS) (Chapter 6).

The total beam current $N = N_0 + N_i + N_e$.

When the focused electron beam is scanned across the specimen the three outputs can be used separately or in conjunction to modulate the brightness of spot on the final image tube. For example, signal N_e/N_0 gives the equivalent of a *dark-field* image in conventional transmitted light or electron microscopy (see page 000). The ratio signal N_e/N_i gives an image whose brightness is proportional to Z. There is no equivalent to this in conventional transmission electron microscopy. The power of the method is diminished, however, if the electron beam has made multiple collisions in passing through the specimen, i.e. for thick specimens.

A subsidiary benefit from this method of image formation is that perturbations which affect both signals are eliminated when one is divided by the other. Thus the effect of fluctuations in beam intensity which sometimes occur in cold field emission sources can be minimised in the processed image.

The final break with tradition pioneered by Crewe in his FESTEM was to construct the entire instrument to operate with ultra-high vacuum. His original instrument was the size of a small suitcase and was easily evacuated to ultra-high vacuum (u.h.v.); it was taken apart to change specimens. This feature, although suitable for a prototype in which the complete microscope was being evaluated, is impractical when the microscope is to be used for examining a number of specimens. In commercial instruments, therefore, only the electron gun is evacuated to u.h.v. and a slightly higher pressure is tolerated in the specimen stage. Specimen exchange takes placed through an airlock chamber. Figure 3.50 shows commercial FESTEM instruments. Compared to a conventional TEM the column is upside-down, with the electron gun at the bottom. This gives the advantage that the specimen stage may be mounted very rigidly on top of the objective lens, which helps to provide the very high mechanical and thermal stabilities necessary when magnifications of millions of times are involved. It also simplifies the mounting of the energy analyser and any other detectors in the vicinity of the specimen. The subtended angle of a single EDX detector can be made as large as 0.3 steradian, which is the highest yet achieved on any electron beam instrument and an effective 0.42 steradian can be achieved by fitting two 0.21 steradian detectors on either side of the specimen (Lyman *et al.*, 1994)

The most spectacular results from the FESTEM have been the imaging of the positions of single heavy atoms in specially prepared specimens. In more mundane microscopy the great advantage of such an instrument has been due to the ease with which intense sub-nanometre electron probes have been available for elemental and crystallographic analyses. Analysis by EELS has been more precise than has been possible on microscopes with thermionic electron sources. The facility for simultaneous comparison of bright- and dark-field images is a valuable one, as well as the *Z*-contrast image produced by dividing one signal by the other. Applications of the FESTEM have extended beyond the biological subjects originally envisaged by Crewe, and it has proved to be a valuable tool in materials science.

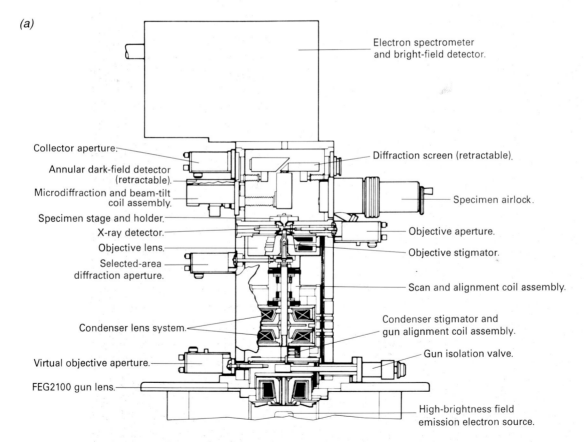

(a)

Electron spectrometer and bright-field detector.

Collector aperture.
Annular dark-field detector (retractable).
Microdiffraction and beam-tilt coil assembly.
Specimen stage and holder.
X-ray detector.
Objective lens.
Selected-area diffraction aperture.

Diffraction screen (retractable).
Specimen airlock.
Objective aperture.
Objective stigmator.
Scan and alignment coil assembly.

Condenser lens system.
Virtual objective aperture.
FEG2100 gun lens.

Condenser stigmator and gun alignment coil assembly.
Gun isolation valve.
High-brightness field emission electron source.

Figure 3.50. (*a*) Cross-section of the VG 601 UX FESTEM, 20–100 kV. The instrument has a lattice-resolving power of less than 0.144 nm and can carry out elemental analysis by EDX and EELS with a 1 nm probe carrying 0.7 nA. (*b*) The photograph shows the VG 603, a 300 kV FESTEM (By courtesy of VG Microscopes Ltd.)

3.5 Combined light and electron microscopes

The light microscope has achieved an honoured place in many laboratories because of its directness of operation and the undoubted advantage of presenting a coloured image. However, its resolving power and depth of field are limiting and it does not have the analytical possibilities of electron microscopes.

(b)

The Topcon (formerly ISI) LEM-2000, combined transmission light and electron microscope illustrated in Figure 3.51, has a horizontal column consisting of four imaging lenses and double condenser illumination, and a light microscope with a rotatable turret of objective lenses. The specimen lies within the vacuum system for both types of examination. The light microscope uses a high-resolution achromat lens system with a magnification range of $\times 50$ to $\times 500$. The TEM optics are designed to get the best out of semi-thin sections in the magnification range $\times 250$ to $\times 45\,000$, using 100 keV electrons. Both LM and TEM images are seen in binoculars under normal room lighting. Specimens can be mounted on standard 3 mm diameter TEM specimen grids or on special 7 mm grids. The specimen stage is motorised and will introduce the specimen into either the LM or TEM observation positions by pressing the appropriate button. The X, Y coordinates of areas of interest in either mode can be memorised by a microprocessor and relocated automatically in the other mode.

3.6 Digital image recording

The SEM has by its nature a sequential image which can be digitised for storage in a computer memory or frame store. The fluorescent final image of a TEM can be scanned with a TV camera and similarly presented as a digital image. Thus all members of the EM family are potential candidates for digital image storage, transmission and processing, and the possibilities of producing prints or putting images directly into reports using desktop publishing techniques are potentially useful.

On a TEM the procedures of operating the microscope and selecting features for examination can be carried out from a TV-rate image of the fluorescent screen. Changes due to specimen movement, focus change, etc., can be seen as they occur ('in real time'). Images can be 'grabbed' and recorded from such a display, but the highest quality (near-photographic) digital image is obtained from a slow-scan CCD camera, which does not produce its image in real time but over a few seconds (Krivanek, 1992). Digital images from such a camera may be analysed using appropriate software to detect imperfections in alignment, astigmatism and focus of the microscope, and automatically feed back correction signals into the instrument (Saxton & Chang, 1988). Such a procedure is essential for microscopy near the resolution limit of the microscope.

More will be said about photographic and digital images in Section 5.3, 'Recording and printing electron micrographs'.

Figure 3.51. The Topcon LEM-2000 combined light–electron microscope. (*a*) (*opposite*) The microscope desk unit. (*b*) (*opposite*) Cross-sectional schematic diagram. (*c*) (*page 134*) Micrographs of a 150 nm section of human kidney in progressively greater detail. ((*a,b*) By courtesy of Topcon (Europe) BV; (*c*) By courtesy of Prof. V. Otaka, Dr S. Suzuki and Dr H. Takizawa, Dept of Pathology, Tokyo Medical College.)

(a)

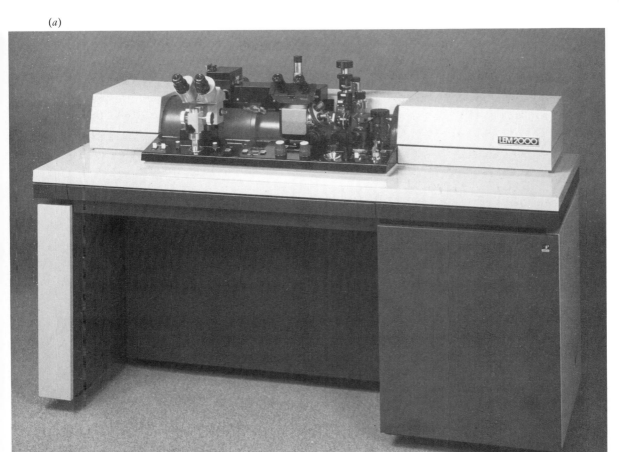

(b)

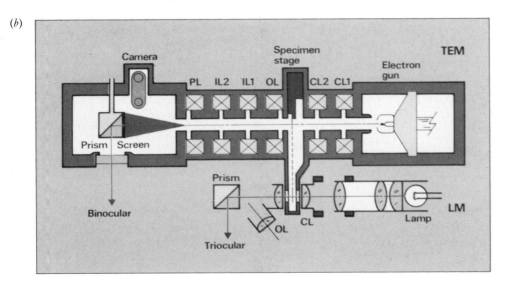

Figure 3.51 (*cont.*)
(*c*)

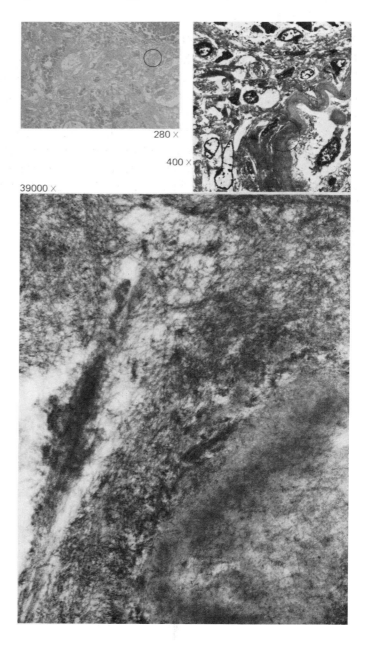

280 ×

400 ×

39000 ×

3.7 Suggested further reading

Following on from the early histories of electron microscopes in Ruska (1980) and Hawkes (1985), the definitive history of the growth and expansion of electron microscopy is given in Mulvey (1996). Descriptions of individual steps in the advancement of instrumentation through the years will be found in the proceedings of the IFSEM conferences (see the main bibliography at the end of the book).

A combined light microscope and SEM is described by Schlüter *et al.* (1979).

The actual procedures involved in operating and maintaining a commercial electron microscope are described and explained by Chescoe & Goodhew (1990) and Chapman (1986).

4 *Specimen preparation for electron microscopy*

4.1 Introduction

The reader has now been introduced to the theoretical background and the essential hardware of electron microscopy. The next step is to see what is involved in preparing actual specimens for examination by the various types of electron microscope. This chapter will give a general introduction to specimen preparation for electron microscopy, followed by an account of the preparation of more difficult subjects, such as moist or living (biological) materials.

The specimen requirements for scanning and transmission microscopy are very different, and preparation for the two techniques will therefore be dealt with separately. The reader is advised to follow both parts, however, as it is only by appreciating the distinctive features of each instrument that a balanced approach to electron microscopy can be obtained and maximum benefits obtained from it.

This book will only give the broad outline of basic procedures; for practical details of how to carry them out the reader should refer to laboratory manuals such as Kay (1965), *Practical Methods in Electron Microscopy*, edited by A. M. Glauert (the Glauert Series) or *Procedures in Electron Microscopy*, edited by A. W. Robards & A. J. Wilson. (See the bibliography, later).

4.2 Specimen requirements for the TEM

The essential conditions to be satisfied by a specimen for the transmission electron microscope are:

(i)   It must transmit sufficient electrons to form an image, with minimum energy loss.

(ii) It must be stable under electron bombardment in a high vacuum.

(iii) It must be of a suitable size to fit the specimen holder of the microscope.

The reasons for (ii) and (iii) are obvious; the requirement of (i) results from the fact that, although some interaction between the electron beam and the specimen is necessary in order to produce image detail, inelastic encounters resulting in energy loss by the electrons degrade the quality of the image because of chromatic aberration in the lenses. For the normal 100 kV microscope this limits specimen thickness to a maximum of about 100 nm, but the value depends on the nature of the specimen and its atomic number. An empirical rule-of-thumb propounded by Cosslett (1956) stated that for amorphous specimens (he had in mind biological materials primarily) the resolution would be limited to one-tenth of the specimen thickness. Objective lenses with lower chromatic aberrations have been introduced since that time; nevertheless, the restriction on specimen thickness for high resolution at normal voltages is still severe. This has provided some of the incentive for higher-voltage microscopy, for the use of STEM techniques and for the development of energy-filtering microscopes.

If a specimen has to be modified to satisfy the above conditions there is a further requirement :

(iv)   Any preparatory treatment should not alter the structure of the specimen at a level which is observable with the microscope.

Most materials are initially in an unsuitable form for transmission microscopy and a number of standard procedures have been developed by which suitable specimens may be prepared for examination. Because of the extreme sensitivity of the microscope, revealing detail on a near-atomic scale, specimen techniques have to be chosen and used with great care. Patience and regular practice are necessary if top quality results are to be obtained consistently. Perhaps because of this the preparation of specimens for the TEM has acquired the reputation of being a difficult and tiresome 'art' (particularly amongst those who have never tried it!).

4.2.1 Specimen support grids

One of the fundamental hardships of preparing and handling specimens for the TEM lies in the small dimensions involved. Materials which are only 10–100 nm thick are very delicate and must be firmly supported in the microscope. They are normally mounted on a circular metallic mesh 3 mm in diameter and about 15 μm thick. This *specimen support grid* is usually made of electrolytic copper but may also be made of other materials, e.g. nickel, gold, palladium, stainless steel, beryllium, or carbon-coated nylon, because of special requirements in the specimen preparation or examination. Specimen support grids are made to a number of designs differing in the size, shape and arrangement of the holes and bars. Some typical patterns are illustrated, enlarged 11 times, in Figure 4.1. It is an intriguing thought that the circular holes of the AEI pattern are only 35 μm across, yet one hole will provide over a thousand different fields of view when photographed at a magnification of ×100 000! The solid metal grid bars are of course necessary to provide support to the specimen, but have the disadvantage that between 25% and 65% (depending on the grid used) of a specimen will not be accessible for examination because it lies over the opaque bars. The same proportion of the electron beam in the microscope is absorbed by the mesh when the illumination is 'spread' over a number of grid openings; hence the preference for copper grids to conduct away the current and heat to the body of the specimen holder.

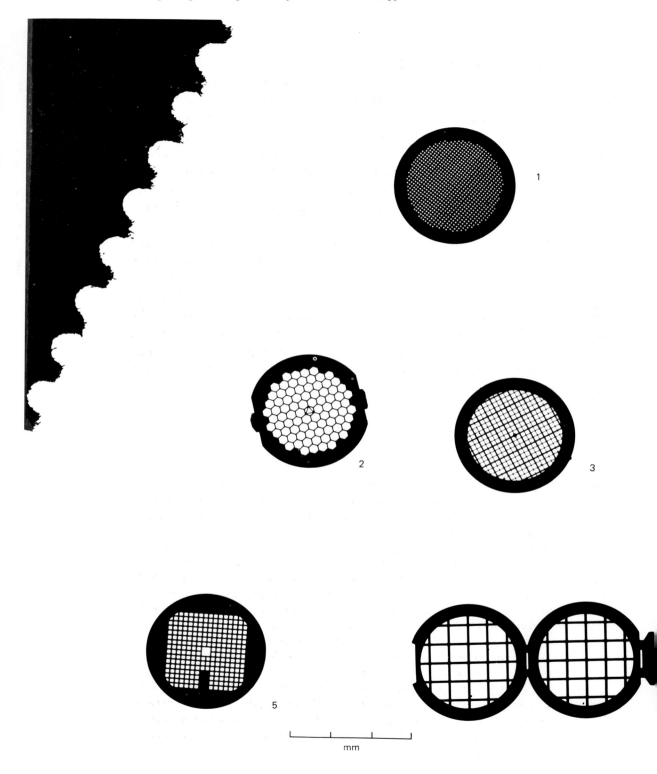

4.2.2 Specimen support films

Self-supporting thin specimens such as wires, fibres, thinned foils or replica films may be mounted directly on to the specimen grids or sandwiched between the halves of a folding grid. More delicate and particulate matter requires support from a thin flat electron-transparent film already stretched across the openings of a specimen grid. Such films may be as thin as 1–2 nm or up to tens of nanometres thick and should ideally be free from microscopic self-structure. Carbon (Bradley, 1954a), silicon monoxide, or plastics such as polyvinyl formal (Formvar) and nitrocellulose (collodion) may be used. The support films are prepared on a flat substrate. e.g. cleaved mica, glass or a liquid surface, and are transferred to the specimen grid.

At magnifications of ×100 000 and above the structure of commonly used support films becomes more obtrusive and the interpretation of specimen detail of 2 nm and smaller can be more difficult. When the study demands an absolutely structureless background it may be possible to mount particulate specimens on perforated support films or micromeshes and to find regions of the specimen spanning holes in the support film. Thin flakes of graphite and crystals of magnesium oxide are also used to provide structureless supports. Further information on the preparation of carbon support films will be found in Appendix 2, and practical details in Kay (1965) or Robards & Wilson (1993).

(*Left*)

Figure 4.1. Specimen support grids of several patterns compared for size with part of a postage stamp. The patterns illustrated are five out of hundreds of available designs, and are mostly intended for specific types of examination. (1) AEI pattern: an array of 35 μm holes spaced at 75 μm centres. Used at high magnifications, with thin support films, when a high degree of specimen support is required. (2) Hexagonal pattern: provides large open areas for examination. May also have the openings identified by letters, allowing specific features to be relocated or identified in both light and electron microscopes. (3) NEW 200: general purpose design due to Cosslett. 200 lines to the inch (8 lines/mm) in both directions, with varied combination of thin and thick grid bars to identify orientation both in and out of the microscope. (4) 'Oyster' folding grid, 50 bars to the inch (2/mm) both ways. Used to 'sandwich' thin metallic foils or other self-supporting specimens. The two halves are locked together by folding back the end tab. (5) Sira 200 grid with open centre, square outer frame and orienting bar. Designed originally to facilitate replication of selected areas of surfaces (Watt, 1964), but open centre and orienting bar provide wider utility.

Although specimen grids were originally punched out of sheets of finely woven wire gauze they are now made by a refined electroplating technique: hence the wide range of patterns which are possible. One face of the disc is always smooth and shiny; the other may be matt or shiny, depending on the plating procedure used in its manufacture. (Patterns 1, 3, 5 made by Smethurst High-Light, 2 by Graticules Ltd and 4 by Mason and Morton.)

4.3 Basic preparatory methods for TEM specimens

4.3.1 Particulate samples

There is a wide variety of techniques for depositing particles in a suitably dispersed state on support films for TEM examination. Dry powders may be dusted directly on to the support film. Particles in liquid suspension may be agitated ultrasonically to break up agglomerates and then deposited on the support in a droplet which is allowed to dry out, or sprayed on using a nebuliser or artist's airbrush. Smoke particles can be deposited by holding a filmed grid in the smoke. There are special techniques for precipitating airborne dust particles and for grading smoke particles for size by centrifugation. More will be said about the study of particles in Chapter 8.

Particles deposited on grids by any of these methods are seen in silhouette in the electron microscope; if they are sufficiently thin in the direction of the beam some internal detail may also be seen. Biological particles may require staining (see pages 153–4) on the grids before their structure becomes clearly visible. Crystalline particles can be studied and identified by electron diffraction (Chapter 6). Information on particle thickness can be obtained by the further process of shadowcasting (Section 4.3.3) or by some form of densitometry on micrographs (Chapter 5).

4.3.2 Surface replication

Most solid surfaces are unsuitable for direct examination by transmission microscopy and surface studies are nowadays carried out in the SEM or analytical TEM. Surfaces which are too valuable, too remote, or too large for the SEM still require replication for TEM or SEM. Fortunately, before the advent of scanning microscopy a number of replica procedures were developed for the study of surface structure in the TEM. These are still widely practised and can reveal surface detail which is invisible to most SEMs.

Surface replication for the TEM involves moulding the surface shape of a solid specimen in a thinner material which is transparent to electrons. The replica may be a positive, exactly duplicating the features of the original surface, or a negative with features which are the reverse of those on the specimen. The former is much to be preferred.

In the early days of transmission microscopy, thick plastic negative replicas were used, supposedly plane on one side and impressed with the surface structure on the other (Agar & Revell, 1956). Viewed by transmission the thickness variations provided information on the surface profile of the specimen. Shadowcasting the replica (see below) emphasised the fine structure of the surface. Thick replicas were found to be unsatisfactory as microscopes and materials improved, however, and nowadays replicas for the TEM are almost invariably made of a layer of carbon some 10–20 nm thick formed by evaporation in vacuo (Bradley, 1954b). The layer may be prepared by direct evaporation on to and subsequent removal from the specimen surface (*single-stage replica*) or by evaporation on to a preliminary plastic negative replica cast on and removed from the specimen (*two-stage replica*). These procedures are illustrated diagrammatically in Figure 4.2 with mention of some of the more commonly used replica materials. An important practical point is that a single-stage replica may be removed from its specimen by dissolving the latter away; this is an important technique in practice, and provides a means of replicating particles or fibres which are enveloped by the evaporated carbon layer.

Some replica processes may involve more than two stages of replication and re-replication, but the overall treatment should be kept as short as possible since there is a risk of losing some detail or introducing artefacts in each new operation.

The absolute limit to the fineness of structure which can be reproduced reliably by replicas is probably below 5–10 nm, and 2 nm resolution has been claimed for direct replication with co-evaporated platinum–carbon (Bradley, 1959).

4.3.3 Shadowcasting
The visibility of detail resolved by replicas is very much bound up with the further process of *shadowcasting* which produces a three-dimensional effect in TEM images. Just as the surface texture of a wall is accentuated by shadows cast under oblique illumination, so the surface undulations on a replica are emphasised if the replica is coated obliquely with a very thin deposit of heavy metal evaporated from a point source. The deposit is thickest on parts of the surface normal to the direction of deposition and least on regions sloping away from the source or in the shadow. Looking normally through the shadowed replica in the transmission microscope, the

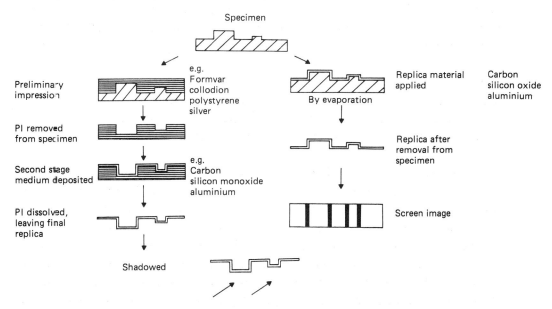

Figure 4.2. Diagrammatic representation of single- and two-stage replica processes (*right* and *left* respectively).

effect of the shadowcasting is to create highlights and shadows in the electron image. The size and form of a surface feature, or thickness of a particle on a support film, can be deduced from their shadows (Figure 4.3).

Shadowcasting may be used quantitatively, since shadow length, shadowcasting angle and specimen height are related, or qualitatively as a means of emphasising surface topography. The technique of conical shadowing provides a qualitative impression free from uninformative empty shadows; the replica is rotated continuously in its own plane during the shadowcasting, tonal differences in the micrograph representing differences in surface inclination. Shadowcasting angles between 45° and 10° to the surface are commonly used. Steeper angles are liable to introduce artefacts.

The choice of the metal to be deposited in shadowcasting is important, as are its thickness and manner of deposition, since these factors control the self-structure of the shadowcasting layer and hence the fineness of detail which can be seen unambiguously in the final micrograph. This is discussed in Appendix 2.

Referring back to Figure 4.2 it will be appreciated that one side of the thin film carbon replica is a positive reproduction of the specimen and the other side is a negative replica. The shadowcasting layer should be applied to the positive side (as shown in the figure), which is equivalent to shadowing the original surface. In some cases it may be possible to shadowcast the specimen itself before applying the evap-

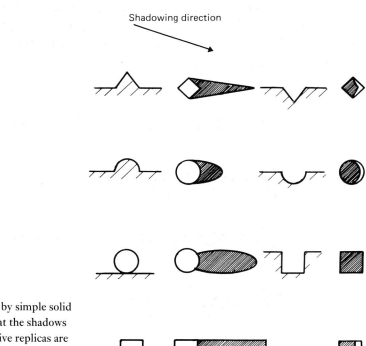

Figure 4.3. Shadows cast by simple solid profiles. It can be seen that the shadows cast in positive and negative replicas are not interchangeable in either shape or tone.

orated carbon replica layer (e.g. in freeze-fracture replication). The combined shadow and replica films are removed together from the specimen and form a *pre-shadowed single-stage replica*. This is the most faithful type of surface replica since the dimensions of the shadows were determined by the actual specimen itself. In 'reading' a micrograph it is necessary to know whether a replica is a positive or negative, and at what stage it was shadowcast, if a correct interpretation is to be obtained. This will be enlarged on later (page 192).

One specialised form of replication is *extraction replication* (Smith & Nutting, 1956), (Figure 4.4), in which certain constituents of a surface are collected for examination and analysis in the microscope, e.g. by electron diffraction or x-ray analysis (Chapter 6). In the study of metals a surface may be polished and etched to reveal certain inclusions. A layer of carbon is evaporated on to the surface which is then etched away further to release the carbon and the inclusions. The extraction replica thus carries with it particles of the second phase whose nature and orientation relative to the parent surface can be studied in the microscope.

A modified form of extraction replication enables the nature of thin, brittle, surface layers to be determined. Running a diamond point lightly across the surface breaks the layer into small chips, which are extracted with a carbon film subsequently evaporated on to the surface.

An extension of the extraction replication technique is *double extraction replication*, reported by a number of users including Czyrska-Filemonowicz *et al.*, (1991). Instead of replicating an etched surface of a bulk sample the sample is mechanically or (electro)chemically thinned to below 0.1 mm thickness before both faces are etched and carbon-coated. The matrix of the sample is gently dissolved away in order to leave

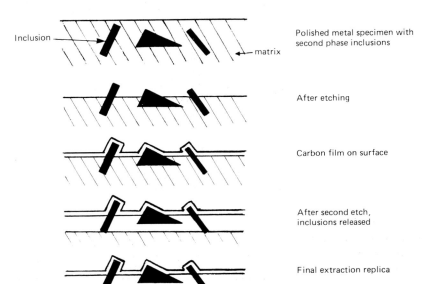

Inclusion — matrix

Polished metal specimen with second phase inclusions

After etching

Carbon film on surface

After second etch, inclusions released

Final extraction replica

Figure 4.4. Schematic diagram of the extraction replica process used in metallurgical studies of alloys.

any precipitates included between the two carbon replica films. The actual positioning and orientation of fine precipitates may not all be retained (as in the single extraction replica) but the number, size distribution and composition of precipitate particles can be determined using TEM or SEM. Very large brittle precipitate particles are preserved, which would almost certainly have been fragmented in a conventional extraction replica. An example of such a particle is seen in Figure 4.5.

4.3.4 Thinned specimens

By its nature the transmission microscope is able to reveal the internal structure of materials, the arrangement of their constituent parts and the degree of crystallinity or atomic order within the material. The TEM is widely used for examinations of biological and industrial materials suitably prepared to allow access to their internal detail. Initially the specimens may not satisfy any of the requirements for TEM specimens and specialised techniques are required to prepare them for examination. This section will explain briefly what the preparation procedures involve. Some further explanation is given in the next chapter but any reader who requires working details of methods and materials is referred to appropriate laboratory manuals on specimen preparation (e.g. Kay, 1965; the Glauert series; or Robards & Wilson, 1993). For convenience the types of prepared specimen will be classified into two separate groups, *ultra-thin sections* and *foils and films*.

Figure 4.5. Chromium-rich $M_{23}C_6$ carbides particles extracted in a double extraction replica from austenitic stainless steel (18Cr–10Ni steel) after ageing at 750 °C. (By courtesy of A. Czyrska-Filemonowicz.)

Ultra-thin sectioning

One way of discovering what structure lies beneath the surface of a specimen is to cut it into slices thin enough to be examined by transmission microscopy. In practice this means that the slices or ultra-thin sections should be 0.02–0.1 μm thick for a 100 kV TEM, but can be up to several μm for the HVEM (Hama & Porter, 1969). The technique of ultra-thin sectioning is most frequently employed in biological examinations, but it is applicable to a wide range of materials. Usable sections may be cut from all but the hardest or most brittle types of material, provided it is suitably prepared for cutting (which implies reducing the size of cross-section which is to be cut).

The ultramicrotome

An ultramicrotome is an instrument in which a suitably shaped block of the material to be sectioned is made to fall past a stationary cutting edge of glass or diamond; a thin slice is sheared from the front face of the block, as shown diagrammatically in Figure 4.6. The specimen face is now advanced towards the cutting edge by a very small but controlled distance and the cutting operation is repeated. If a series of consecutive uniformly thin sections is required the ultramicrotome must be capable of making a succession of identical small advances in specimen position between cuts. Specimen advances adjustable between 5 and 100 nm are usually provided. Several different methods are used for this, including a geared-down mechanical advance and elongation of the arm holding the specimen by controlled thermal expansion or magnetostriction. Because of the extremely small distances involved in sectioning it is important that the specimen and ultramicrotome be insulated from mechanical vibrations and variations in temperature due to draughts, heat from lamps and the operator's breath. In this context it is worth bearing in mind that a specimen block

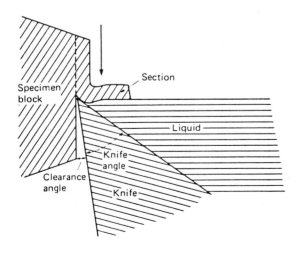

Figure 4.6. Schematic cross-section of the process of ultra-thin sectioning. The specimen falls (or is pulled) past the cutting edge of a glass or diamond knife. The thin section is sheared from the parent block and floats on a liquid bath behind the cutting edge. Important parameters, which have to be optimised for each specimen/knife combination, are the angle of the cutting edge, the clearance angle between the cut face and the knife, and the speed of cutting. Because of the forces involved the sections always suffer some degree of compression, being shorter than the length of face from which they were cut but thicker than the specimen advance. (N.B. The thickness of section in relation to the length of the block face has been greatly exaggerated in this drawing; in reality the ratio would be nearer 1:5000.)

projecting 5 mm from its holder on the microtome will change in length by about 20 nm for a one degree change in temperature.

The important features in an ultramicrotome are illustrated schematically and described in Figure 4.7(*a*). A practical layout of the component parts is shown in the photograph of a commercial ultramicrotome (Figure 4.7(*b*). A more detailed description of the possible variations in ultramicrotome design and construction is to be found in Reid & Beesley, (1990).

A close–up view of an actual cutting arrangement is shown schematically in Figure 4.8. The specimen, encapsulated in an 8 mm diameter cylindrical block of resin or hard plastic, is clamped in a collet on the end of the arm of the ultramicrotome. The end of the block has been trimmed to form a truncated pyramid, so that sections cut from it will be between 0.5 and 1 mm square.

It has been found in practice that neat ribbons of sections can be formed if the block face is trimmed to a trapezium, the section being cut from the longer parallel side to

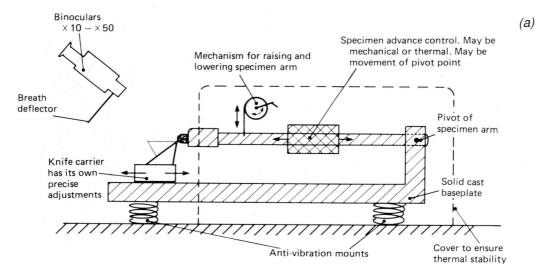

Figure 4.7. (*a*) Salient features of an ultramicrotome. The specimen is clamped firmly at one end of a rigid arm which is pivoted at the other end relative to the massive base plate of the instrument. The arm is raised and lowered by a mechanism which may be manually or motor driven. The cutting stroke may be under gravity or may be power-assisted. The speed of the cutting stroke is adjustable over a wide range, typically 0.1–20 mm s^{-1}. The angular positions of the knife edge and specimen block are both adjustable so that the operator has complete control over the speed, direction and frequency of the cutting operation, observing it through a stereozoom binocular microscope at magnifications of between 10 and 50 times. (*b*) A commercial ultramicrotome (Leica Ultracut UCT) with mechanical specimen advance. The specimen advance is stepping-motor controlled via a separate electrical control unit (left). Length of cutting stroke, cutting speed and specimen advance are all controlled from this box. Up to five combinations of cutting parameters can be retained in memory, for use on different specimen types. For initial setting-up, and for coarse sectioning, the knife advance is controlled electrically via a wheel on the control unit. On the right side is a hand wheel which can be used for manually raising and lowering the specimen arm. (Photograph by courtesy of Leica Ltd.)

the shorter one, as illustrated. The cutting edge is a freshly fractured corner of an accurately broken triangular piece of plate glass. Cut sections float on the surface of a bath of liquid, frequently water, contained in a sheet-metal reservoir waxed to the knife behind the cutting edge. Selected sections are picked up from the surface of the liquid with a specimen grid held in fine forceps. A thin plastic or carbon support film may be deposited on the specimen grid beforehand to provide additional support. Specimen

(b)

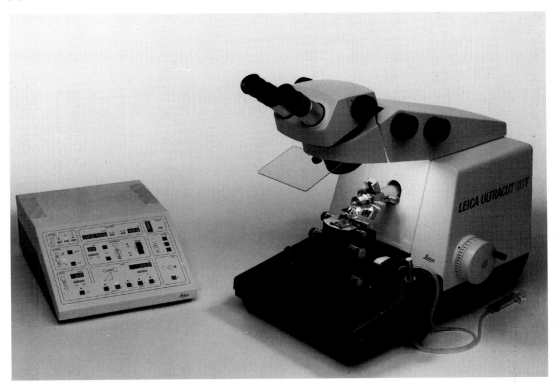

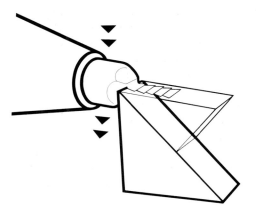

Figure 4.8. An artist's impression of section cutting in practice. A specially trimmed specimen is sectioned on to a bath of liquid, usually water, behind the cutting edge. (N.B. The size of the sections in relation to the knife edge has been exaggerated.) (By courtesy of Leica Ltd.)

grids may have the usual square mesh or, if a series of sections is to be mounted on a single grid, may have a slot aperture, in which case a support film will almost certainly be required. Some microscopes with side-entry rod-type holders have provision for extra-long grids (e.g. 10×4 mm) on which a string of sections may be mounted and viewed without interruption by opaque grid bars.

Embedding procedures

A wide range of specimen types are sectioned by ultramicrotomy for examination by transmission microscopy. The majority are of a biological nature since there is still no more satisfactory method of studying the reactions of human and animal tissue to treatment with drugs, etc., than by cutting sections and comparing treated and untreated material.

Freshly prepared biological materials destined for sectioning are first of all stabilised or 'fixed' either chemically with a reagent such as osmium tetroxide or glutaraldehyde or by rapid freezing to preserve its structure (cryo-preparations are described later in the chapter).

After chemical fixation the materials are encapsulated in a resin or other plastic material to give mechanical support during sectioning and subsequent handling. Biological specimens usually contain a proportion of water, and this is progressively replaced by a dehydrating agent such as ethanol, which is in turn replaced by increasing concentrations of the liquid resin. The impregnated tissue, etc is placed in a plastic mould which is topped up with neat resin. The liquid is then polymerised by heat treatment or UV irradiation for a number of hours to form a solid block of embedded specimen with suitable mechanical properties for section cutting. The resin must also satisfy a number of other requirements. It must show good penetration into the specimen (i.e. not be too viscous), good adhesion to the specimen during and after sectioning, should undergo as nearly zero dimensional changes on hardening as possible and should be stable under electron bombardment in the high vacuum of the microscope.

Finally, in biological work the cut section must be readily penetrated by staining solutions which are used to enhance the limited natural contrast of the section, or in immunocytochemistry to mark the sites of specific antibodies. Particular types of Epon® and Araldite® epoxy resins are favoured by many workers, and kits of resin, hardener, plasticiser, etc., are available from electron microscope equipment suppliers for this purpose (Glauert, 1991). Acrylic resins are also favoured by many workers. Low-viscosity resins are available to assist penetration into microporous specimens.

Modifications of structure which may accompany dehydration can be reduced by dehydrating and embedding at reduced temperatures, e.g. below -20 °C. Embedding resins are used which are hardened by exposure to UV light instead of heat. One method of working which has been found satisfactory is known as PLT

(Progressive Lowering of Temperature). The replacement of water by resin commences at room temperature, and as the concentration of resin increases, the specimen temperature is progressively reduced. The embedded sample is returned to room temperature for hardening by UV (Armbruster *et al.*, 1982).

Procedures for embedding non-biological materials are similar except that the controlled dehydration is not usually necessary. Powders and other fine, granular materials may be dried in an oven, stirred into the liquid resin and then poured into a plastic mould. More substantial specimens are placed directly in the bottom of the mould which is then filled with liquid resin.

Degassing under vacuum for a short while helps towards a more thorough penetration of the resin into porous specimens; polymerisation by heating or UV irradiation is carried out as before. The hardened resin provides mechanical support to the specimen during the sectioning process and facilitates handling of the ultra-thin section afterwards. A micrograph of an ultra-thin section nominally 50 nm thick, cut from a resin-embedded granule of bone charcoal, is shown in Figure 4.9. The importance of achieving complete penetration of the resin into the internal cavities of specimens of this kind can be appreciated, since the fine internal structure would otherwise be unsupported and disintegrate when the section was cut.

Cryo-ultramicrotomy

A technique originally developed for cutting elastic and plastic materials (Cobbold & Mendelson, 1971), and now increasingly used in biological work (Appleton, 1977; Echlin, 1992), is cryo-ultramicrotomy. In this the specimen and knife are cooled to sub-zero temperatures, when soft materials become hard and suitable for sectioning and frozen-hydrated materials can be sectioned. In the case of a number of rubbers and plastics the *glass transition temperature* lies between -50 and $-150\ °C$ and sections of suitably shaped specimens may be cut without a preliminary embedding treatment. Cryogenic attachments are available for a number of commercial ultra-microtomes (e.g. Figure 4.10(*a*)) and allow the operator to control temperatures of knife and specimen within close limits. One practical problem in cryo-ultramicrotomy is to provide a suitable liquid for the bath behind the cutting edge, since water can no longer be used. Often, no liquid at all is used. Another problem is due to electrostatic charging of the cut sections – cured with a deionising attachment to the ultramicrotome. Figure 4.10(*b*) is a transmission micrograph through a nominal 0.03 µm section cut from red rubber tubing at $-145\ °C$.

Knife materials

Freshly fractured plate glass provides readily prepared cutting edges for softer materials such as resin-embedded biological tissue, but the edges are dulled after a comparatively small number of sections have been cut, and harder materials cannot be cut at all. Fernandez-Moran (1953) showed that diamond could be sharpened for use as an ultramicrotome knife material. The length of cutting edge is limited in comparison with glass, and its cost is much greater, but its hardness allows a very wide range of materials to be cut over an extended period of time. In some laboratories, therefore, diamond knives are used for all ultra-thin sectioning, but generally glass is still used to cut softer materials and diamond is kept in reserve for harder specimens.

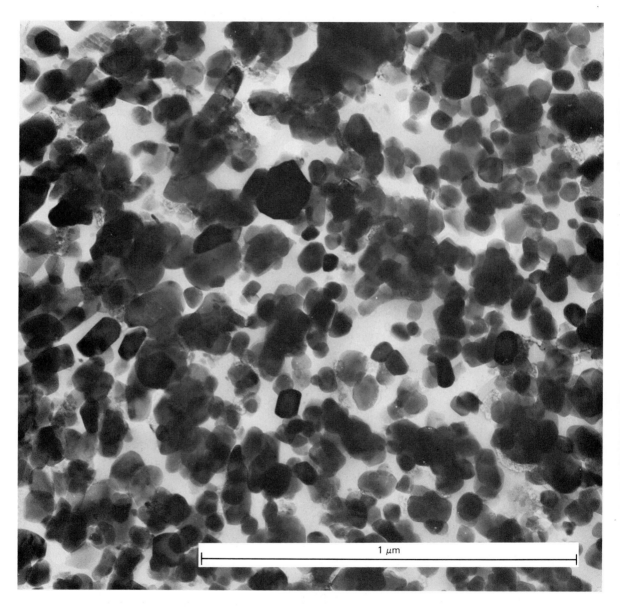

Figure 4.9. 100 kV transmission micrograph of a 50 nm section of resin–embedded bone charcoal.

Using diamond it is possible to cut sections of hard metals, although the shearing forces during cutting introduce so much new damage into the metal section that it is unsuitable for studies of lattice defects although still satisfactory for studies of composition. Therefore, thin metallic specimens are usually prepared by chemical or electrochemical processes, (see, for example, Kay, 1965; Goodhew, 1985; or Hirsch *et al.*, 1965), or by ion–beam thinning. These will be outlined in a later section.

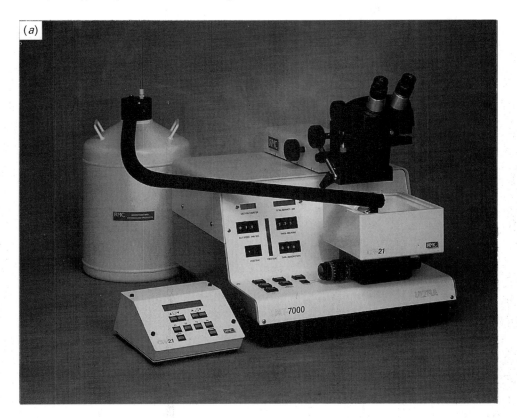

Figure 4.10. (*a*) Adaptation of ultramicrotome for cryo–ultramicrotomy. Accessories are available from the major manufacturers to enable their ultramicrotomes to be used at temperatures approaching that of liquid nitrogen ($-196\,^\circ$C). This fluid is used as coolant for the specimen and the knife, and its effect is balanced against electrical heating coils until the desired temperatures are achieved. The specimen is normally cooled to a lower temperature than the knife edge. The illustration shows a cryo–chamber fitted to the RMC MT 7000 ultramicrotome. One hazard of cryo–ultramicrotomy is frost formation of the specimen and the cutting edge, and the cryo–accessories provide a closed environment around the cooled components to prevent this. The MT 7000/CR21 cryo–accessory enables the temperatures of the knife, specimen and environment to be independently controlled down to $-150\,^\circ$C. (By courtesy of RMC Europe Ltd.) (*b*) (*overpage*) Thin section of red rubber tubing cut at $-145\,^\circ$C. The various filler materials in the rubber can be seen. Stereoscopic microscopy revealed that the rod-like structures are in fact inclined platelets which have been trimmed at both ends in the cutting of the section. Taken at 100 kV. (Preparation and micrograph by A. M. Allnutt, Sira Ltd.)

An important parameter in ultramicrotomy is the angle of the cutting edge. Smaller angles aid the cutting process, but the mechanical strength of the edge is reduced, and it can be damaged by hard materials. Diamond knives are therefore produced with a range of cutting angles, from 35° for cutting soft, resin-embedded tissue, to 45° for general use and 55° for hard metals.

These angles can be set up in the mechanical polishing operation to produce the cutting edge. The angle of a glass knife, formed by a controlled fracture of a piece of plate glass, is less controllable, and knives of a particular cutting angle may need to be selected from a number of samples after fracturing.

Figure 4.10 (*cont.*)

(b)

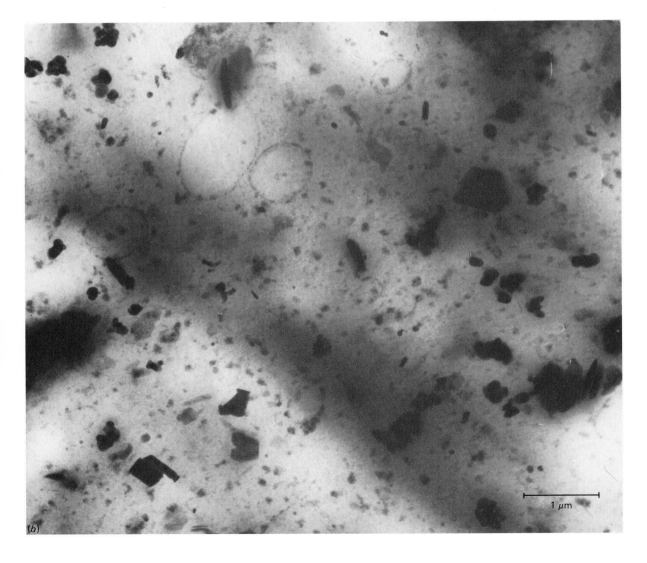

(b)

1 μm

Several attempts have been reported to prolong the life of glass knives, including surface coating the freshly fractured edge with a thin layer of hard metal such as molybdenum. Sapphire knives of hardness/durability/cost intermediate between glass and diamond have been marketed but have not found wide acceptance.

Staining

In light microscopy the selective absorption of coloured dyes by biological structures enables the detail of a thin section to be seen more clearly. In electron microscopy the equivalent technique uses the absorption of heavy metal atoms to produce differential electron scattering and hence increased contrast in ultra-thin sections. Organic materials consist mainly of elements of low atomic number, e.g. hydrogen ($Z=1$), carbon (6), nitrogen (7), oxygen (8), phosphorus (15), sulphur (16), etc., which have a low electron-scattering power. Hence the contrast in an image of untreated material is low. Staining by soaking in solutions containing tungsten (74), osmium (76), lead (82), or uranium (92) results in sites of high electron-scattering power, and consequently images with more clearly visible specimen detail.

If the specimen is particulate, e.g. viruses, macromolecules or cellular fragments, there is an alternative staining technique, *negative staining*. The difference between the two techniques is illustrated diagrammatically in Figure 4.11. In positive staining the heavy metal is absorbed by the tissue or particle itself, whereas in negative staining it is the surroundings and background which are stained, leaving the particle itself unstained and clearly visible against them. This is a very good way of revealing fine surface structure on particles.

Ultra-thin sections are most conveniently stained by immersion in, or floating on, a drop of staining solution for an appropriate length of time. The stain is absorbed into the cut surface of the specimen and fortunately does not affect the embedding medium. The process of staining is not fully understood or predictable, and the effect of different stains on a particular type of specimen has to be found by practical trials. Commonly used materials are osmium tetroxide, uranyl acetate, lead, and phosphotungstic acid, all at about 1% concentration at an appropriate pH. Figure 4.12 illustrates the need for, and the effect of, positive staining in an ultra-thin section of biological material.

Negative stains commonly used include phosphotungstate, silicotungstate, uranyl acetate and ammonium molybdate, at concentrations of a few per cent. Trace

Positive stain

Negative stain

Figure 4.11. Schematic diagram illustrating difference between positive and negative staining processes for biological particles.

additives, e.g. bovine plasma albumen, may be used to improve the wetting property of the stain on the support films.

Solutions of the selected stain may be applied to a particle dispersion already on a plastic- or carbon-filmed specimen support grid, using a spray gun or micropipette, or the stain may be mixed with the particle dispersion and then applied to the grid. In either case the liquid is quickly dried off to leave the specimen incorporated within a film of stain. The micrographs of Figure 4.13 show the dramatic increase in visibility of detail in adenovirus particles with use of a sufficient concentration of negative stain. Further examples of stained biological preparations are given in Chapter 8. Harris & Horne (1994) critically assess the utility of negative stains.

The selection of suitable electron stains is a specialised subject, and the appropriate handbooks (e.g. Kay, 1965; Lewis & Knight, 1992; Weakley, 1981; Hayat, 1989a; and Robards & Wilson, 1993) should be consulted for practical details. Staining is not

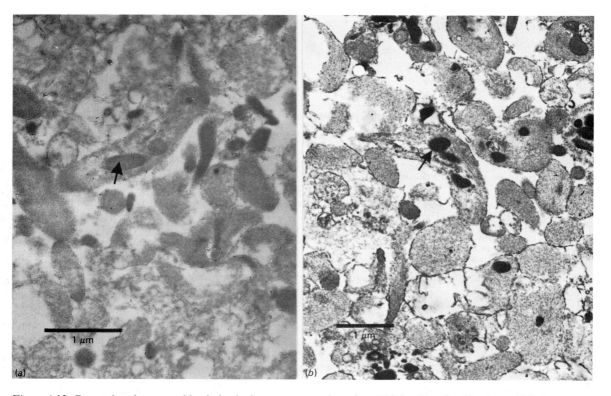

Figure 4.12. Comparison between a blood platelet homogenate section when (*a*) it has been fixed in glutaraldehyde followed by osmium without further staining, and (*b*) the same material has been stained further with uranyl acetate and lead citrate. Osmium used during fixation also stains membranes, which provides some contrast, but this is very poor compared with the general contrast introduced by lead ions and the ability of uranyl ions to bind with nucleic acids. The granules (arrows) contain nucleic acids and the different contrast seen in (*b*) is very clear. (Micrographs and commentary by G. R. Bullock, Ciba–Geigy Pharmaceutical Ltd.)

Figure 4.13. Negative staining of adenovirus particles. (*a*) In the absence of any stain the particles show no recognisable structure. (*b*) A weak (0.3%) solution of phosphotungstic acid (PTA) begins to reveal a central structure within the particle, but the contrast is insufficient to show further detail. (*c*) 3% PTA solution produces a strong outer deposit of stain and reveals the characteristic morphology of the virus particle within this. (Micrographs and commentary by A. M. Field, Central Public Health Laboratory, London)

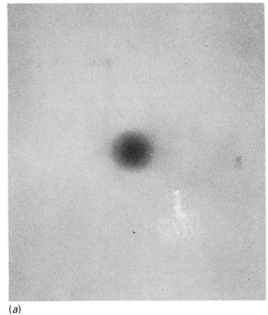

(*a*)

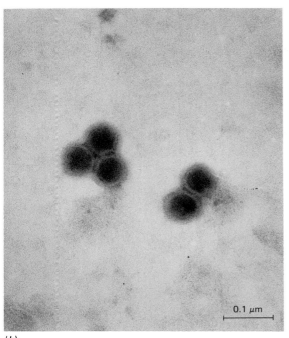

0.1 μm

(*b*)

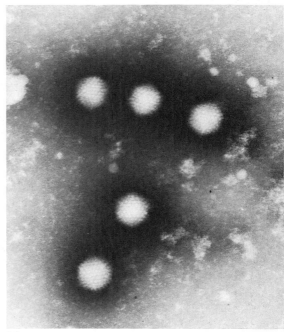

(*c*)

limited to biological specimens and can be used with advantage in the examination of industrial materials such as papers and plastics.

Autoradiography

Autoradiography is effectively an extension of the principle of selective staining. It can be used in conjunction with electron microscopy although its spatial resolution is much poorer than that of the microscope itself. Its operation relies on the fact that natural processes do not distinguish between elements and their radioactive isotopes. If a living organism is treated with a solution containing a proportion of radioactive compounds, any site in the organism which absorbs the solution will become temporarily radioactive. The distribution of the absorbed material is studied in ultrathin sections cut from the specimen. A section, supported on a filmed grid, is coated with a monolayer of silver halide crystals in a gelatin binder. β-particles (high-energy electrons) emitted by the radioactive isotope expose halide crystals immediately adjacent to that part of the specimen. After a suitable duration of exposure to the β-radiation, over a period of days or weeks in darkness, the halide emulsion is developed to metallic silver at the exposed grains, and fixed out so that all unexposed halide is removed. The silver deposit can then be seen superimposed on the structural detail of the section by transmission microscopy. Although of limited applicability this technique can produce invaluable information on processes occurring within cells and tissues (Baker, 1989). An example of its use is shown in Figure 4.14.

4.3.5 Thin foils and films

The distinction between these two types of specimen, which are nearly always non-biological and frequently metallic, is simply that a thin foil is a piece of bulk material which has been thinned for examination whereas a thin film is a specimen specially grown in a thin form, and may be from zero thickness up to the practical limit of the TEM or HVEM.

Preparation of thin foils

Non-biological materials such as metals, alloys and ceramics are in general too hard and in some cases too brittle for ultra-thin sectioning on an ultramicrotome unless a certain amount of structural disturbance can be tolerated, and a very small section is adequate. Uniform mechanical thinning and polishing of large areas ceases to be feasible below a specimen thickness of about 40–50 μm and other means have therefore to be found if the internal structure and crystallinity of such materials are to be accessible to electron microscopy.

A thin foil, fifty to several hundred nanometres thick, is prepared by eroding away material from a sample perhaps tens of micrometres thick, which may itself have been cut or spark-machined from a larger block. Since the final thinned foil has to be accommodated in a holder for 3 mm grids it may be convenient to produce the thin region at the centre of a 3 mm disc punched out of a sheet or cut from a rod. Alternatively a larger area of sheet may be thinned down and the thinnest areas, perhaps surrounding a region thinned to perforation, cut out and sandwiched between the two halves of a folding grid.

The actual process of erosion may be carried out by electropolishing in a liquid or by a gaseous sputtering process (ion-beam thinning). For most metals and alloys the former method has been used satisfactorily over a number of years and commercial apparatus for specimen thinning is readily obtainable. In the case of 'difficult' metals and alloys, and solid specimens of other types, e.g. ceramics, paint films, bones and teeth, ion-beam thinning is satisfactory and indeed it may be the only practical method. Neither process results in a parallel-sided specimen of ideal thickness, but rather in a hole surrounded by a thinned region of tapered section. Under these circumstances the area available for study will be greater for the Intermediate and High

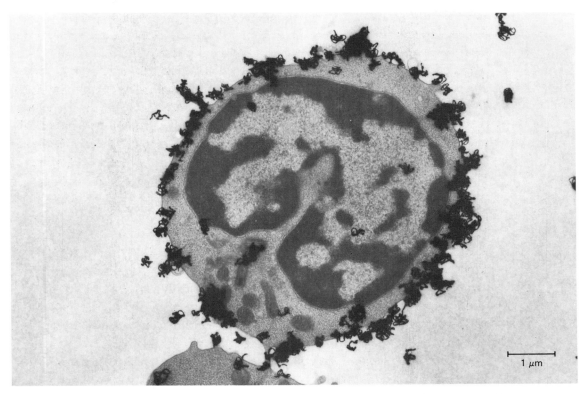

Figure 4.14. Electron microscope autoradiograph of peripheral blood lymphocytes incubated with $Na^{125}I$ in the presence of the glucose–glucose oxidase/lactoperoxidase reaction. The cell and cell membrane appear well fixed and most of the silver grains are associated with the cell membrane. (Micrograph and commentary by J. Baker, Ciba–Geigy Pharmaceuticals Ltd.)

Voltage TEMs and for STEM than for the 100 kV TEM. In practice there is evidence for some types of specimen that the reduction of thickness to, say, 0.1 μm has resulted in the total loss of certain gross specimen features, and the thin foil is therefore not truly representative of the total bulk material from which it was prepared. This is an aspect which should be considered and, if possible, checked on when using any form of specimen preparation for the electron microscope.

(i) Electrochemical thinning ('electropolishing'): A graph of current passed against voltage applied between two electrodes in an appropriate electrolyte has the form of two steep sections joined by a plateau. If the surface of the positive electrode is examined it will be found to have become etched and pitted in the lower steep region and smooth and polished on the plateau. *Electropolishing* results in preferential removal of surface irregularities and overall progressive removal of material to give a thin foil which is eventually thinned locally to perforation. The hole or holes are surrounded by more or less uniformly thin areas of metal which are suitable for structural examination in the TEM. The perforations can be encouraged to occur at a desired point, e.g. the centre of a 3 mm disc of material, by suitably shaping the other electrode, or by squirting the electrolyte through a fine jet directed at the centre of the disc. The temperature and voltage are closely controlled and thinning is stopped as soon as perforation occurs, to maximise the thinned area.

Recommendations have been published for electropolishing many metals and alloys (see, for example, Goodhew, 1985). Equipment is commercially available to enable TEM specimens to be obtained routinely, usually starting from mechanically polished thin discs punched out of sheet metal 25–250 μm thick. Great care must be taken in washing, drying and storing the freshly polished foils, since any traces of electrolyte remaining will cause the foil to corrode in storage.

(ii) Ion bombardment thinning ('*ion-beam thinning*' or '*ion milling*'): A collimated beam of energetic positive ions (argon$^+$, Ar$^+$ is usually employed, since it is inert and heavy) directed at any angle at a solid surface in a vacuum will cause atoms to be removed from the surface by the process of sputtering. Ion energies up to 10 keV are used. The number of atoms removed for each normally incident ion is called the 'sputtering yield' and varies from element to element. If the ion beam is incident at a glancing angle the rate of sputtering will be lower, but the bombardment tends to smooth out irregularities in the target surface by preferentially removing the high spots. In an ion-beam thinning apparatus a 3 mm disc of material preferably $\leqslant 50$ μm thick is bombarded from both sides by narrow beams of argon ions, and is simultaneously rotated in its own plane about an axis through its centre. Initially the ion beams can be incident at 45°–90° to the surface, for a high thinning rate, but final thinning to perforation is carried out at near-glancing incidence, e.g. 10° to the surface, in order to produce a large area of thinned foil around the perforation. The cross-

section of the thinned foil will be wedge-shaped, with angle twice the angle of thinning; as low a thinning angle as possible is thus desirable. The limit is usually imposed by the thickness of the clamping ring used to hold the disc in the thinning machine. Thinning from one side only allows smaller thinning angles to be used, but with reduced thinning rates.

High ion energy and normal incidence can result in permanent damage to the crystal lattice of the specimen, which is visible in the microscope afterwards. The specimens can also become very hot whilst being thinned. If thinning is allowed to continue beyond perforation the thinnest regions around the hole are rapidly sputtered away by the ion beams, leaving a truncated wedge section of reduced transparency. Commercial ion-beam thinning equipment is therefore fitted with a sensitive detector to cut off the ion beams automatically when the first hole appears. The final stage of thinning of some materials, e.g. heavy metals and their alloys, may take hours if beam damage artefacts are to be avoided, and the equipment may be left running unattended in the knowledge that the power will be shut off automatically when necessary.

The amount of material to be removed by ion milling can be reduced considerably if initial thinning is carried out by a *dimpling* procedure. This can be achieved by grinding against a small spherical ball, but a similar effect may be achieved by other mechanical arrangements. The small circular crater may achieve a specimen thickness as little as 5 μm at its centre, surrounded by a thicker region to aid handling. Dimpling on its own may be sufficient without further ion milling to allow examination of some semiconductor materials in an HVEM.

Ion-beam thinning can be applied to a variety of difficult materials including ceramics, for which there would be no other means of thinning. Fine powders, e.g. lunar dust particles, have also been thinned after being suitably mounted in epoxy resin. It is a useful non-contact technique for specimen preparation, when used with discretion, and is ideal for preparing large thin areas of clean sample for high-voltage microscopy. The technique and its characteristics are reviewed by Barber (1970, 1993).

Several variations on ion-beam thinning use chemical attack in addition to normal ion sputtering, in order to reduce the preparation time of certain types of specimen, principally semiconductors.

In reactive ion-beam etching (RIBE) the inert gas used for sputtering is replaced by a gas which is chemically active in relation to the material to be thinned, e.g. O^{++} is used in place of Ar^+.

In chemically assisted ion-beam etching (CAIBE) inert gas ions are used for sputtering, but the specimen surface is exposed to a reactive gas directed through a jet assembly. In thinning the compound semiconductor InP a jet of sublimed iodine is used to prevent the formation of indium islands on the surface during normal thinning with Ar^+ ions (Alani & Swann, 1993).

(iii) Focused ion-beam machining: Although this book has concentrated on the focusing and manipulation of electron beams (electron optics) the principles can be applied to high-energy beams of other charged particles and ions. Focused beams of 30 keV ions can be used for localised sputtering at a high rate, and micromachining is possible. Since this also results in ionisation and electron emission, a form of inspection by scanning ion microscopy is possible if the ion beam is scanned in a raster and the secondary electrons are detected. Used particularly for the examination of the detailed structures of microelectronics devices, focused ion beams (FIBs) can be used to isolate a chosen area of specimen and then to thin it more gently for TEM examination. Useful reviews of the production and properties of FIBs are given by Orloff (1993) and Prewett & Mair (1991).

(iv) Wedge cleaving: The ion bombardment techniques outlined above can give rise to preparation artefacts such as loss of crystallinity (surface amorphisation) and inter-diffusion, caused by the temperature rise under bombardment. A non-bombardment technique which is both quick and gives a thin region which is free from such artefacts, although very limited in extent, is available for some semiconductors which have well-defined cleavage planes.

Thin films

Thin films, formed by vacuum- or chemical deposition on to solid substrates, have been studied extensively since the TEM was first developed. In this case the thin film is either deposited directly on to an electron-transparent substrate, e.g. a film of carbon or silicon monoxide or a flake of molybdenum sulphide on a specimen support grid, or removed from its substrate by a mechanical or chemical stripping process and transferred to a specimen grid. Needless to say, where the structure being studied includes isolated clusters of atoms, it is essential to confirm that the specimen put into the microscope is representative of the starting material. One way of doing this is to compare the structures revealed by surface replication or high-resolution scanning microscopy of the thin film deposit before and after change of substrate.

Much useful information on the structure and growth of thin films has been obtained from *in-situ* depositions within transmission electron microscopes. With care a suitable substrate and evaporator/sputtering source may be built into the specimen stage, and stages of growth recorded on cine film or video tape.

4.4 Specimen requirements for the SEM

These are very different from those for TEM specimens. Except in the transmission mode of operation the restrictions on specimen thickness and size are much relaxed, and will be limited only by the capacity of the specimen stage and chamber of the particular microscope being used. Features to be studied must be accessible both to the scanning electron beam and to the detector in whichever mode the microscope is operating.

Specimen stability under examination is essential in any form of microscopy; operating current and voltage of the SEM sometimes need to be modified to suit specimens of a more delicate nature. Care must be taken that any necessary specimen preparation does not alter the features to be studied.

The essential condition peculiar to the SEM is that the specimen must remain at a constant (zero or earth) potential during examination. Specimen preparation is mainly concerned with satisfying this last requirement.

4.4.1 Specimen size

The specimen in transmission microscopy is allowed to occupy a few cubic millimetres of space within the objective lens and hence is restricted to 3 mm across and tens of micrometres thickness inclusive of support grid. In most SEMs the specimen position lies well outside the lens system and specimens may therefore be more massive; it is only the outside surface which is being examined so the thickness is immaterial as far as the process of examination is concerned. The function of the specimen stage in this type of instrument is merely to hold and manipulate the specimen so that the region of interest can be brought under the electron probe. The size of specimen could be semi-infinite provided that the specimen chamber were large enough to accommodate it and the specimen stage able to hold it and translate, tilt or rotate it in a smooth and steady manner. In practice absolute rigidity and large traverses are expensive to provide and although most microscopes nowadays will accept specimens several centimetres across, the majority of scanning microscopy is probably carried out with specimens mounted on stubs like the 12.5 mm diameter 'mushroom' originally introduced with the Cambridge Stereoscan (Figure 4.15). Nearly all current SEMs will permit examination of the entire surface of specimens up to 12.5 mm diameter and 3 mm thick with complete 360° rotation and up to 75° or 90° of tilt from the flat, square-on position. Larger and thicker specimens can usually be accommodated, but it may not be possible to examine the whole surface from all angles because of limited X, Y and Z stage movements. If a very limited range of specimen movements (or none at all) is acceptable, the largest specimen which can be examined in a scanning microscope is limited eventually by the dimensions of the access port into the specimen chamber. Several currently available SEMs specify a maximum specimen size about 75 mm diameter $\times$50 mm deep.

The critical dimension (CD) and wafer-inspection SEMs for semiconductor wafers up to 150 mm diameter are invaluable production tools in today's world, but are too specialised for further mention here. Large specimens such as aero-engine turbine blades are the exception rather than the rule, however, and descriptions in the following pages will assume that the specimen is mounted on the upper surface of a 12.5 mm Stereoscan 'stub' like a flat-topped aluminium mushroom, whose stem plugs into a 3 mm socket on the specimen stage. Other manufacturers use specimen mounts differing in detail of size, shape or material, but all have a flat upper surface for specimen mounting. To save specimen exchange time, clusters of 12.5 mm mounts may be mounted on a single large stub and put into the microscope together (see Figure 3.27).

4.5 Preparatory methods for SEM specimens

4.5.1 Mounting

Whatever their electrical properties specimens have to be suitably mounted on stubs before examination. There are many different ways of doing this; those to be mentioned have been found satisfactory in everyday use.

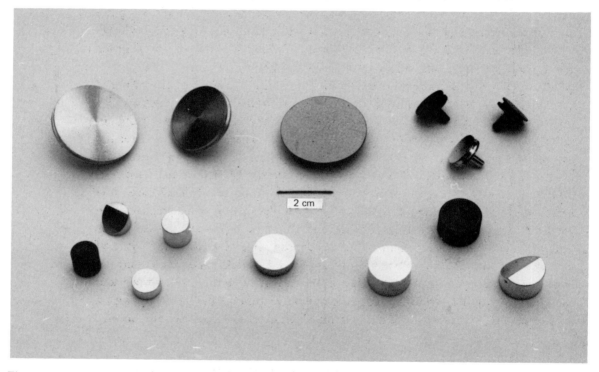

Figure 4.15. Examples of specimen mounts used in scanning microscopes. Two basic geometries of mount are in common use: the pin-type mushroom, used by Amray, Gresham–CamScan, LEO (Leica-Cambridge & Zeiss), Philips and others, and the plain cylinder used by Hitachi, Topcon (ISI) and JEOL. Cylindrical mounts with chamfered tops are used to obtain high tilt angles in specimen stages with a restricted range of tilt angles. The mounts illustrated are: *back row, left to right*: 32 mm and 25 mm diameter pin-type mounts (aluminium); 32 mm carbon for microanalysis stage; 12.5 mm pin-type mounts in carbon and Al. *Front row, left to right*: 10 mm diameter JEOL mounts – 10 mm high carbon, angled Al, 5 mm and 10 mm high Al; 15 mm diameter Hitachi; 15 mm Topcon (ISI) mounts – Al and carbon, 45° angled Al. Standard stubs machined from Al may be re-used indefinitely after carefully dismounting the specimens. For low-backgound microanalysis the same geometry of mount is made from carbon, which requires gentler handling and is best considered as expendable. Alternatively, thin discs of carbon or beryllium may be glued to the top surface of metal stubs for low-background use. (Specimen mounts kindly supplied by Agar Scientific Ltd.)

The main division of specimen types is between bulk and particulate materials. The former can include lumps of metal, rock or inorganic material as well as open structures such as paper and woven fabrics. Particulate materials can have a size range covering four orders of magnitude down from a centimetre or so.

Mounting bulk materials is straightforward in most cases, an appropriately sized portion being cut or broken-off and stuck on the surface of the stub with a suitable adhesive. The latter should set firmly and not volatilise in the high vacuum of the microscope. One useful material is colloidal silver paste (e.g. silver 'dag') which performs the dual function of bonding the specimen to the stub mechanically and electrically. Sheet materials and lightweight specimens are conveniently held flat on the stub with double-sided sticky tape, although this does not conduct electricity and may 'creep' if it becomes warm in the microscope. A conducting stripe of silver paste is painted over the edge of the sticky tape from specimen to the stub surface. 'Stubs' with mechanical clamping jaws can also be used satisfactorily, if the requirements for stability and electrical earthing under the beam are met.

The method of mounting particulate matter depends very much on the nature of the particles and what sort of information is required. Large solid particles and dry powders can simply be sprinkled on the surface of a very thin adhesive layer. Again, silver paste is a convenient medium, provided that the specimen particles are readily distinguishable from the flakes of silver paste and are unaffected by the solvents in the paste. Swollen gelatin, colloidal graphite and Apiezon W vacuum wax have been used as mounting media. Particles dispersed in liquids are conveniently deposited from a spray-gun or airbrush or in a liquid droplet placed to dry on a piece of clean glass microscope slide. It is helpful if the glass substrate is pre-coated with a conducting film of gold or aluminium by vacuum deposition. The glass substrate is itself mounted flat on a stub with silver paste or double-sided sticky tape with a conducting stripe of paste painted over the edge of the glass between the upper surface and the stub. Identifying arrows or numbers can be written on the substrate with a writing diamond or a special conducting ink which is visible in the electron beam.

4.5.2 Coating

The primary electron beam, amounting to between 10^{-12} and 10^{-6} amperes depending on the spot diameter, is brought to rest in the specimen in the scanning microscope. A certain fraction of this current is emitted in the opposite direction as secondary or backscattered electron current. The net difference between the two currents must be enabled to leak away to earth via the stub, or the specimen will charge up as a capacitor. The change in specimen potential can be such as to influence the scanning beam and adversely affect the quality and resolution of the image. If the specimen does not possess adequate inherent electrical conductivity to discharge itself it should be provided during specimen preparation with a thin conducting 'skin' of metal or carbon to conduct the charge away to the stub. *Specimen coating*, usually by vacuum deposition (Appendix 2), is normally carried out on all insulating and semi-insulating materials before they are examined in the SEM. The only exceptions are when the specimen is to be examined at very low voltage, $\simeq 1$ kV, where most materials have a secondary emission coefficient $\delta = 1$, and may be examined without charging-up, and in a low-vacuum or environmental SEM when specimen charge is neutralised by ionised molecules of the chamber atmosphere.

Certain fixation treatments for biological materials have the effect of giving the specimen sufficient inherent conductivity to make further treatment by coating unnecessary (see Section 4.8.2).

The basic requirement of a coating is that it provides a continuous electrical path between the stub surface and any point on the specimen likely to be exposed to the electron beam. The degree of conductivity required is not critical and indeed may be relatively low. In general coatings are applied on an empirical basis; probably they will err on the generous side, especially with smooth, flat, specimens, because the effects of putting on an excess of material are usually less noticeable than the reverse.

Similar materials tend to be used for SEM coatings as for shadowcasting and support films in transmission microscopy. This is partly because the techniques and equipment for deposition are well known, and the materials are known to conduct in very thin layers, but also because they are materials which do not readily form insulating oxide layers during storage in air between the coating process and the examination. Thus, carbon, gold/palladium, platinum and several other metals and alloys are deposited on to specimens after they have been mounted on stubs. Unlike shadowcasting procedures, however, which demand evaporation from a point source to form sharply defined shadows, the requirement in coating for SEM is for an all-over, even, coating penetrating into all nooks and crannies of the specimen.

If the coating material comes from a localised source, as in evaporation or low-pressure sputtering, it is necessary to rotate the specimen stub during the deposition and to rock it so that all points on the specimen surface receive evaporant. This is shown schematically and as commercially realised in a planetary work holder in Figure 4.16. Adequate all-over coatings can be applied using this technique, although insulating particulate specimens require more metal to be applied than do more continuous structures. Any slight vibration in the rotary mechanism is liable to slow down the formation of conducting paths from particles to the stub.

A popular and easy technique for applying all-over conductive coatings is d.c. sputtering. This method of deposition, which can operate with a poorer vacuum than the evaporation process, provides a diffuse stream of coating material and enables specimens to be coated for the SEM without the need for a rotating jig, and in much less time. Gold, chromium and platinum are three materials which are often applied by sputtering. The principles of sputtering and thin-film formation are described in Appendix 2.

As well as providing a conducting film the coating layer also performs the useful

Figure 4 16. The rotary coater for SEM specimen preparation. In order to achieve a uniformly deposited layer of conducting material over the surface of the mounted specimen the stub is spun about the axis of its stalk and rocked so that the deposition angle from the evaporation source varies between 90° to the stub surface and a lesser angle, ideally parallel to the stub surface. This is shown schematically and from the viewpoint of the evaporation source in the planetary workholder of a commercial coating unit. (Photograph by courtesy of Edwards High Vacuum International.)

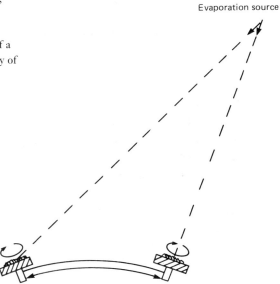

Evaporation source

function of giving enhanced electron emission from the specimen surface. In some instances, therefore, metal coating may be carried out with advantage on specimens which would be able to dissipate the charge sufficiently well but have poor electron-emitting properties. This applies particularly to low atomic number materials such as carbon but even metallic specimens, e.g. fracture faces or polished metallurgical specimens, may have sufficiently poor surface characteristics to benefit from an added coating.

The thickness of coating layers is not readily determined on practical specimens and is likely to vary considerably between the horizontal surfaces on a stub, which are continuously facing the evaporation source or sputtering target, and vertical faces or re-entrant features. Hence the tendency, already noted, of coating rather generously. Estimates of thickness, e.g. 20 nm, which have appeared in the literature are doubtless true for some portions of any specimen, and will give good electrical conduction for any coating material. The important thing is that the coating thickness at any point to be examined should not fall below the value at which the coating material conducts electricity, e.g. about 4 nm for gold or 1–2 nm for its alloys and platinum. This aspect is dealt with in greater detail in Appendix 2.

Coatings for high-resolution SEM

Users of SEMs with resolutions of 2 nm or less are faced with similar problems to TEM operators and shadowcast replicas; i.e. at magnifications of $\times 100\,000$ and more the self-structure of their metal coatings becomes visible and can be mistaken for genuine specimen structure. However, they have the advantage over TEM users that they are not restricted to high atomic number elements.

No hard and fast rules can be given – it may be necessary to experiment to find a suitable coating procedure for a specific specimen type. Oily or greasy surfaces, and specimen heating during coating must be avoided; the minimum of material should be deposited. Chromium has been reported as satisfactory by some workers. Refractory metals such as tungsten, molybdenum or platinum form continuous layers when only 1–2 nm thick, and should be worth trying. Carbon gives structure-free conducting coatings at a nanometre thickness but has very low electron emission at normal gun voltages. A combined coating of carbon followed by a nanometre thick discontinuous coating of gold or gold/palladium alloy to improve electron emission would be hard to detect.

Specimen coating for x-ray microanalysis

Metallic coatings which are ideal for scanning microscopy in the secondary or backscattered emissive modes can be an embarrassment if any form of electron-excited x-ray microanalysis is undertaken on the same specimen. This is partly because the characteristic emission lines of the coating material may obscure the elements being sought in the analysis and partly because the coating absorbs some of the x-rays emitted from the specimen; this latter effect is more pronounced with higher atomic number coatings such as gold and platinum. The emission and absorption of x-radiation by the conducting coating can be minimised if the coating thickness is reduced, and platinum films have an advantage in conducting electricity at less than one-half the thickness of gold. However, it may be undesirable to have any interference at all from the coating material. In this case coating with evaporated carbon is the normal practice. Adequate electrical conduction is provided by a comparatively thin layer (in terms of mass/unit area, which determines the electron and x-ray absorption) and the low energy x-radiation from carbon ($CK\alpha=0.277$ keV) is only detected by energy-dispersive detectors in the window-less mode.

4.5.3 Replication

Replicas are needed in scanning microscopy in order to deal with specimens which may be too large for the available specimen stage, too valuable to cut up or even coat for examination, or part of a system which must be examined *in situ*. SEM replicas can be solid bodies and not the fragile thin carbon films prepared for transmission microscopy, but must be electrically conducting with good secondary emission and able to withstand electron bombardment in the microscope. Since specimens for the SEM may be very irregular the initial replica material and subsequent processing should be able to reproduce rough surfaces, even to the extent of re-entrant surface features. Correct interpretation of negative replicas is difficult, so a positive process is desirable.

Two stage replicas have been prepared using solvent-softened cellulose acetate sheet or a suitable grade of silicone rubber to form the primary impression, which is then reproduced by a further process and coated to give a robust metal-surfaced final replica suitable for the scanning microscope. The SEM has an advantage over the TEM in that the fidelity of the replication process can be checked by direct comparison of a replica with an original specimen which is small enough for insertion into the microscope. Figure 4.17 shows such a comparison on a specimen which contains both coarse and fine detail.

4.5.4 Ultramicrotomed cross-sectioning

In some cases an ultramicrotome may be used, not to cut an ultra-thin section for transmission microscopy, but to produce a flat-surfaced cross-section for examination either by light microscopy or by scanning electron microscopy or microanalysis. The technique is useful with soft and porous materials such as coated papers, for which conventional metallurgical grinding and polishing methods would be unsuitable. Materials may be cross-sectioned after encapsulation in resin or, if appropriate, when frozen. An example of the latter is shown in Figure 4.18.

4.5.5 *Freezing*

One of the most useful accessories to a scanning microscope is the cryo-stage, which enables materials to be examined at temperatures down to about $-150\ °C$. Before it can be examined in a cold stage the specimen must be prepared in a cryo-preparation unit, which may be a separate unit or an accessory bolted on to the specimen chamber of the microscope.

The specimen holder for a cryo-stage is much smaller than the 13 mm Stereoscan stub. The specimen is mounted on it with a paste, e.g. colloidal graphite, which freezes solid, gripping the stub and the specimen, when the stub is plunged into a freezing bath of, e.g. liquid nitrogen 'slush' at $-210\ °C$. The specimen must be preserved at this temperature throughout its further preparation and examination,

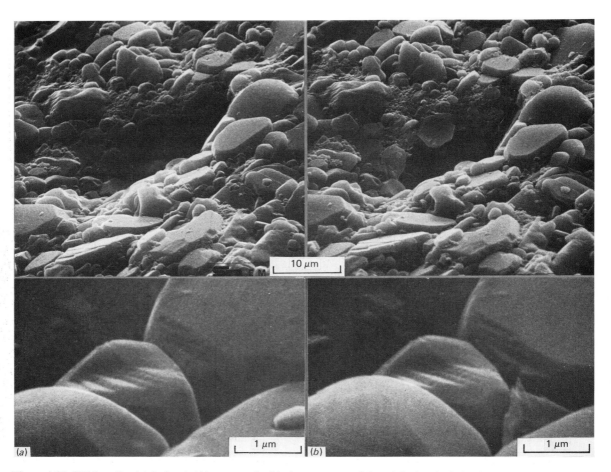

Figure 4.17. SEM replica (*right-hand side*) compared with the same area of the original etched alumina surface after replication. Both coarse and fine structural detail have been well reproduced. Specimen stub tilt angle 45°. (Replica procedure and micrographs by N. A. Wraight for Sira EMTG; reproduced with permission from Sira Ltd.)

except that controlled warming up may be necessary to sublime away any superficial ice crystals (frost) which may have been formed on the surface of the specimen during freezing. The specimen is transferred into the microscope column under vacuum and after inspection at low voltage to check whether the surface is frost-free it is given a conducting coating so that it can be examined under higher resolution conditions.

Commercial cryo-preparation units have facilities for mechanically fracturing frozen specimens if internal structures are to be studied. Controlled sublimation to reveal the non-aqueous structure is carried out before the fracture surface is carbon- or metal-coated.

Specimens do not need to contain water to benefit from low-temperature examination. Waxes, paint films and polymers can be mounted, coated and examined in cryo-stages.

4.6 Preparation of moist specimens

A large proportion of biological specimens and some materials of industrial interest contain water or other liquids in their normal state. Preparation of these specimens for examination by the various forms of electron microscopy and analysis presents a number of problems (unless an Environmental or Low-Vacuum SEM can be used), the major ones being concerned with the preservation of specimen size, microstructure and composition unchanged through the various stages of treatment. A fundamental difficulty in assessing the suitability or otherwise of the various treatments of biological material is that there is no knowledge of what the

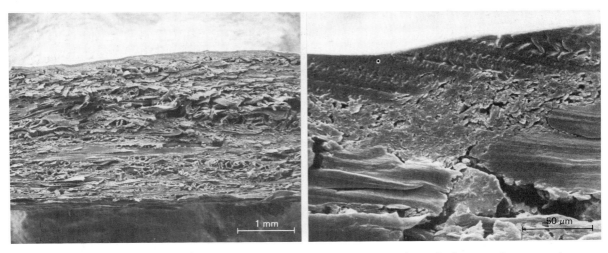

Figure 4.18. Cold-cut cross-section of coated cardboard. A thin sliver of board was frozen in the cryo-microtome and cut across with a cooled glass knife. Although not perfectly cut, in so far as some of the fibres appear to have been torn rather than sectioned, the cross-section enabled the thickness and penetration of the surface coating to be examined, as seen in the more detailed view.

'correct' answer is, so that any changes from the results given by the time-honoured traditional techniques may be leading either closer to or further from the genuine 'live' structure. Certainly there must be structural modifications introduced by chemical fixation and resin impregnation, but the extent of these is still a matter of conjecture.

The traditional approach of progressive dehydration and impregnation with embedding resin has already been outlined. It is a routine which can be conveniently automated. However, the increasing use of scanning microscopy for the study of biological materials in a form other than ultra-thin sections, and of x-ray microanalysis on SEM and TEM has stimulated the search for other methods of preparation which are less likely to disrupt delicate structures and to remove or replace dissolved salts during processing.

4.6.1 Drying biological specimens

The oldest and simplest method for drying a material is natural or forced evaporation of the water in air or in an oven. In some cases, e.g. when a rigid body is being dried, this may be quite satisfactory, but when the moisture provides much of the volume and 'stiffening', or has to be extracted from fine cells or capillaries, natural drying irreversibly modifies the internal structure. A major disruptive force is surface tension associated with the passage of the liquid/vapour boundary through the specimen as it dries.

The technique of critical-point drying is one of the less disruptive ways of extracting water from materials.

Critical-point drying.

In this process no phase boundaries pass through the specimen – there is no freezing and no evaporation. Use is made of the instantaneous transition from liquid to vapour which occurs at the critical point of the liquid. A system of liquid and vapour phases in equilibrium has a critical temperature T_c above which the system is always gaseous, no matter how much it is compressed. The transition from liquid to vapour at the critical point takes place without an interface because the densities of gas and liquid are equal at this point. If a wet specimen is totally immersed in a liquid below its critical point and is then taken to a temperature and pressure above the critical values it is then immersed in a gas (i.e. it has been dried) without having been exposed to surface tension forces.

This procedure has been used for a number of years (Anderson, 1951), and commercial apparatus is available to enable it to be carried out. In practice, the critical temperature of water itself is inconveniently high and other liquids with lower critical points are substituted for water in the specimen. Two such liquids are carbon dioxide and Freon 13; their critical temperatures and pressures are 31.0 °C, 74 bar (1080 psi) and 28.8 °C, 38 bar (560 psi), respectively. The deciding factor between these is usually cost; in England liquid carbon dioxide is considerably cheaper and more readily available than Freon, and is therefore more widely used.

The procedure in critical-point drying is, briefly, to substitute a liquid miscible with liquid carbon dioxide for the water in the specimen. This may take the route water → acetone → carbon dioxide or: water → ethanol → amyl acetate → carbon dioxide.

The specimen, immersed in the substitution liquid, is placed in a sealed pressure vessel which is flushed through thoroughly with liquid carbon dioxide, which in turn is allowed to substitute for the previous liquid. With the chamber part-filled with liquid CO_2, the specimen being submerged, the valves are closed and the temperature raised slowly up to and above the critical value. A gauge enables the internal pressure to be followed at the same time. At the critical point the meniscus between liquid and vapour blurs out and disappears; the specimen is now dry and in a gaseous environment, which is de-pressurised before removal of the specimen. A commercial critical-point dryer is seen in Figure 4.19(*a*). Although used primarily on biological specimens there is no reason why critical-point drying should not be used on other types of material when the special attributes of the processes can be turned to advantage. For example, in the preparation of particulate specimens for examination the surface tension forces involved during evaporation of liquid from droplets of suspension can cause suspended particles to become strongly aggregated, making it difficult to see and measure the size of individual particles. Critical-point drying can be of assistance here in keeping the particles separate. Critical-point drying can be a help in preventing a wet fibrous structure from collapsing as the liquid is dried out. In the examination of lubricating greases, for example, it is common practice to remove the oily constituent with an organic solvent, and there is a possibility of distorting the fine soap fibres as the solvent evaporates to dryness. Substitution of liquid carbon dioxide and its subsequent removal by the critical-point method can eliminate the possibility of such distortion occurring (Figure 4.19(*b*)).

The other techniques for the removal of water from specimens are all based in one way or another on cryogenics and start by rapidly freezing the specimens. The essential treatments, and in some cases the examinations, are carried out at low temperatures. Figure 4.20 shows the variety of cryogenic procedures which are currently being used. Some of these are still being evaluated in laboratories in various countries; details of reliable techniques may now be found in textbooks but the latest thinking is best looked for in the proceedings of international conferences and the appropriate journals.

Freeze- and molecular distillation drying.
The vapour pressure of ice (Table 4.1) is sufficiently high for direct sublimation from solid to vapour to take place into an enclosure at a sufficiently low pressure. Moist materials can therefore be dried without a liquid/vapour interface passing through

Figure 4.19. Critical point drying. (*a*) Polaron E 3100 Critical Point Dryer. The cylindrical chamber contains the specimen to be dried. After part-filling the chamber with liquid CO_2, its temperature and pressure are raised by passing warm water through a jacket surrounding the chamber. Changes of pressure and temperature inside the vessel are followed on the dial gauges, whilst the liquid/vapour meniscus can be observed through the glass window of the chamber. (By courtesy of VG Microtech Ltd.) (*b*) Scanning micrographs show the different textures of fibres from a lithium-based grease after solvent extraction of the oil and: (*left*) critical point drying, (*right*) air drying. (preparation by D. E. Webber; micrographs by N. A. Wraight, both of Sira Ltd.)

them if the water is first of all frozen and then sublimed away under reduced pressure. The specimen is quickly frozen (see below) and cooled to the temperature of liquid nitrogen (-196 °C). In freeze-drying the frozen specimen is placed on a cold surface at about -80 °C in a chamber which is then evacuated to a high vacuum. The specimen is kept cold until all the ice has sublimed away and been removed from the chamber as vapour. This may take several hours, depending on the size and water content of the specimen and the temperature at which it is held. Freeze-drying is an

(b)

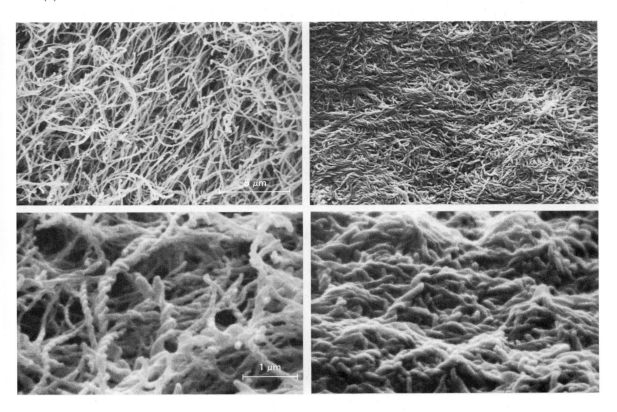

effective and easy procedure for drying aqueous particle suspensions without the risk of forming agglomerates in the drying process.

In molecular distillation drying the frozen specimen is held at a lower temperature, down to $-196\,°C$, when the vapour pressure of ice is so low that an ultra-high vacuum ($\leqslant 10^{-7}$ mbar) is necessary for sublimation of the ice to take place. Drying takes place over a period of days, the temperature being raised slowly in stages during this time to minimise the damage due to transformation of ice from amorphous to crystalline forms (Livesey *et al.*, 1991).

The subsequent treatment depends on the method of examination to be used. The cold, dry specimen may be impregnated with a special low-temperature embedding resin and then warmed up for sectioning; it may be warmed up before embedding, replication, shadowcasting, etc., are carried out, or it may be examined directly by scanning microscopy. A shadowcast freeze-dried preparation of tobacco mosaic virus is shown in Figure 4.21.

Table 4.1. *Variation of vapour pressure of ice with temperature*

(Data reprinted, with permission, from Lide, 1990.)

Temperature (°C)	Vapour pressure (mbar)
0	6.11
−10	2.6
−20	1.03
−30	3.8×10^{-1}
−40	1.3×10^{-1}
−50	3.9×10^{-2}
−60	1.1×10^{-2}
−70	0.3×10^{-2}
−80	0.1×10^{-2}

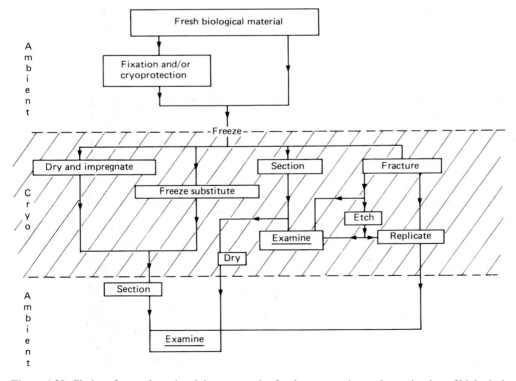

Figure 4.20. Choice of procedures involving cryogenics for the preparation and examination of biological materials containing fluids. Routes including impregnation and sectioning are principally for TEM examination.

4.6.2 Freeze-substitution

Some biological specimens suffer damage when the embedding resin penetrates into a freeze-dried structure, and a modified procedure known as freeze-substitution can be used in such cases. The specimen is quickly frozen, as for freeze-drying, but instead of the ice being removed by sublimation it is dissolved away by a cooled liquid dehydrating agent such as methanol or acetone. The frozen specimen is left to stand for long periods of time in several changes of liquid until the water has been substituted first by the dehydrating agent, and then by the desired embedding medium, which may be methyl methacrylate or epoxy resin. The process of substitution proceeds slowly and several weeks may be necessary at temperatures of $-80\ °C$ and below; methanol is reported to be faster than acetone at such temperatures, but there are doubts about the use of either of these liquids if the specimen is being prepared for x-ray microanalysis (Pålsgård *et al.*, 1994). It may be desirable to add low concentrations of fixatives, e.g. osmium tetroxide, uranyl salts or glutaraldehyde, to the substituting liquid to stabilise particular constituents of the specimen (e.g. proteins, lipids).

4.6.3 Freeze-fracturing and freeze-etching

These are techniques which enable quite unique information to be obtained about the internal structure of water-containing biological structures, without any possibility of modification by chemical fixatives, stains, dehydrating agents or the

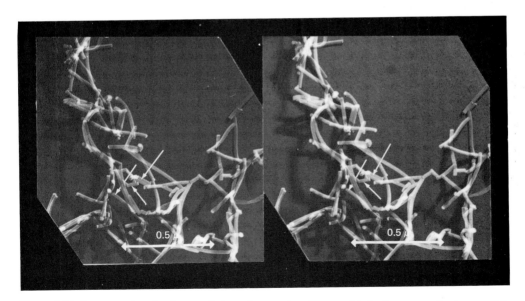

Figure 4.21. Stereoscopic pair of transmission micrographs of freeze-dried tobacco mosaic virus particles showing protein subunits. (By courtesy of R. L. Steere.)

substitution of constituents. They are a step closer to the goal of examining living material; the potential in this respect was demonstrated by Moor & Mühlethaler (1963), who studied the surfaces of yeast cells by freeze-etching and found that they were subsequently capable of resuming normal functioning and reproduction.

Essentially, in freeze-fracturing a water-containing specimen is quickly frozen and then split into two pieces. The frozen fracture faces contain topographical features such as cell walls, membranes, pores, etc., which are examined either directly in the SEM or indirectly by means of replicas in the TEM. In freeze-etching a carefully controlled thickness of ice is allowed to sublime from the fracture faces before they are examined or replicated. In either case replication takes place in the same chamber as fracturing, without breaking the vacuum in between.

There are two principal ways of producing the initial fracture faces, illustrated schematically in Figure 4.22. The method of Steere (who originated the technique of freeze-fracturing in 1957) produces two complementary faces by fracturing the specimen under tension in a two-part hinged holder (Steere, 1957, 1981). The technique of Moor produces the fracture face by chipping away successive layers of brittle, frozen material with a cooled cutting edge, often a modified razor blade. In either case the specimen is held at a temperature which may be as low as $-196\ ^{\circ}C$ (using liquid nitrogen). For the TEM the fracture face or faces are shadowcast with a fine-grained film of platinum/carbon or refractory metal, and then coated with carbon to provide the replica film. After being warmed to room temperature and removed from the vacuum chamber the replica is removed from the specimen by dissolving the latter in chromic acid or a bleach solution appropriate to the specimen material. The replica film is washed and mounted on a specimen grid for examination in the TEM. If complementary replicas are available they may be mounted together on a grid so that identical areas on the opposing fracture faces can be identified more readily.

Freeze-fracturing and freeze-etching of biological material reveal structures which would probably not be exposed by any other method. The plane of fracture follows the line of least resistance, and there are several schools of thought as to whether this is along outer surfaces or within membranes, etc., or a combination of each. There is a risk of artefact formation because of ice crystal growth in the initial freezing stage; this may be minimised by soaking specimens in glycerol cryo-protectant prior to freezing. One way of detecting the presence of artefacts is to compare the structures of the two halves of a complementary replica, which should be a good match for one another if there has been no modification. An example of a side-by-side positive/negative comparison is given by the arrangement of stereoscopic pairs of micrographs in Figure 4.23(*a*). Examples of freeze-etching are shown in Figures 4.23(*b*) and (*c*).

There is of course no reason why the technique of freeze-etching should be confined to biological materials, provided the specimen has a structure which is conveniently fractured whilst its water content is frozen. Emulsions and pastes could be studied in this way.

The term freeze-fracturing is also applied to a procedure commonly used for the preparation of certain types of dry material for examination in the scanning microscope. This will now be described.

Freeze-fracturing of dry materials.

We saw earlier that rubbery and elastic materials may be sectioned on a cryo-ultra-microtome which freezes them below their glass transition temperature, so that they become hard enough to cut. It is also possible to induce brittle fracture into such

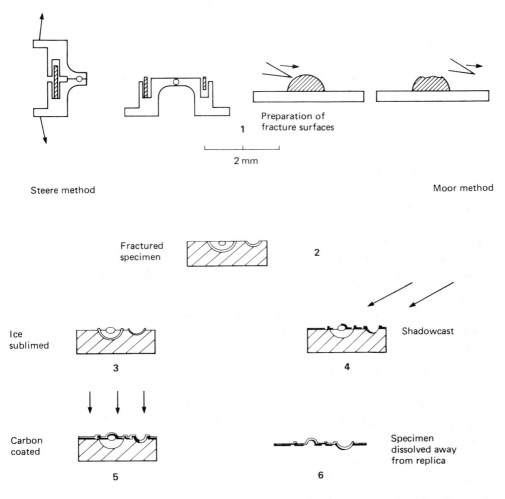

Figure 4.22. Schematic diagram of the freeze-etching process. The first stage may be either the complementary fracture method of Steere or fracture with a cooled knife, as practised by Moor. From then onwards the procedure for controlled sublimation of ice to form an etched surface and the preparation of a pre-shadowed carbon replica of this surface are common to both schools.

glassy materials so that internal structures are revealed. There is a direct parallel with biological freeze-fracturing in so far as the fracture plane takes the path of least resistance within the material and hence follows grain boundaries and similar types of feature which are significant in relation to the structure of the material. Information on the construction of surface coatings, etc., can be obtained in the scanning microscope from the study of fracture faces prepared in this and other ways. Plastics and coated papers are particularly suitable for this type of study.

Although it is possible to construct a cooled mechanical jig for clamping and fracturing samples it is convenient and easy to carry out the fracturing by holding the specimen in two pairs of forceps, placed either side of the point at which the fracture is to take place, under the surface of liquid nitrogen in a beaker or insulated container. If the specimen is in the form of a thin strip some 0.5–1.0 cm wide it is helpful to cut a notch out of the strip to provide a line of weakness at which the fracture will commence. A scanning micrograph of a typical freeze-fractured specimen is shown in Figure 4.24.

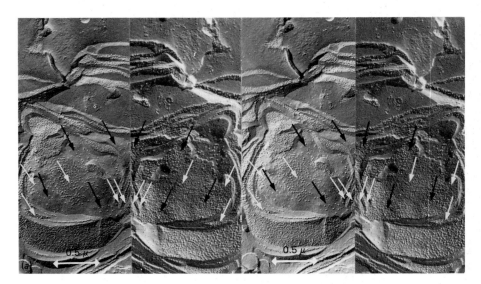

Figure 4.23. (*a*) Complementary stereo electron micrographs of freeze-fractured *N. rustica* chloroplast showing thylakoid faces (white arrows) connected by stroma membrane (black arrows). Chloroplast within cell in 25% sucrose, 25% glycerol in 0.2M phosphate buffer, pH 7.0 (×25 000). (By courtesy of R. L. Steere, J. M. Moseley and E. F. Erbe.) (*b*) Stereoscopic pair of micrographs of a replica of a true three-dimensional crystal of tobacco mosaic virus within an infected tobacco cell. Specimen frozen in phosphate buffer only and etched at −98 °C for two minutes after fracture, before replication (×115 000). (By courtesy of R. L. Steere.) (*c*) (*p. 180*) Micrograph of a freeze-etched replica of a dividing yeast cell (*Saccharomyces cereviciae*). The different components are identified as follows: C: cell wall; PM: plasma membrane; N_1: nucleus, surface view; N_2: nucleus, cross-fractured; M, mitochondrion; V_1: vacuole, surface view; V_2: vacuole, cross-fractured; L: lipid granula. (By courtesy of Bal-Tec AG.) Compare with micrographs by cryo-SEM in Figure 4.26.

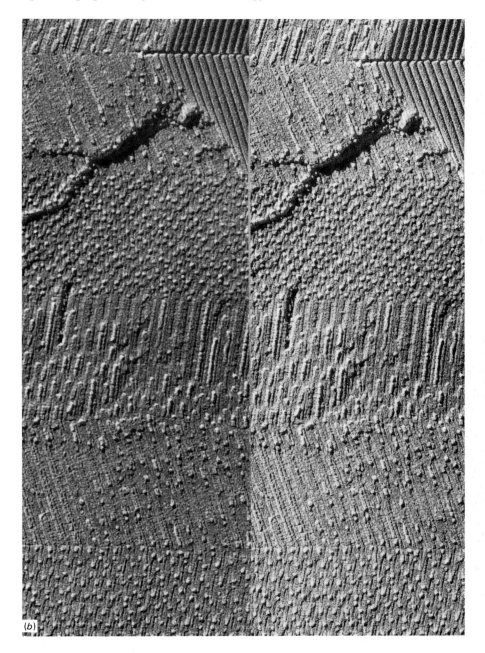

4.6.4 *Cryo-microscopy*

The closest approach to the goal of examining fresh, untreated, biological material is provided on electron microscopes equipped with specimen stages able to accept and examine frozen specimens. Thus, frozen sections cut on a cryo-ultramicrotome can be transferred to the TEM and examined whilst still hydrated. Frozen, fractured or etched bulk specimens can be examined in an SEM with a cryo-stage. There are considerable practical difficulties in transferring frozen specimens into microscopes adequately cold and free from frost, but this is now possible into certain models of SEM and TEM. The results represent the norm against which other preparations involving drying and substitution are judged. Figure 4.25 is an example of cryo-SEM

Figure 4.23 (*cont.*)

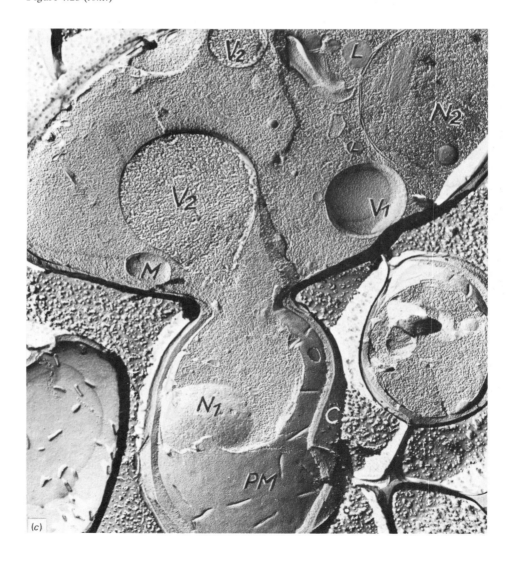

(c)

on a botanical specimen which would be difficult, if not impossible, to examine in such detail by any other means. Spores are seen emerging from stomata on a leaf surface which can be seen to have a waxy coating – usually the first detail to be lost in normal dehydration procedures.

Using the well-characterised test specimen bakers yeast, Walther *et al.* (1995) showed that frozen specimens in the cryo-SEM are very beam-sensitive in the uncoated state, but that high-resolution examination in an SEM with a cryo-stage was possible when the fractured frozen specimens were thinly shadowcast with heavy metal and then coated all-over with carbon. (Effectively TEM freeze-fracture replication but examined in the SEM without dissolving away the specimen.) The TEM analogy was carried further by using backscattered electrons for imaging the heavy-metal layer, using the carbon only as a stabilising layer. Figure 4.26 shows the fracture surface of the frozen yeast suspension imaged by backscattered electrons at 5 kV (*a*) and 30 kV (*b*). Micrographs of similar areas of nuclear envelope by

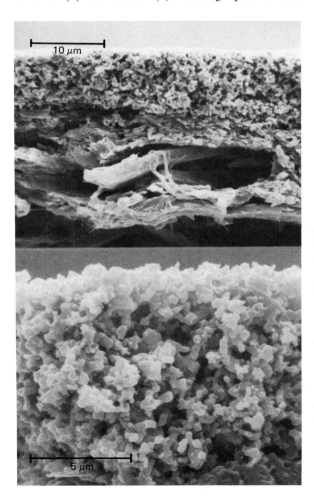

Figure 4.24. Scanning micrographs of a zinc-oxide-coated document copying paper frozen and fractured under liquid nitrogen. The evenness of coating thickness and the variations of porosity within the coating can be seen from examination of such a fracture face.

cryo–SEM at 20 kV and by a conventional freeze etch replica in the TEM, are shown in (*c*) and (*d*), respectively. These last two micrographs show that the double-layer cryo–SEM method is a reliable one, with the added advantages over the TEM that more extended areas may be examined and there are no problems caused by incomplete removal of the specimen from the replica. The bulk material is also available for electron probe microanalysis. On the other hand, removal and storage of a frozen specimen for re-examination at a future date is not as convenient as filing a 3 mm TEM grid in a grid storage box.

4.6.5 Freezing processes

A considerable amount of effort is being devoted to establishing the most suitable way to perform the freezing operation which precedes the variety of cryogenic treatments we have met in the previous paragraphs. If freezing is carried out too slowly, or if frozen material is kept at temperatures above about $-80\ °C$ large ice crystals are formed which cause structural damage to the specimen and provide artificial structures (artefacts) in the micrographs. Glycerol and other anti-freezing agents ('*cryoprotectants*') are often introduced into specimens to inhibit the process of crystallisation. Increasing the rate of freezing results in smaller ice crystals being formed, but the ideal is to cool the water so rapidly that it freezes in an amorphous vitreous (glass-like) state. For this to occur it is postulated that specimen cooling

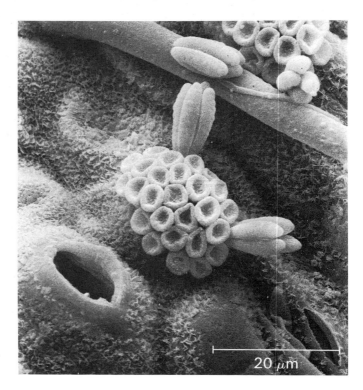

Figure 4.25. Cryo–SEM. An unusual Basidiomycete genus, *Microstroma*, on the giant sensitive plant *Mimosa pigra*. The fruiting structures are emerging from stomata on the lower leaf surface. $\sim -160\ °C$, gold coated, 15 kV. (By courtesy of G. F. Godwin and H. C. Evans, C.A.B.I.)

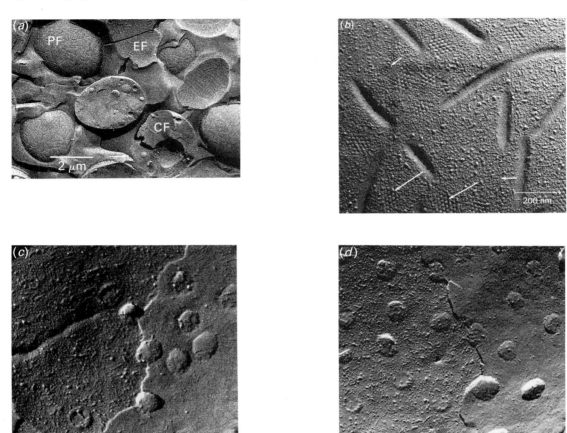

Figure 4.26. Direct examination in high-resolution (*in-lens*) cryo-SEM of bakers yeast *Saccharomyces cerevisiae*. An aqueous suspension between copper plates was rapidly frozen and fractured at −150 °C (123 K). Fracture surfaces were coated at 45° with 1.5–3 nm of Pt/C and cone-shadowed at 80° with 5–7 nm carbon. (*a*), (*b*), (*c*) are backscattered-electron images. (*a*) General view (5 kV) with cross-fractured cells (CF), protoplasmic fracture faces (PF) and extra plasmatic fracture faces (EF) (*b*) Protoplasmic fracture face from same sample at maximum resolution (30kV). Large and small arrows show particle grouping as seen previously on high-resolution TEM replicas. (*c*) Detail of nuclear envelope in a cross-fractured cell by SEM at 20 kV. (*d*) Transmission micrograph of similar detail from a shadowcast freeze-fracture replica, tone-reversed to have dark shadows. (Micrographs by courtesy of P. Walther, E. Wehrli, R. Hermann and M. Müller, 1995.)

rates must be in the range 10^5–10^6 degrees per second. Since biological materials are, in general, poor conductors of heat it is obvious that the outside of a specimen will cool much faster than the interior, and a frozen specimen will have only a limited thickness for which the cooling rate was sufficiently high for vitreous ice to have formed; beyond this the specimen may be unusable because of crystalline ice formation. The vitreous region is estimated to range between 5 and 20 μm in thickness for a number of cooling procedures being investigated and used. It has been found that amorphous ice can crystallise into cubic ice at temperatures as low as -160 °C (113 K) (Dowell & Rinfret, 1960). By -90 °C the crystals will have transformed into the hexagonal form.

The earliest attempts at rapid cooling involved plunging the specimen into a bath of liquid nitrogen (-196 °C, 77 K). Experience showed that this was an inefficient technique since an insulating sheath of gaseous nitrogen formed around the specimen and restricted the heat transfer from it. A number of other liquids have been used, including nitrogen slush (liquid nitrogen partly frozen under reduced pressure) at -210 °C, (63 K), and various Freons; it is widely accepted that plunging into liquid-nitrogen-cooled liquid propane is a simple and effective method of cooling, achieving 5–10 μm of vitreous ice. Another method of using liquid propane is to spray it on to the sample, or in the case of a sample to be freeze-fractured, on to a pair of copper plates with the sample between them (Costello & McIntosh, 1981).

Another very successful technique ('slamming') is to bring the specimen rapidly into contact with a highly polished plate of pure copper cooled to the temperature of liquid helium (-271 °C, 2 K). Rapid cooling results in a vitreous ice layer of 15–20 μm being formed, with the added attraction that pressing the specimen against the metal mirror results in a plane homogeneous block face for thin sectioning.

A method which has proved successful for freezing thicker (0.5–1.0 mm) specimens is high-pressure freezing (Moor, 1987). The technique has been shown to work very satisfactorily on botanical tissues, and commercial equipment for its use is now available. The method makes use of the fact that under an applied pressure hydrated specimens may be (super)cooled to a much lower temperature (-90 °C instead of -40 °C) before ice crystals begin to form, and the rate of ice-crystal growth is also much reduced. Cooling by liquid nitrogen at pressures of 2000–2500 bar is carried out under carefully controlled conditions. The deleterious effect of high pressures on specimens is minimised by carrying out the process quickly.

It is likely that the requirements and the 'best' technique will differ slightly for every different type of sample. However, the successful use of the various techniques described above will make possible a great step forward in the study of the structure and composition of biological materials.

4.7 Cytochemical preparations

4.7.1 *Immunocytochemistry and colloidal gold*

Antibodies labelled with (=attached to) dyes or small metal particles are used to locate and mark the position of certain types of tissue constituent (antigens). For light microscopy fluorescent labels may be excited by blue- or UV-radiation, as in epifluorescence confocal microscopy. For electron microscopy the use of antibodies labelled with small particles of colloidal gold has proved successful (Beesley, 1989). Because of its high atomic number gold has a high visibility by both transmission and scanning electron microscopy. A range of sizes is available, from tens of nanometres down to 1–2 nm diameter. The smaller the gold particle, the more readily it can be inserted into cells, and the more efficient its use becomes; in addition there are more particles for a given mass of metal.

Small gold particles may be increased in size and hence made more visible by 'enhancement' with silver, to give visibility at the light microscope level if necessary. For SEM and TEM stereoscopic microscopy is a great help in enabling the exact location of gold particles to be seen.

It is normal to apply the labelled antibody after the specimen has been prepared for the electron microscope, ie after the necessary fixation, embedding, sectioning or drying processes have been carried out. One of the major problems is that very often these pre-treatments weaken or destroy the immune response. Chemical fixation, particularly with glutaraldehyde and osmium tetroxide, is destructive of the reaction. Milder fixatives such as formaldehyde are less destructive, but the ideal treatment is cryo-fixation, in which no chemicals (apart from cryoprotectants to restrain ice crystal growth) are used, but only rapid freezing. Once frozen, the troubles are not over, since embedding resins can also interfere with antibody reactions. The PLT procedure appears not to interfere in some specimen types, but preliminary experiments should always be carried out to make sure that there is still an immune response to be labelled at the end of the specimen preparation.

Although there is a strong desire to use colloidal gold decoration quantitatively to measure antigen concentrations, Hayat (1992) considers that this is not yet justified.

Gold labelling for TEM

After the pre-treatment, ultra-thin sections of resin embedded material are labelled by placing a drop of the gold-labelled marker on the cut surface, leaving it for a set time before rinsing off the surplus. Gold particles down to 1 nm can be seen in micrographs.

Gold labelling for SEM

For normal SEM examination 20 nm gold particles are the smallest that can be detected reliably. If smaller gold particles are to be used either a field emission SEM can be used to detect them or the particle size can be increased by silver enhancement. It is necessary to use backscattered electrons for imaging colloidal metal particles, since the topographical contrast of small surface particles is low, and sub-surface particles would not be detectable.

4.8 Other preparatory techniques

4.8.1 Microwaves in specimen preparation

Some of the stages of biological specimen preparation can be speeded up by irradiation with microwaves (2.45 GHz, 12.24 cm). In a suitably controlled microwave oven tissue specimens can be heated rapidly to a temperature of 45–55 °C at which deterioration is arrested without the use of chemical fixatives such as aldehydes. Staining reactions are speeded up considerably, whilst immunocytochemical reactions are retained. It is important not to exceed the temperature range mentioned, or changes in tissue structure are liable to occur. Subsequent conventional fixation, e.g. with osmium tetroxide, is necessary to obtain permanent stability of tissues. The theory and practice of microwave processing are fully described by Kok & Boon (1992).

4.8.2 Conductive fixation

Certain fixation treatments for biological materials have the effect of giving the specimen sufficient electrical conductivity to make futher treatment by metal coating unnecessary before examination in the SEM. Two such treatments are known by the initials OTOTO and TAO, representing the different stages in the processes. Before applying either of these, the blocks of tissue are fixed in glutaraldehyde. Then, for OTOTO the tissue is immersed successively in solutions (1%) of OsO_4 (O), Thiocarbohydrazide (TCH), OsO_4 (O), TCH and finally in OsO_4 again, prior to drying and mounting on a stub for examination by SEM. The function of the TCH is to increase the uptake of OsO_4, with consequent improvement in bulk conductivity.

In the TAO preparatory treatment, more appropriate with collagenous tissues, the stages of treatment involve Tannic acid (T), followed by Arginine (A) and finally OsO_4 (O).

One advantage of these non-coating procedures is that the same specimen may be embedded for sectioning and examination by TEM without the complication imposed by metal coating layers. Details of these treatments are given by Murphy (1978) and Jongebloed (1990).

4.9 Specimen storage

One aspect of electron microscopy which tends to be overlooked (because it's always there and taken for granted) is specimen storage. Unless the freshly prepared specimen is rushed straight to a waiting electron microscope for examination it usually has to wait its turn somewhere. Similarly, after examination the specimen is usually retained at least until the work has been written up and no further examination is required.

One of the greatest nightmares in microscopy is that specimens have been mixed up, and possibly the reported conclusions really refer to another sample. It is highly desirable, therefore, that immediately after preparation and examination the grid or stub be placed in a uniquely labelled storage place. Furthermore, since conclusions are being drawn from a microscopic amount of sample, the storage conditions should be such that no further changes are forced on the specimen. A cool, dry, neutral environment, free from vibration, is usually adequate, but storage *in vacuo* may be necessary if specimens and/or coatings are reactive. Thin sputtered or evaporated metal coatings will change their structure and resistance with time, and should be examined as soon after coating as possible. The stability of elemental distributions in freeze-dried cryo-sections stored under vacuum has been questioned by Pelc & Žižka (1993).

In cryo-microscopy, of course, the requirements are exceptional, and arrangements must be made for storage of the specimen frozen until there is no chance of it being re-examined.

Figure 4.27 shows simple but adequate storage units for routine TEM grids and SEM stubs.

4.10 Suggested further reading

The most commonly used aspects of biological preparations, i.e. embedding, sectioning and staining, are dealt with in detail in Robards & Wilson (1993), and in Glauert (1974), Reid and Beesley (1990) and Lewis & Knight (1992) of the Glauert series. These procedures are also described by Weakley (1981), Meek (1977), Hunter (1993) and Dykstra (1993). Fixation is discussed by Glauert (1991), Hayat (1989a),

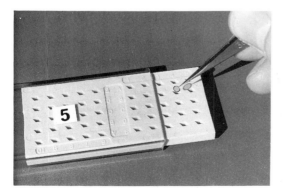

Figure 4.27. Storage boxes for TEM grids and SEM stubs. Numerous designs are available; those shown are simple but effective in preserving the prepared specimen from physical damage, and in an identifiable place.

and Bullock (1984). Other aspects of preparation, such as drying, various cryo-techniques and autoradiography, are described by Hayat (1970), Köhler (1973), Robards & Sleytr (1985), Baker (1989), Roos & Morgan (1990), and Morel (1993). Techniques in immunocytochemistry are covered in Bullock & Petrusz (1982-9).

A historical review of ultra-thin sectioning is given by Villiger & Bremer (1990). There is a great deal of practical advice on freeze-fracturing and freeze-etching procedures and on stereoscopic microscopy in Steere (1981). Cryo-techniques are described in Steinbrecht & Zierold (1987), Echlin, (1992) and Robards & Wilson (1993). A number of papers on freeze-substitution will be found in Gilkey (1993). An illustrated review of cryo-preparation techniques, with numerous references, will be found in Quintana (1994).

A very useful collection of techniques for analytical and quantitative microscopy, principally in the TEM, is contained in Meek & Elder (1977). Biologists should be aware of potential artefacts such as those described by Crang & Klomparens (1988). Beesley (1989) and Hayat (1989b and 1992) summarise the state of gold labelling in cytochemistry.

The biologist using scanning electron microscopy should be interested in Hayat (1974).

The non-biologist will find specimen preparation described in Kay (1965), Goodhew (1973, 1985), and Robards & Wilson (1993).

Specimen and general techniques for scanning microscopy are contained in Wells *et al.* (1974), Goldstein *et al.* (1992), and Lyman *et al.* (1990).

Microwaves in specimen preparation are comprehensively dealt with by Kok & Boon (1992), whilst two special microwave issues of *The Histochemical Journal* were published: **20**, 311–404, 1988 and **22**, 311–93, 1990.

The interpretation and analysis of micrographs

5.1 Interpretation of transmission micrographs

The specimen has been prepared and put into the microscope. The beam is turned on and the image focused on the fluorescent screen. How should we now interpret the light and shadow in relation to the original specimen which was provided?

With most types of specimen working out the meaning of a TEM image is straightforward once a few basic principles have been appreciated. The image we are considering is the electron image on the fluorescent screen, or the finished photographic or video print which can be examined more conveniently than the screen image. The silver image on the processed photographic film, which is strictly speaking the true micrograph, will be reversed in tone from the other two, bright areas on the screen giving dark areas on the film and vice versa.

In the absence of a specimen the image field would be uniformly bright. It is darkened when the specimen is inserted because electrons have been scattered out of the illuminating beam by passage close to the atoms in the specimen (Figure 5.1). Detail in the electron image results from localised variations in the scattering power of the specimen. More electrons may be scattered, i.e. the image will be darker, if the specimen is thicker or composed of heavier atoms, or both. An important distinction between light- and electron microscopy lies in the fact that image contrast in the former is caused by differential *absorption* of illumination whereas in the latter the operative mechanism is *scattering* without absorption. A TEM specimen which absorbs electrons is too thick.

In the fluorescent image, therefore, the brightest areas are those from which scattering is least. So in a micrograph of dispersed particles the dark regions are the par-

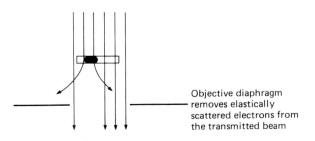

Objective diaphragm removes elastically scattered electrons from the transmitted beam

Figure 5.1. Strongly deflected elastically scattered electrons are removed from the image by the objective aperture giving *amplitude contrast* to the final image. This is the main source of image detail at all but the highest magnifications, when scattered electrons and the undeflected beam interfere to produce *phase contrast* detail in the image (1 nm and less in size). Inelastically scattered electrons are deflected through smaller angles and pass through the imaging system to focus in a slightly different plane because of the effect of chromatic aberration of the lenses.

ticles seen against a background of a lighter 'translucent' support film. Variations of density within the borders of the particles represent internal structure. In a micrograph of a parallel-sided microtomed ultra-thin section the presence of density variations shows that there are local changes in atomic arrangements, which may be accentuated by the presence of local deposits of heavy atoms in the stain. In the directly exposed electron micrograph the optical density of the developed silver image at any point is proportional to the electron exposure at that point; this can be made into a quantitative measurement technique, as will be described in Section 5.5.3.

5.1.1 Amorphous and crystalline materials

Figure 5.2(*a*) gives an example of an image which can be interpreted along the lines outlined above. Light and shade are produced by *amplitude contrast* which is the main mechanism in the microscopy of amorphous materials, except at the highest magnifications. This will suffice for the correct interpretation of most biological transmission micrographs. Materials science transmission micrographs of crystalline specimens will show mainly *diffraction contrast* in which light and shade is determined by local orientations relative to the electron beam, rather than mass–thickness changes. The difference in behaviour between amorphous and crystalline specimens is illustrated schematically in Figure 5.2(*b*). Considering the simple example of two rectangular blocks of material, one amorphous, one crystalline, we see that altering the orientation of the former produces only a geometrical change in the image, whereas slightly tilting the latter makes all the difference

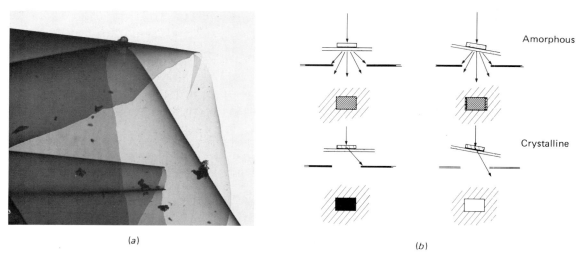

(*a*)   (*b*)

Figure 5.2. (*a*) Amplitude contrast demonstrated with a micrograph of overlapping folded shreds of carbon support film. Up to five superimposed thicknesses can be identified; folds and curls in the layers can be inferred from density variations in the image. Local additions of density are provided by isolated particles on the support film. (*b*) Schematic diagram showing the different effects of specimen tilt on image contrast in amorphous and crystalline specimens.

between light and dark tones in the image, as a consequence of the Bragg condition being satisfied or not (Chapter 2). If the specimen has many small crystals randomly orientated the overall effect may average out to a 'grey' tone, but the individual black and white patches will be seen separately if the magnification is increased sufficiently. This effect is the one which restricts the usefulness of some materials as shadow-casting films in electron microscopy, since the black/white effect in the micrographs is dependent on the chance orientation of the individual crystallites rather than on the controlled variation of film thickness due to substrate topography.

Transmission electron micrographs of crystalline materials show a number of characteristic features, which we will consider briefly (Figure 5.3 (i–iv)). Extended thin crystals will show the effect of light and dark bands running across the crystal. If the specimen is tilted or if it distorts because it is heated by the electron beam the bands will move to other positions. These bands are *bend contours* which join up regions with the same orientation relative to the electron beam. They have no physical existence and do not belong to any particular region of the specimen. Parallel light and dark

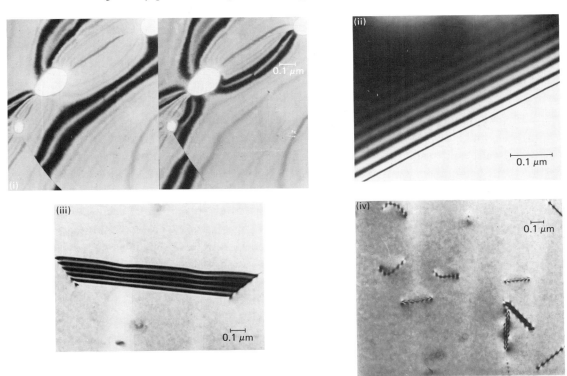

Figure 5.3. Crystallographic image effects. (i) Bend contours in Al foil move to different positions when the foil is tilted or buckled. (ii) Equal thickness contours in Si indicate that the specimen tapers in thickness towards an edge. (iii) A stacking fault in GaAs. (iv) Dislocations in GaAs. A practical specimen may exhibit all of these effects, and others, simultaneously; they will be familiar to the electron metallographer. (Micrographs by Philips Electron Optics, Cambridge.)

fringes which may be seen parallel to the thin edge of a thick crystal are *equal thickness contours*, and indicate that the crystal is tapering towards the edge. The change in thickness from one contour to the next depends on the nature of the crystal (i.e. its *d*-spacing) and how it is orientated relative to the electron beam. Other characteristic effects are *stacking faults* and *dislocations* (sometimes dislocation loops) which indicate the position of local defects in the atomic arrangement in the crystal; these may be caused by the presence of an impurity or second phase in the crystalline matrix.

The full story of electron–crystal interaction and image formation is outside the scope of this book, and readers whose major interest is in the *electron microscopy of thin crystals* should refer to the book of this title by Hirsch *et al.* (1965) or other books on electron metallography.

The particular circumstances of near-atomic resolution have their own requirements, and these are addressed in publications such as Buseck, Cowley & Eyring (1988).

5.1.2 Shadowcasting

Shadowcasting is applied to electron microscope specimens to reveal added information on the third dimension. This can be either qualitative or quantitative, but in order to extract it correctly it is necessary to know the angle and direction of the shadowcasting and at what stage in the specimen preparation it was applied.

The height or thickness of particles spread and shadowed on a plane substrate can be determined from the shadow length when the shadowcasting angle and the magnification are known. For this purpose the expression of shadowcasting angles as $\cot^{-1}x$ rather than so many degrees is more helpful in quick calculations. This is seen in theory and practice in Figure 5.4.

The original main use of shadowing, however, was to give a meaningful appearance to carbon replicas. These have little inherent contrast except at boundaries where the carbon film is seen edge-on. Figure 5.5(*a*) shows a carbon replica of a rough surface. Figure 5.5(*b*) is a replica of a similar area shadowcast at 45° ($\cot^{-1}1$) to the plane of the grid. The impression of surface relief makes it clear that there are pits and peaks on the replica. Knowing that shadowcasting was directed from the top of the picture enables the correct interpretation to be made.

The above micrograph was tone-reversed to give dark shadows and light highlights. Without this reversal process the fluorescent image and the final print are negatives in which 'shadows' are reproduced white and 'highlights' dark. In this case an approximation to correct interpretation (pseudo-relief) is obtained by orientating the print so that the source of shadowcasting lies at the bottom (Figure 5.5(*c*)). White shadows and dark highlights now combine to give a mental impression of a surface with correct peaks and depressions, but illuminated from the upper edge and having dark shadows.

Unfortunately this pseudo-relief is inaccurate in detail, since it replaces shadows

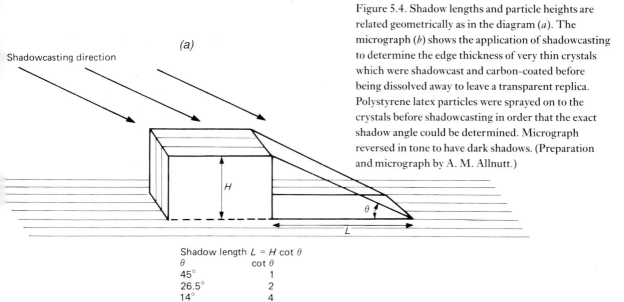

(a)

Shadowcasting direction

H

θ

L

Shadow length $L = H \cot \theta$

θ	$\cot \theta$
45°	1
26.5°	2
14°	4

Figure 5.4. Shadow lengths and particle heights are related geometrically as in the diagram (*a*). The micrograph (*b*) shows the application of shadowcasting to determine the edge thickness of very thin crystals which were shadowcast and carbon–coated before being dissolved away to leave a transparent replica. Polystyrene latex particles were sprayed on to the crystals before shadowcasting in order that the exact shadow angle could be determined. Micrograph reversed in tone to have dark shadows. (Preparation and micrograph by A. M. Allnutt.)

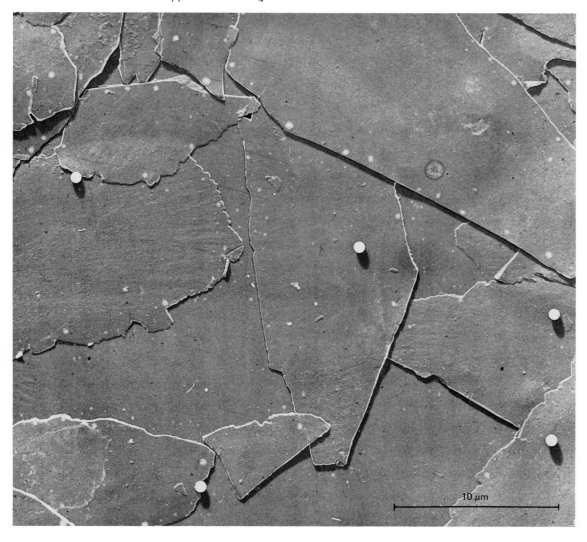

10 μm

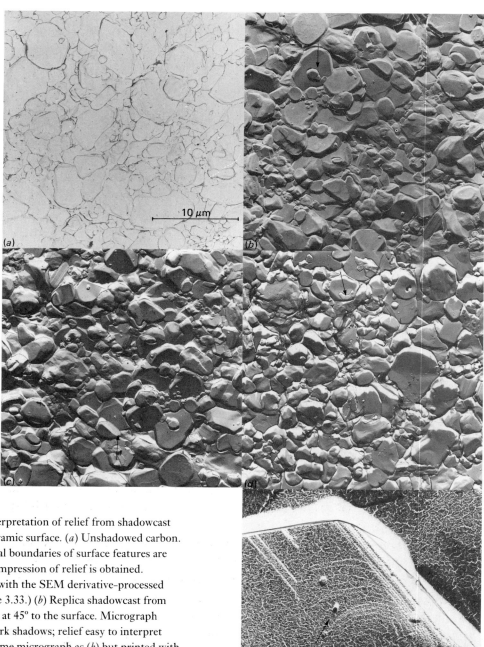

Figure 5.5. Interpretation of relief from shadowcast replicas of a ceramic surface. (*a*) Unshadowed carbon. Only the vertical boundaries of surface features are visible and no impression of relief is obtained. (Compare this with the SEM derivative-processed image in Figure 3.33.) (*b*) Replica shadowcast from top of the field, at 45° to the surface. Micrograph printed with dark shadows; relief easy to interpret correctly. (*c*) Same micrograph as (*b*) but printed with white shadows. Correct pseudo-relief when shadowing is from the bottom of the page. (*d*) Carbon replica shadowed on wrong surface, as a negative. Light and shade will not give true impression of the surface relief. (*e*) Direction of shadowcasting deduced from tapered shadow behind speck on surface. Direction of shadowcasting shown by arrow on each micrograph.

by highlights which rarely have the same form. On microscopes with digital or video images reversal of contrast to give dark shadows is only a matter of turning a switch or pressing a button.

If the direction of shadowcasting is not indicated on a micrograph it can generally be inferred by examining the shape of the shadows (compare, for example, the shape of shadows within a pit and behind a protrusion in Figure 4.3) or by looking for the typical tapered shadows behind pimples or specks of dust on the replica surface (Figure 5.5(*d*)).

Shadowcasting the negative replica in a two-stage process will never give a true surface impression, whether with black or white shadows, but pseudo-relief in the correct direction will be seen best when the print is orientated so that the direction of shadowcasting is from the top of the print (Figure 5.5(*e*)) with white shadows.

5.1.3 Making measurements from micrographs

Since a transmission micrograph is an enlarged plan view of the specimen it is possible to lay a ruler across the print and make measurements of features of interest. The precision with which these measurements may be related to actual dimensions on the specimen depends on the accuracy of calibration of magnification of the microscope and the reproducibility of this figure from one micrograph to another. Most microscopes nowadays give a digital readout of magnification which changes as the magnification knob, switch or button is operated. This readout will probably be within 10% of the actual magnification and the exact value within about 1% may be determined by calibration using a suitable specimen, as in Figure 5.6. However, the calibration is only meaningful if special precautions are taken in operating the microscope, to achieve a reproducibility of magnification also of about 1%. There are two possible sources of error – magnetic hysteresis in the lenses and variability of specimen plane along the microscope axis.

The magnification readout is linked to the current in the projector lens coils; however, the magnetic strength of the lenses (and hence the magnification) is not uniquely determined by this current, but differs according to whether the current was decreased from a higher value or increased from a lower one. The operator of an older microscope can deal with this situation by 'normalising' the lenses after each change in magnification. A normalising button may be provided which takes the lens currents to maximum and then drops them to their desired values each time; or the lenses may be simply switched off and on several times. On computer-controlled microscopes the necessary routine may have been automatically programmed into the magnification changing process.

For the magnification calibration to be of practical use all specimens must occupy the same plane in the specimen stage as the calibration grid – within tens of micrometres if the 1% reproducibility is to be attained. Top-entry specimen cartridges can differ in length and are best calibrated individually; they must also seat reproducibly in the stage. Specimen grids must be flat and undistorted. The magnitude of the objective lens current at focus is a sensitive test of the location of the specimen plane. The side-entry eucentric goniometer specimen stage is free from these uncertainties, as the specimen plane can be adjusted precisely to the eucentric point, even for individual grid squares, and absolute reproducibility of the specimen plane is therefore possible.

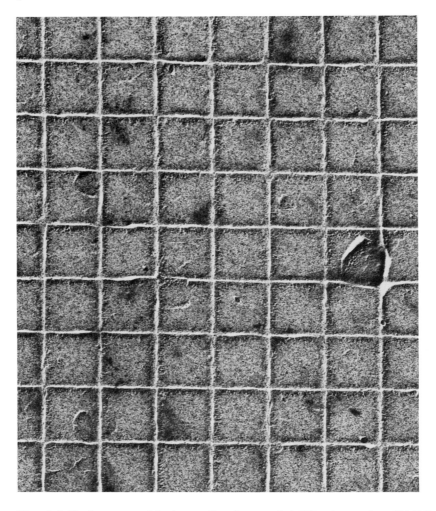

Figue 5.6. Shadowcast metal/carbon replica of cross-ruled diffraction grating of 2160 lines/mm (Oster & Skillman, 1962). With this replica the microscope can be calibrated to an accuracy of about 1% up to about 150 000 times. Above this figure the lattice plane spacings of crystalline materials are used, e.g. beef catalase (8.76 nm), copper and platinum phthalocyanines (1.437 and 1.2 nm respectively) and graphite (0.34 nm). In the figure one line space is enlarged to 14.86 mm and the magnification is therefore $14.86 \times 2160 = 32\ 100$. At lower magnifications the crossed grating enables image distortion to be observed and measured. This specimen is also suitable for magnification calibration and determination of distortion in the STEM mode of the ATEM.

Having ensured that the magnification from specimen to micrograph is accurately known the size change on to the final print, whether by photographic enlargement or video printer, must be known equally precisely.

Dimensions in the depth of the specimen can also be measured from micrographs by using the technique of stereoscopic microscopy, to be described in Section 5.4. Size measurements of bumps and pits from single-stage post-shadowed replicas will show some error because of the thickness of the carbon layer. This is illustrated in Figure 5.7.

5.1.4 Instrumental effects

The operator of a microscope naturally tries to produce the best possible micrograph from the specimen under the conditions prevailing at the time of examination. The design of some microscopes makes it easier to achieve the optimum on them than on others, but even so there is an element of personal judgement involved and different operators on a single microscope may well produce noticeably different results. It is as well, therefore, to see what variations are possible, so that they can be detected and allowed for in interpreting the micrographs.

Focusing

These are mainly due to the presence of Fresnel fringes around image features. These are interference effects and vary with the difference between the plane of the specimen and the plane of focus of the objective lens. The Fresnel fringe is most clearly demonstrated with a specimen consisting of a thin carbon film with circular holes (the microscopist's '*holey carbon film*'). When the objective lens of the microscope is focused exactly on the plane of the carbon film there are no visible fringe effects. Reducing the objective lens current moves the plane of focus beyond the carbon film and produces a bright fringe within the holes, the breadth of the fringe

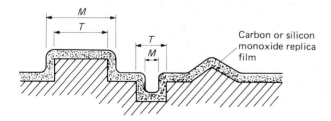

Figure 5.7. A thin-film replica, especially of carbon, forms a comparatively uniform layer over the surface on which it is deposited. Steep-sided surface features will therefore appear larger when they protrude from the surface, or smaller if they are recessed, when measured over the replica film. The difference between true (T) and measured (M) dimensions may be between 20 and 40 nm in a practical replica, which may be a significant proportion of the size of a sub-micron feature. Size errors can be avoided if the specimen surface is shadowcast directly before the replica film is applied, or by making a two-stage replica.

increasing the further the objective is taken from focus. In this condition the lens is said to be *underfocused*. When the objective is more strongly excited than exact focus (*overfocused*) there is a bright fringe produced within the image of the carbon film itself, the breadth again increasing with the degree of overfocus. The change of effect from underfocus to exact focus and overfocus can be seen clearly in Figure 5.8.

The holey carbon film provides a very useful test specimen for microscopists and service engineers but few practical specimens provide the same clear-cut edges. However, similar Fresnel fringe effects occur at every piece of edge detail in a transmission specimen. True focus is a condition of minimum image contrast, whereas underfocus gives an apparent sharpening up of specimen features, and the overfocus fringe results in an apparent blurring or loss of sharpness. Consequently most micrographs are recorded with the objective slightly underfocused. The very slight drop in resolution is more than compensated by the enhanced visibility of detail. However, if the underfocus is overdone the loss of resolution is noticeable, and the outlining becomes obtrusive. Figure 5.9 shows the different focusing effects on an ultra-thin section.

Microscopes fitted with wobbler focusing devices actually lead the operator to the point of exact focus; the operator can then introduce what he considers to be the desirable degree of underfocus before recording the micrograph. Fresnel fringes provide a very real help towards focusing images in the TEM; correct focus is more

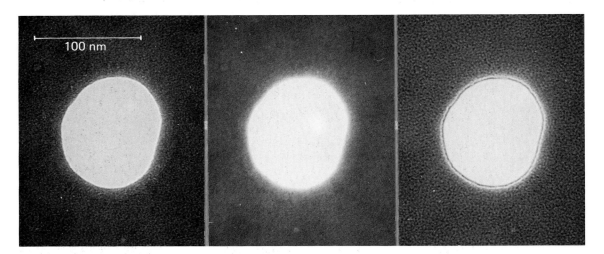

Figure 5.8. Fresnel fringes are an electron interference phenomenon. Their change in appearance with the focus of the objective lens can be seen from the accompanying micrographs. It should be noted that both under- and over-focused fringes (*left* and *right*, respectively) are accompanied by enlarged and obtrusive structure of the carbon film background. Although only the first and brightest Fresnel fringe is clearly visible the fringes continue with alternate light and dark bands of decreasing intensity. Strongly defocused illumination or use of a pointed filament will result in more fringes, as shown in Figure 2.22.

difficult to judge at low magnifications (e.g. ×10 000 or below) when the fringe width is so small as to be invisible unless the microscope is a long way from focus. Finding true focus in dark-field microscopy (Chapter 7) is difficult since Fresnel fringes are bright-field phenomena.

Astigmatism

The light microscopist can do nothing about the astigmatism of his objective lens. If there is any he must put up with it, but at least he knows it will remain constant during the lifetime of the lens. In an electron microscope, however, astigmatism is a variable and varying quantity which can have a profound influence on the resolution of the micrograph. Most microscopes have a stigmator which enables the astigmatism to be either eliminated **or made worse** by the operator. Automatic astigmatism compensation is possible on a microscope under computer control with digital imaging and appropriate image analysis software, but for most microscopes the compensation of astigmatism is a procedure whose success depends entirely on the operator's experience. It is also only met at relatively high magnifications, where the effect of resolution limitation becomes detectable in the micrograph, above about 50 000 times on the TEM.

The nature of the defect of astigmatism, as we saw earlier, is such that the lens has a different focus for different orientations of image detail, with the direction of sharpest focus turning through 90° as the plane of best focus is passed through. It can be regarded as a combination of a spherical lens with a cylindrical lens and compensation is by superimposing a second cylindrical lens at right angles to the first, of equal and opposite effect.

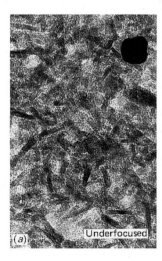 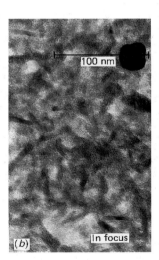 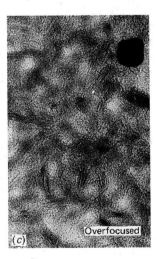

Figure 5.9. Focusing effect on a thin section in the TEM. Micrographs of a thin section of resin-embedded microcrystalline alumina: (*a*) underfocused, (*b*) in focus and (*c*) overfocused. In practice, for enhanced visibility of fine detail, micrographs would be recorded with focus settings between (*a*) and (*b*), as seen in Figure 5.10.

Astigmatism in electron microscopes can be caused by several mechanisms. There is a constant instrumental defect resulting from the practical difficulty of making electromagnetic lenses with completely symmetrical magnetic properties. There is also a variable astigmatism due to electrostatic charges on insulating films on apertures and lens bores in the electron microscope column. These charges are continually changing during operation of the microscope, and compensation of the resultant astigmatism is therefore a continuing requirement until the effect increases beyond the limit of compensation. The microscope or its components must then be dismantled and given a thorough clean. The specimen itself may provide a third source of astigmatism, if it shows local variations of potential under examination. Unless the specimen preparation has been inadequate this effect is usually smaller than the other two, but in work near the resolution limit of the microscope it may be necessary to recompensate the astigmatism for each field of view. The 'customer', however, is presented with the finished micrograph and in his own interests should be able to determine whether or not there is any residual astigmatism which would result in the resolution being poorer than he might have expected.

Astigmatism shows as a directional effect which rotates through 90° as focus is passed through. The simultaneous presence of over- and underfocus Fresnel fringes in astigmatic TEM images near focus is shown in Figure 5.10. The asymmetric fringes result in a directional effect which can make true structure difficult to identify.

One final point, sometimes overlooked, is that the correct elimination of astigmatism depends on the microscopist's eyesight being itself free from directional defect; persistent image defects in micrographs from a particular operator could well be due to a deficiency of the operator's eyesight.

Mechanical effects

Image features drawn out in one direction can also be the result of mechanical instabilities of the microscope or specimen. Nowadays the latter is much more likely to be the case than the former, although small particles of dust between specimen cartridges and their seatings are well-known sources of drift in top-entry TEM specimen stages. Dispersed particles in the TEM may move on their support films during examination and support films or replicas may likewise drift across the metal bars of the specimen grid. Often these mechanical movements of the specimen are not detected during setting up of the exposure, and are first seen in the micrograph. The effect of slight drift is similar in appearance to astigmatism, and has a disastrous effect on resolution. The only remedy is to reinsert the specimen and examine the image for stability over a period of time. Astigmatism can be looked for by going through focus and looking for changes in directional effects. If found, the astigmatism should be recompensated. Examples of the effect of mechanical instability are shown in Figure 5.11.

Contamination

Improvements in column design and in vacuum systems have combined to make contamination much less of a problem than it used to be. However, certain types of specimen, careless 'housekeeping', and older designs of microscopes all give rise to the characteristic appearance of contamination deposits, which will be described and illustrated.

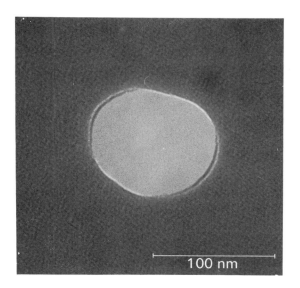

Figure 5.10. Astigmatism in the TEM. Astigmatism can result in under- and overfocus fringes being present simultaneously, in directions at right angles to one another. This is seen readily with an ideal subject like a holey carbon film. The disastrous effect on the detail in a thin section of resin-embedded alumina is seen by comparing the other two micrographs.

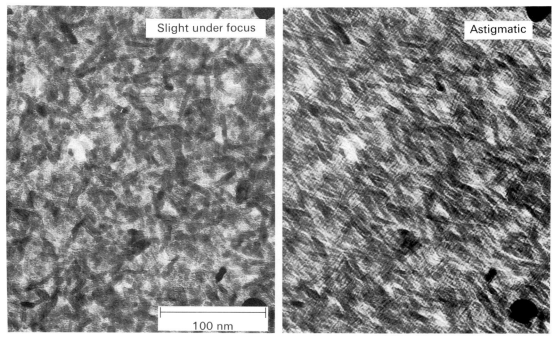

Slight under focus

Astigmatic

The residual atmosphere inside the column of an electron microscope will contain the normal atmospheric gases, oxygen, nitrogen, carbon dioxide and water vapour, as well as vapours of hydrocarbon compounds. These may have originated from vacuum pumping fluids, vacuum sealing gaskets ('O' rings) and any removable parts, e.g. specimen holders, plate- or film holders, which are handled directly by the operator when they are removed from the microscope. Under normal conditions there will be a thin layer of hydrocarbon adsorbed on all surfaces within the column, including the specimen. Electron bombardment polymerises the surface film and permanently fixes it in place. The depleted surface layer of hydrocarbon is replenished by continued condensation of vapour and by diffusion of fresh condensed material inwards across the surface from outside the electron-bombarded area. Thus, a permanent deposit grows on the specimen surface whenever it is under examination. The deposit is referred to as *contamination*, and its rate of formation as the *contamination rate*.

Contamination builds up over the whole of the illuminated field of view, and becomes noticeable if the magnification is reduced and the previously illuminated area forms part of the now larger field of view. This is illustrated in Figure 5.12. The progress of contamination can be followed, and its rate measured, with a holey carbon film as specimen. The polymerised layer builds up as a blanket over the whole field and also within the holes, so that these reduce in size and can eventually fill up with deposit. As the same time the diameter of isolated particles in the specimen will be increased. The two effects are illustrated in Figure 5.13. It is found that the contamination rate increases with the electron current density, i.e. with increasing magnification. The actual size of the illuminated area also has an effect on the contamination rate, which is greater for small spots, even at the same current density.

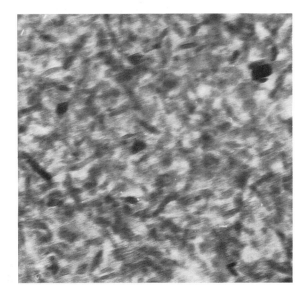

Figure 5.11. Visible effects of mechanical instability in electron microscopy. In the TEM slow movement of the specimen across its support grid, of the specimen holder in its seating, or of the stage itself gives the effect of drawing out the image in the direction of movement. When the drift is considerable in relation to specimen detail, as in the illustration, it is totally destructive of image resolution. Careful observation of the image on the screen before making the micrograph will usually show whether there is any mechanical instability present. If it is very slight it may be possible to reduce its effect to a tolerable value by shortening the exposure time, but only if the increased electron beam intensity needed does not result in an increased rate of drift.

The methods for minimising the effects of contamination are based on reduction of the amount of adsorbed hydrocarbon available for polymerisation and reduction of the time of exposure of the specimen to the electron beam. The column and specimen stage of the microscope should be maintained as clean as possible, and components and specimens should not be touched with bare hands. The rate of build-up of adsorbed films on specimens will be reduced if the latter are heated (up to several hundred degrees above ambient) and the rate of surface diffusion of contaminant is reduced if the surface is cooled below its normal temperature. In any case placing liquid-nitrogen-cooled metallic surfaces in the vicinity of the specimen reduces the vapour pressure of the contaminant to a much lower value. It is also advisable to focus the microscope and take the micrograph as rapidly as possible after the field of view has been brought into the electron beam.

5.1.5 Specimen preparation artefacts

An apparent microscopic structure which results from the processes of preparation or examination and is not a property of the original specimen is termed an *artefact*. There are well-defined types of artefact in transmission microscopy caused by the use of imperfect techniques for specimen preparation. In some cases the false result is obvious and the remedy is to repeat the preparation. In other cases the error is the use of unsuitable materials or techniques and the falsity of result may well go unsuspected unless some alternative method is used which produces a different result. Artefacts can occur in the preparation of replicas and in the examination of beam-sensitive materials. Loss of crystallinity because of beam-heated dehydration in the microscope, and change of crystalline structure of thin films on removal from their substrates, are all artefacts which may pass for genuine structures on first acquaintance but which can be recognised as false on more critical examination. A number of these effects are illustrated in Figure 5.14.

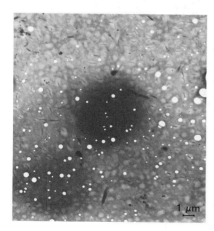

1 μm

Figure 5.12. The visible results of contamination seen when magnification is reduced.

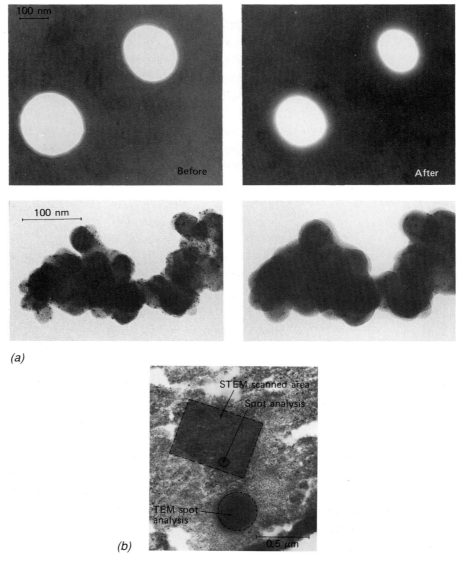

Figure 5.13. (*a*)(*Upper*). Part of the field of Figure 5.12, showing holes in a carbon film reduced in diameter by several minutes' examination with intense focused illumination. (*Lower*) A featureless 'skin' of contamination formed on the outside of particles after a period of detailed examination. The small black specks, which are metal particle on the outer surface of carbon black particles in a catalyst, can be seen to have become encapsulated by the layer of contamination. (*b*) Contamination effect on a thin section due to localised spot analysis in the TEM and STEM modes. In this case the contamination was a characteristic of the sample itself and not of the microscope.

5.2 Interpretation of scanning micrographs

The most commonly used imaging mode of the scanning electron microscope is the secondary electron emissive mode. This is used for imaging surface topography using secondary and backscattered electrons. Interpretation of this mode will be described, before the other modes using cathodoluminescence (CL), absorbed current and x-radiation.

Micrographs made in the SEM at magnifications up to a few thousand, are usually found easy to interpret as three-dimensional, 'real', subjects even if the exact reasons for light and shadow are not appreciated by the observer. One reason for this is that the specimen in the SEM is frequently tilted relative to the scanning electron beam. The lower end of the specimen (which appears at the top of the micrograph) is further from the final lens than the top end (see Figure 5.15(*a*)), and its image has a lower magnification. Consequently, there is a genuine perspective effect in the image (Figure 5.15(*b,c*)), and it looks three-dimensional to the observer. In addition, the top and bottom of the pictures are usually the first to show unsharpness because of depth of field limitations, and this augments the impression of distance in the mind of the observer. The explanation of the light and shade in a scanning micrograph is more complex, and is worth considering in some detail.

We learned earlier that the specimen is seen from the viewpoint of an observer in the bore of the final lens, i.e. coincident with the source of illumination. It might be expected, therefore, that the image would always be flatly lit and lacking in contrast, but this is not usually the case (except, of course, in completely flat subjects); in practice, scanning micrographs made in the secondary emission mode appear to have directional lighting from the top of the picture with strong rim lighting of isolated objects. (N.B. As far as the writer is aware the arguments and effects reported here apply to all makes of microscope when used in their SE mode using an Everhart–Thornley detector, and all accessories (scan rotation, dynamic focus, tilt correction, derivative processing, etc.) switched *off*. Images from different models may not be identical, for reasons which will be discussed later.)

To understand this apparent discrepancy between the lighting conditions and the appearance of the micrograph, we must recall that the secondary electron emission from a surface varies with the angle of incidence of the electron beam on the surface, being least when it is irradiated at normal incidence and greatest at grazing incidence; our flat lighting will therefore result in maximum signal from surfaces nearly parallel with the incident beam. It will also be higher from thin regions and edges where secondary electrons can be collected from two surfaces of the specimen simultaneously. This explains the appearance of rim lighting, but not the apparent directional nature, from the top of the picture. To see the reason for this, we recall that in Figure 5.15(*b*) the electron collector was shown situated beyond the bottom of the specimen, i.e. above the top of the image on the screen. Electrons emitted from

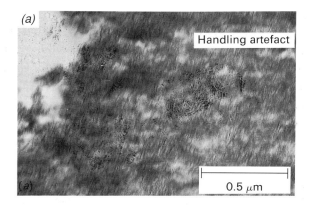

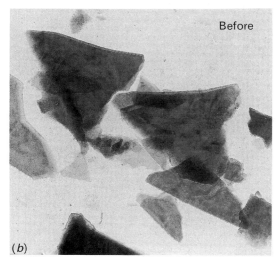

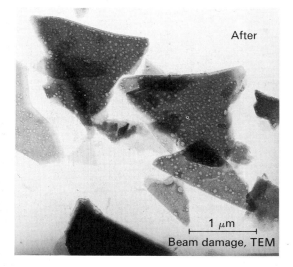

regions of the specimen which are facing the collector are readily collected by it, whereas electrons emitted by surfaces hidden from the collector may not reach it. The brightness of tone at any point on the image is therefore determined by the combined effects of surface inclination to the scanning beam and the efficiency of collection of the electrons from it.

An example of this is given in Figure 5.16, using the simple shape of a record-player stylus as subject; this is seen at a tilt angle of about 40°. Regions designated A are bright because high secondary electron emission is combined with a high efficiency of collection. At B, the secondary emission is low because of near-normal

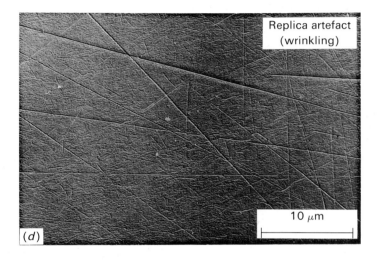

Figure 5.14. Common (recognised) artefacts in transmission electron microscopy. There are probably more still being accepted as true structures. (*a*) Handling artefact. A puzzling combination of blurred featureless detail in a thin section with a sharply rendered particulate deposit turns out to be the operator's fingerprints (×1) on the photographic film exposed to a micrograph (2 seconds at × 30 000) of a drifting specimen. (*b*) Beam damage. Talc particles before and after electron-irradiation in the TEM. Many types of crystal suffer damage in the beam, sometimes so rapidly that it is impossible to photograph an undamaged one. The blotchy appearance may be due to loss of water from the crystal; its appearance is acompanied by loss or modification of the diffraction pattern. Beam damage may be reduced by using a cooled specimen stage, a minimum-dose focusing and exposing device, or by using the HVEM or STEM. (*c*) Replica artefacts. There are various artefacts associated with replication, frequently due to the inability of the replica material to conform to all the fine structure of the specimen. Shadowcast replicas of an abraded stainless steel surface show different renderings of the same features. The right-hand one is most likely to be correct, since it is a pre-shadowed replica (page 143), in which the surface was directly shadowcast before applying the carbon replica film.
(*d*) The commonly used collodion/carbon two-stage process may also suffer from wrinkling which may be associated with electron bombardment during the carbon-coating procedure. The topography of this pitch-polished glass surface could have been accepted as genuine if this artefact had not been identified previously. (Micrograph (*d*) by A. M. Allnutt, Sira Ltd.)

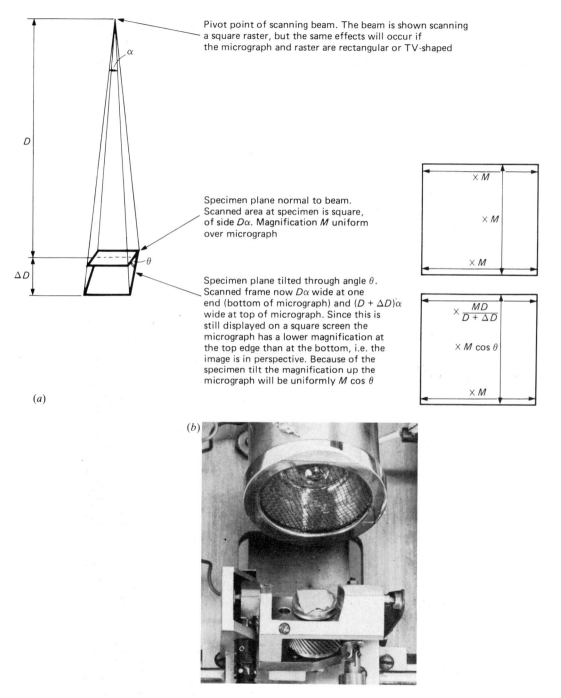

(a)

Pivot point of scanning beam. The beam is shown scanning a square raster, but the same effects will occur if the micrograph and raster are rectangular or TV-shaped

Specimen plane normal to beam. Scanned area at specimen is square, of side $D\alpha$. Magnification M uniform over micrograph

Specimen plane tilted through angle θ. Scanned frame now $D\alpha$ wide at one end (bottom of micrograph) and $(D + \Delta D)\alpha$ wide at top of micrograph. Since this is still displayed on a square screen the micrograph has a lower magnification at the top edge than at the bottom, i.e. the image is in perspective. Because of the specimen tilt the magnification up the micrograph will be uniformly $M \cos \theta$

$\times M$

$\times M$

$\times M$

$\times \dfrac{MD}{D + \Delta D}$

$\times M \cos \theta$

$\times M$

(b)

Figure 5.15. Origin of perspective in scanning micrographs of tilted specimens. Scanning micrographs of tilted specimens have a 'real' appearance because they have genuine perpective. This is shown schematically in (a). (b) Specimen in its tilted stage as it appears from the viewpoint in the final lens of the column. Note that the electron collector is situated above the top of the picture. (c) Illustration of the perspective effect using the parallel grooves of a gramophone record, tilted in the microscope.

probe incidence and, in addition, the electrons are screened from the collector by the body of the stylus. In regions C, the secondary emission would be expected to be high because of the surface inclination to the scanning beam but the final effect is of low brightness, because the emitted and backscattered electrons will be almost immediately reabsorbed by the surrounding surfaces.

5.2.1 *Effects of tilt*

The combined effects of changes in secondary emission and perspective when a specimen is tilted can be observed in Figure 5.17(*a*), in which the same particle is seen at normal incidence and tilted 45° and 70° from the normal. The apparent shadow under the tilted particle and the outlining effect at the edges should be noted. The use of medium and high tilt angles has the effect of showing up small surface irregularities more clearly, but the advantages are mitigated if depth of field limitations unduly restrict the field of view which can be studied with the increased sensitivity to detail.

(*c*)

A further example of the effect of tilting the specimen in the SEM is given in the sequence of micrographs in Figure 5.17(*b*). The same field of view is shown in each case, and the change of magnification with tilt angle is evident as well as change in viewpoint. The micrograph at zero tilt should be compared with the transmission micrograph in Figure 5.5(*b*) of a shadowcast replica of a different part of the same specimen.

5.2.2 *Depth of field effects and dynamic focusing*

In spite of the greater depth of field of electron microscopes, the act of tilting the specimen frequently takes the upper and lower edges of scanning micrographs out of the zone of sharp focus. If greater depth of field is essential, the final probe forming lens can be 'stopped down' by inserting smaller aperture discs into the beam. The increased depth of field resulting from using apertures of decreasing sizes (200, 100 and 50 µm diameter) is illustrated by the three micrographs in Figure 5.18(*a*). Since part of the electron probe current to the specimen is removed when a smaller aperture is used the image-forming signal is reduced relative to random background 'noise' in the detection system. The micrograph appears more granular and is said to have become more 'noisy'; a slower scanning speed and/or frame averaging will be necessary to overcome this effect. Depth of field with a given aperture is greatest at long working distances, in line with the similar effect which the reader may have met in photography with a camera.

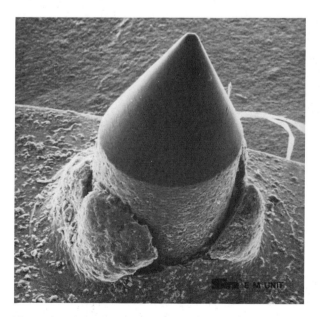
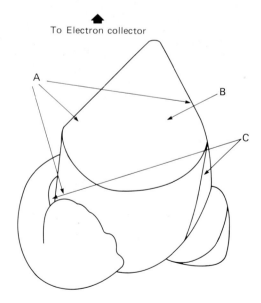

To Electron collector

A

B

C

Figure 5.16. Scanning micrograph of a worn record–player stylus. The accompanying diagram indicates regions of light and shade which are explained in the text. Specimen tilted approximately 40°.

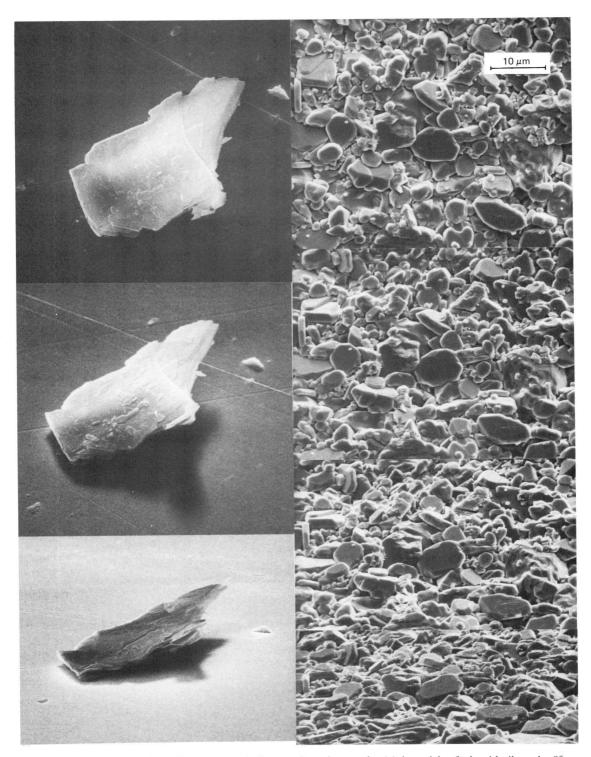

Figure 5.17. Effect of specimen tilt on perspective in scanning micrographs. (*a*) A particle of talc with tilt angles 0°, 45°, 70°. Dispersed on gold–coated glass and coated with Au/Pd. (*b*) The surface of an alumina ceramic at tilt angles of 0°, 30°, 50°, 70°. Coated with Au/Pd; beam energy 10 keV.

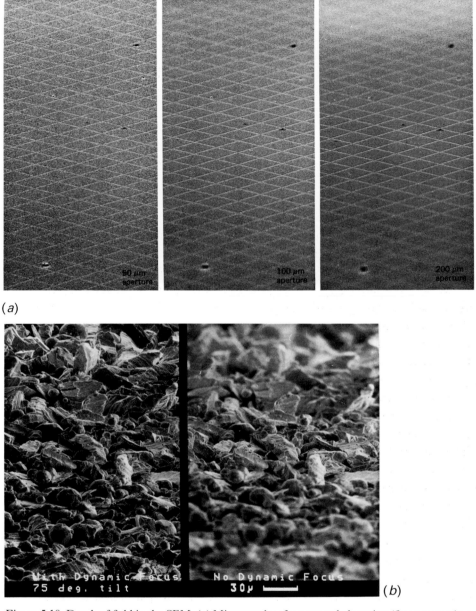

(*a*)

(*b*)

Figure 5.18. Depth of field in the SEM. (*a*) Micrographs of a cross-ruled grating (fine grooves in a flat surface) tilted to nominally 70° to show the effect of final aperture diameter on depth of field. With a 200 μm aperture the extent of sharply focused image is only about three squares (about 210 μm in the plane of the specimen), doubling at each halving of the aperture diameter. Clearly noticeable on the original prints is the progressive increase in granularity of the image as the probe angle was reduced. (*b*) Micrographs of a layer of powder particles on a substrate tilted 75° taken with (*left*) and without (*right*) dynamic focusing of the electron beam. (By courtesy of Gresham–CamScan Ltd.)

The need for increased depth of field in a particular micrograph may result from the rugged nature of the specimen or use of a large angle of specimen tilt. The former can only be compensated for by use of a smaller final aperture in the probe-forming lens. In the special case of tilted flat specimens the whole of the field can be retained in focus, even with the largest aperture, if the microscope has a *dynamic focusing* facility as described earlier (Chapter 3, page 110). The effect of this is seen in the micrographs of Figure 5.18(*b*).

5.2.3 Electron energy effects

As already explained in Section 2.1 the emissive mode signal in the SEM is made up of a combination of secondary and backscattered electrons not only from the point directly impacted by the electron probe but from a micrometre or so in all distances from this point. There may also be secondary electrons from the underside of the final lens pole-piece excited by the backscattered electrons from the specimen.

Two ways of increasing the ratio of surface to bulk information are to reduce the electron penetration and to increase the electron emission from the surface itself. The two series of micrographs in Figure 5.19 show the effect of reducing the electron gun voltage on a ceramic specimen coated (*a*) with carbon, (which removes surface charge but contributes very little secondary emission), and (*b*) with an evaporated film of gold/palladium, with higher electron emission. It is seen that in the latter case surface detail remains visible at higher voltages, and hence the microscope may usually be operated with higher gun brightness and better resolution than in the carbon-coated case.

A further effect of reducing the gun voltage is to decrease the energy given to the specimen, usually as heat, when the primary electron beam is brought to rest within it. The heating effect is frequently sufficient to cause permanent damage to the specimen; typical damage to paper fibres is illustrated in Figure 5.20. Beam damage results from an excessively high combination of electron current and gun voltage and it is of advantage to reduce both of these. Reduction of beam current results in a noisier picture, however, and reduction of voltage, e.g. from 30 to 15 kV or lower, is therefore the more usual practice unless the consequent sacrifice of resolving power cannot be tolerated. When examining organic and polymeric materials it is usually wise to restrict electron gun voltages to 10 kV and to refrain from prolonged examinations at high magnifications, which produce very localised stresses in the specimen. Generalisations are unwise, however, since the nature of the metal coating can have a strong influence on the stability of the specimen under electron bombardment.

Microscopes with a range of operating voltages enable the microscopist to experiment with different working conditions in order to find the best compromise between revelation of the detail he is interested in and the onset of beam damage in the specimen.

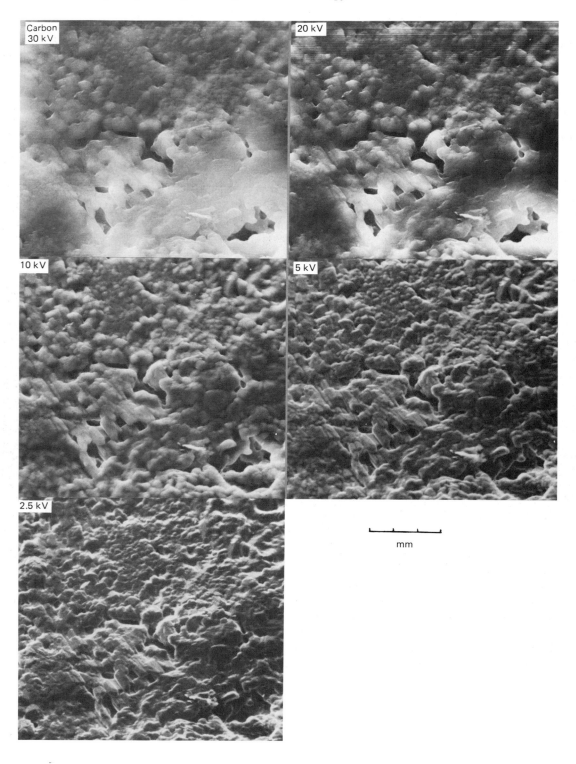

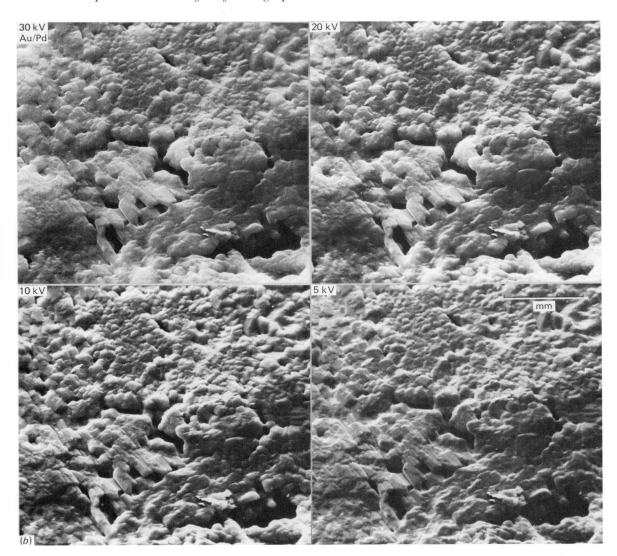

Figure 5.19. Scanning micrographs of a porous ceramic body coated with thin conducting layers of (*a*) carbon and (*b*) gold/palladium and examined with a range of electron beam energies. The smooth, glazed, appearance at 30 kV results from the electron emission variations due to the surface roughness being swamped by the 'featureless' emission due to electrons which have penetrated and been scattered below the surface. The proportion of surface information in the picture is increased by cutting down the electron penetration into the specimen and improving the emission from the surface itself. In the case of a specimen of low atomic number (e.g. organic materials, polymers, ceramics) a thin coating of heavy metal is in general more effective than reduction of beam energy, and enables the higher brightness and better resolution of higher operating voltages to be retained.

If the gun voltage is reduced to very low values of around 1 kV there is an advantage to be gained from the enhanced secondary electron signal obtainable from most specimens at voltages of this level. Referring back to Figure 2.7 (Chapter 2, page 36), we see that at low kilovoltages $\delta = 1$, in which case as many electrons leave the specimen as arrive at its surface. Hence, even if the specimen is an uncoated insulator, it will not charge up under examination in the SEM. Many materials can therefore be studied without coatings if the voltage range of the microscope includes the $\delta = 1$ region. All microscopes need to have their probe diameters increased to give enough emission for a good picture at 1 kV. The resolution is thus degraded and finer detail would be seen if the specimen were metal-coated and examined at a higher voltage.

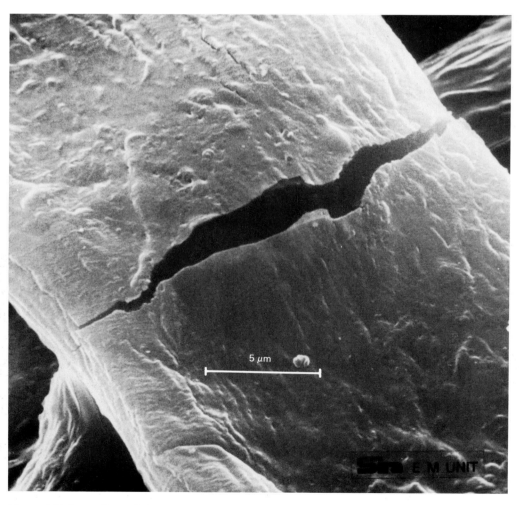

5 μm

Figure 5.20. Paper fibre split by exposure to electron bombardment at 30 kV. Damage is most likely to occur at higher magnifications; a common time is when critical focusing and astigmatism correction are being undertaken with a reduced raster before recording a micrograph.

The ability to provide high resolution at low accelerating voltages is a valuable asset of microscopes with brighter electron emitters in their guns: lanthanum hexaboride (LaB_6) to some extent, and field emitters in particular. (For a more detailed description of electron guns and brighter emitters see Appendix 4.)

It should perhaps be mentioned here that the presence of electrostatic and/or magnetic fields at the surface of a specimen will have an effect on the emissive mode micrographs. There are ways of using the SEM in order to maximise this type of surface information, leading to micrographs with *voltage* or *magnetic contrast*. (See Chapter 7 and the specialised books on scanning microscopy, e.g. Wells *et al.* (1974); Goldstein *et al.* (1992); Newbury *et al.* (1986); or Reimer (1985)).

5.2.4 Making measurements from scanning micrographs

Dimensional measurements accurate to about 1% can usually be made from scanning micrographs after suitable calibration of the instrumental magnification and specimen stage tilt scale. Because of any distortions, which may differ between the visual and recording cathode-ray tubes, magnifications can vary over the image field. Magnification, stage tilt and image distortion can be calibrated accurately by means of a suitable test specimen (Figure 5.21) (Watt & Wraight, 1971). Digital readout of magnification and electronically generated micron bars on micrographs require to be confirmed by calibration against a standard just as much as do analogue meter readings. Reproducibility of magnification appears to be less affected by magnetic hysteresis than in the TEM, although errors of several percent are possible unless precautions are taken. However, size measurements are not, in general,

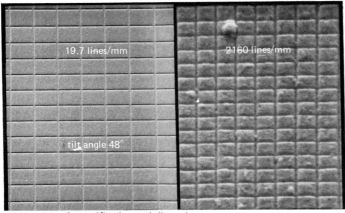

Calibration of magnification and tilt angle

Detection and measurement of distortion

Figure 5.21. Magnification calibration specimens for scanning microscopes comprise replicas of two cross-ruled gratings, with 19.7 and 2160 lines/mm in both directions. These provide over 60 mm² of flat test specimen with which magnification calibration can be checked to 1% accuracy over a wide range ($\times 20$–$\times 40\,000$). Specimen stage tilt and image distortion can be checked and measured readily with the specimens. (Specimens developed for Sira EMTG. Sets of specimens made by the Sira procedures are available from EM suppliers.)

Table 5.1. *Ratios of vertical to horizontal magnifications on scanning micrographs of plane specimens tilted at an angle of θ° (to 1%)*

θ (°)	0	5	10	15	20	25	30
$\cos \theta$	1.00	1.00	0.98	0.97	0.94	0.91	0.87
θ (°)	35	40	45	50	55	60	65
$\cos \theta$	0.82	0.77	0.71	0.64	0.57	0.50	0.42
θ (°)	70	75	80	85	90		
$\cos \theta$	0.34	0.26	0.17	0.09	0		

as straightforward as in the case of transmission micrographs because of the effect of specimen tilt. If M is the image magnification at zero tilt, at a tilt angle of θ degrees the magnification perpendicular to the tilt axis is reduced to $M\cos\theta$ but remains unchanged parallel to the tilt axis. This axis is usually parallel to the base of the micrograph but rotates from this direction as the working distance is changed unless compensated by a scan rotation unit. In other directions the magnification is intermediate between the two values. Values of $\cos\theta$ at 5° intervals are given in Table 5.1.

The $M\cos\theta$ relationship only applies to features in the plane of the specimen stub surface. The true magnification in a specimen plane which is neither the stub plane nor the zero-tilt plane can only be guessed by way of an estimated tilt angle.

The $M\cos\theta$ variation can be counteracted, at the cost of distorting the perspective of the image, by using a 'tilt correction' unit. This causes the scan generator to cram the scan lines closer together when the specimen is tilted, so as to maintain them a constant distance apart on the specimen surface, over a wide range of specimen tilt angles. M is then the same in all directions on the micrograph – for features in the plane of the stub. This is illustrated in Figure 5.22(*a*). However, the procedure can introduce dramatic and possibly unsuspected distortions to features in any other plane, as can be seen in Figure 5.22(*b*).

Indicating the dimensional relationship between the specimen and the final print, which may be a different size from the original fluorescent image, may introduce another possible source of inaccuracy. If the microscope display includes an electronic marker bar signifying, for example, the magnified length of one micrometre on the specimen, this bar will change size with the rest of the image and (subject to being shown itself to be accurate) is a good way of indicating scale.

Another method is to specify the horizontal field width (HFW), being the distance on the specimen contained in the full width of the micrograph. This should be reproduction-proof provided no uncompensated cropping of the image takes place.

Dimensions of coated specimens

Some caution should be exercised in the measurements of overall size of features on carbon- or metal-coated specimens. Particles, fibres and similar discrete features will be increased in size by twice the thickness of the coating at the points measured. As was stated earlier this dimension is difficult to establish, so the error introduced by neglecting it is equally uncertain. If the coating thickness is 10 nm the error on a diameter would be 20 nm, which constitutes more than 1% for features appearing to be 2 μm or less across. For a feature 0.1 μm across the error is 25%, which might be serious.

Size errors due to beam diameter

A further possible source of error in dimensional measurement, which would be significant only on sub-micrometre objects, occurs because the electron probe has a finite diameter. Referring to Figure 5.23 we see that when the probe of diameter d scans across a feature of width W the overall width of the feature on the line scan is $W + d$. If the probe diameter d is 10 nm there would be 1% error on a feature width of 1 μm and 11% on 0.1 μm.

5.2.5 Other modes of operation of the SEM

Briefly mentioned in an earlier section (Chapter 2, page 32) these are examination methods in which the detector responds to something other than the total secondary and reflected electron emission from the specimen. The modes are seldom used in isolation, and micrographs recorded in several modes enable a more complete picture to be built up concerning the specimen.

Backscattered emissive mode

The type of image obtained in the backscattered mode will depend on the detection system on the microscope (Figure 5.24(*a*). With a straightforward Everhart–Thornley scintillator detector situated to one side of the specimen, with the latter tilted towards it, biasing-off the secondary electrons (SE) will emphasise the roughness of the specimen by introducing dark, empty shadows. In appropriate specimens the effect of atomic number will make heavier elements appear lighter-toned (Figure 5.24(*b*)). An annular solid-state detector or a Robinson detector positioned symmetrically above the specimen will collect a much higher signal than the Everhart–Thornley detector, and will respond only to backscattered electrons with energies above several keV. Because of its symmetrical position over the specimen (which should be examined untilted) there is no directional lighting effect in the image, and atomic number (Z) contrast is emphasised. Adjacent elements in the Periodic Table can be distinguished by their difference in brightness in the backscattered image.

If the solid-state detector is split into two halves (or four quadrants) (Figure

5.24(*c*)) and the signals from two portions on opposite sides of the specimen are combined, the resultant image is the equivalent of the previous one, showing maximum *Z* contrast. If the difference between the two signals is taken the image will show primarily topographical detail.

Backscattered electron imaging is very effective for examining the surfaces of metals and alloys. Applying a metal coating to an insulating specimen will obviously diminish any atomic number effects from the specimen, but not the geometrical one. Since much of the backscattered signal originates below the surface of the specimen there can be considerable differences between the BSE and SE images, especially at high primary electron energies. One particularly useful feature is that BS electrons are much less susceptible to surface charging effects than are secondary electrons; BSE images may be obtainable on occasions when SE imaging is impossible. Backscattered electrons are also used in the formation of channelling patterns in the study of crystalline specimens by scanning microscopy. This technique will be described in the next chapter (page 270).

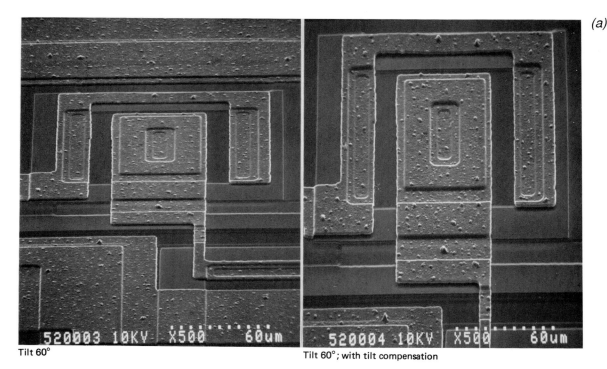

(a)

Tilt 60°

Tilt 60°; with tilt compensation

Figure 5.22. Correct and incorrect application of 'tilt correction'. (*a*) Micrograph of a tilted plane specimen with the scanned frame tilt-compensated to give a uniform magnification over the whole field. (*b*) A sphere should appear round from all angles. A micrograph of round copper particles at 80° tilt (*upper*) shows this. Turning on the 'tilt correction' (*lower micrograph*) results in a spectacular example of the distortion which can be introduced by this procedure. (Micrographs (*a*) by courtesy of Hitachi Scientific Instruments; (*b*) by A. Stubbs, Johnson Matthey Technology Centre.)

(*b*)

Cathodoluminescence mode

Light emitted by the specimen when bombarded by electrons is detected and amplified to provide the final image. Useful in geological examinations, the micrographs
are limited to line-of-sight effects and are therefore most appropriate to flat, polished
sections of material (although not exclusively, as the micrographs of phosphor particles show, Figure 5.25). Light areas on the micrographs correspond to regions of
specimen which fluoresce under electron impact. Placing coloured filters over the

detector enables fluorescences of different wavelength to be distinguished. A mirror or lens can usefully be employed to increase the efficiency of collection of the luminescence emission. If three separate detectors are used, sensitive to the primary colours Red, Green, Blue, separate CL signals may be amplified and recombined on a colour monitor to provide micrographs of luminescence in the actual colours, (Saparin & Obyden, 1988). Using three image stores enables the comparatively weak signal from a high-resolution probe to be frame-averaged for better results. However, there are effects of intensification and bleaching of cathodoluminescence emissions which have to be watched out for.

Specimen current mode

Primarily of use in studies of semiconductor materials, this mode can also be used to provide information on other types of specimen. Variations in the net current flowing to earth from the specimen under examination are used to modulate the brightness of the enlarged image. Changes in electron emission from the surface and generation of electron–hole pairs within the specimen are recorded. In the latter case the mode is described as the electron beam induced current (EBIC) mode. The effect of applied potentials on the various electrodes of semiconductor devices can be studied in this mode (Figure 5.26). For non-semiconductors the specimen current micrograph is similar to the emissive mode micrograph but with reversed contrast. If the microscope does not have an efficient detector for backscattered electrons it may be preferable to image channelling patterns (Chapter 6 pages 270–1) in the specimen current mode.

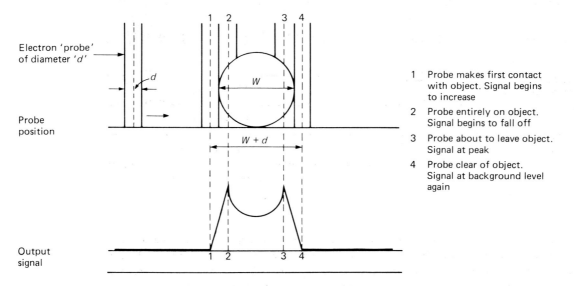

Figure 5.23. Diagrammatic illustration of possible errors in size determination of sub-micron objects due to the finite size of the scanning electron probe.

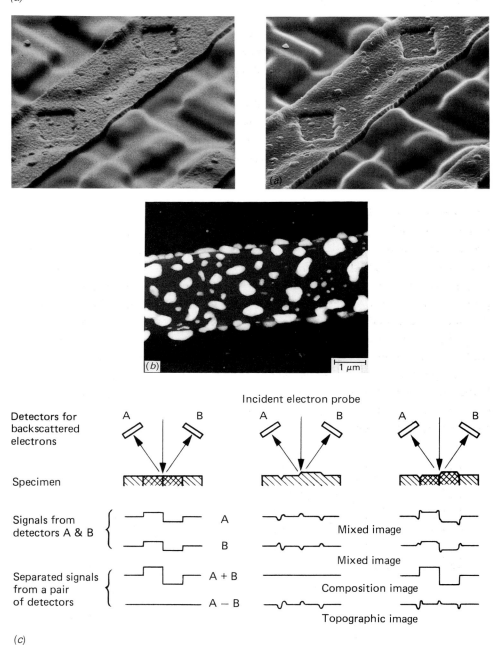

(a)

Incident electron probe

Detectors for
backscattered
electrons

Specimen

Signals from
detectors A & B

Separated signals
from a pair
of detectors

A

B

A + B

A − B

Mixed image

Mixed image

Composition image

Topographic image

(c)

Figure 5.24. Backscattered emissive mode. (*a*) Part of a microcircuit imaged by backscattered and secondary electrons (*right*) and backscattered electrons alone (*left*) using a single Everhart–Thornley scintillator detector. Note the greater prominence of topographical features and the empty shadows in the backscattered image. (*b*) Strong atomic number contrast between islands of Pt and an Al_2O_3 fibre is obtained using a solid-state backscattered electron detector positioned directly above the specimen. (*c*) Schematic diagram of (A−B) and (A−B) compositional and topographical backscattered imaging using solid-state detectors positioned on opposite sides of the specimen.

(*continued on next page*)

Transmitted electron mode

Specimens prepared and mounted similarly to those for TEM can be examined in the SEM if the electron detector is placed on the opposite side of the specimen from the incident electron beam. Useful, high-contrast, images can be obtained by this form of scanning transmission electron microscopy, with effective electron penetration through non-crystalline materials equivalent to a TEM operating at a higher

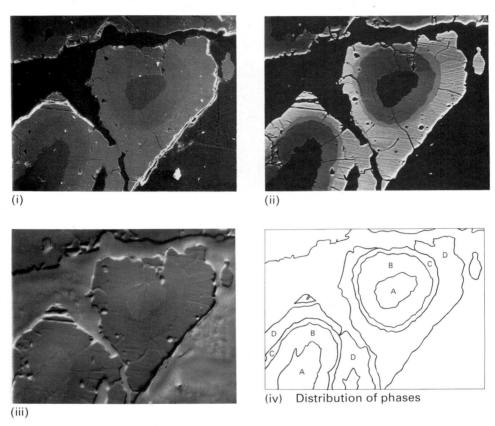

(i)

(ii)

(iii)

(iv)   Distribution of phases

	Composition	Mean atomic number (Z)
A	SiO_2	10.76
B	$CaSiO_3$	13.45
C	$Ca_3Si_2O_7$	14.02
D	Ca_2SiO_4	14.40

Figure 5.24. (*c*) (*cont.*) (i) Secondary, (ii) compositional and (iii) topographical images of a polished mineral section lead to (iv) a composition map. Note how atomic number differences of less than unity are clearly differentiated, and black spots in the compound D are shown by the (A−B) signal to be holes and not finely divided inclusions. ((*c*) by courtesy of JEOL Ltd.)

gun voltage, (Figure 5.27). Although, unlike the TEM, the image is not affected by electron energy losses in passing through the specimen, there is a *top/bottom effect* whereby detail is resolved better at the electron input side than the exit side because of the increase in beam size by multiple scattering.

The interpretation of images in this mode is very similar to that already described

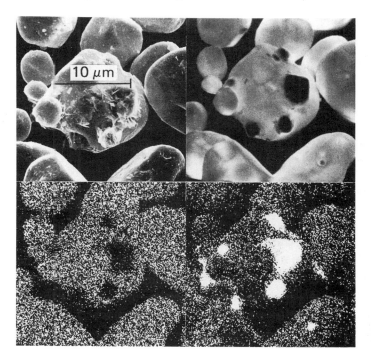

Figure 5.25. Cathodoluminescent and x-ray distribution modes. Phosphor particles are imaged using secondary electrons (*upper left*), cathodoluminescence (*upper right*), Zn Kα and Mg Kα x-radiations (*lower left* and *right*, respectively). (By courtesy of LEO Electron Microscopy Ltd.)

Figure 5.26. Specimen current (EBIC) mode. A split-screen micrograph compares part of an integrated circuit in the electron beam induced current and the secondary emission modes. (By courtesy of Gresham–CamScan Ltd.)

for the TEM. STEM images may readily be produced by either bright- or dark-field techniques (Chapter 7, p. 303). There is no difficulty in producing high-contrast images – only in reproducing them satisfactorily on paper afterwards.

X-ray emissive mode

In this mode information about the elemental composition of the specimen can be presented as a spot elemental anlysis or an elemental distribution map. These will be described in more detail in Chapter 6.

Application of several examination modes to the same specimen normally produces much complementary information, although it may not be straightforward to explain all the observed features in detail.

5.2.6 Instrumental effects to be seen in micrographs

Focusing

Focusing the scanning microscope is the operation of bringing the electron probe to its smallest diameter exactly at the specimen surface. At any other lens excitation the probe is larger and the resolution poorer. Running through focus on a flat specimen square to the beam (zero tilt) merely results in the image going to maximum sharpness and out again. On a tilted specimen a correctly focused micrograph will have its sharpest definition running parallel to the tilt axis and passing through the centre of the image. Under- or overfocus results in the plane of sharpest focus being towards the top or bottom of the frame, respectively, or even out of the field completely (Figure 5.28). The effect is only one of decreasing sharpness; there are no Fresnel fringes as in the TEM.

Astigmatism

The effect of astigmatism is only met at magnifications of about $\times 20\,000$ and above on the SEM. The SEM image will show a directional sharpness (Figure 5.29) which

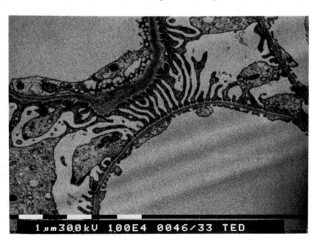

Figure 5.27. Scanning transmission mode. Micrograph of a thin section of rat kidney made using the transmitted electron detector in a scanning microscope. (By courtesy of Philips Electron Optics.)

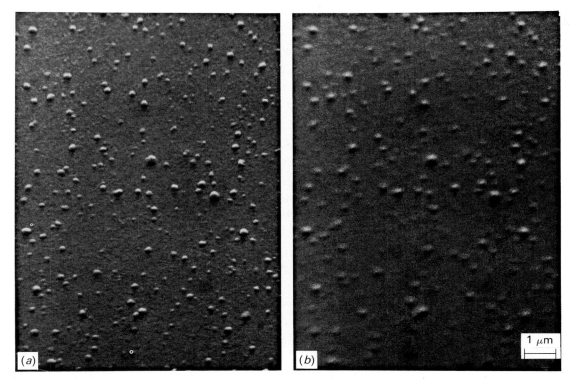

Figure 5.28. Effect of defocusing. Between (*a*) and (*b*) the final lens was defocused slightly, resulting in the plane of focus going out of the top of the picture (the specimen plane was tilted). The resolution is worsened since the electron probe at the specimen is now larger in diameter than before.

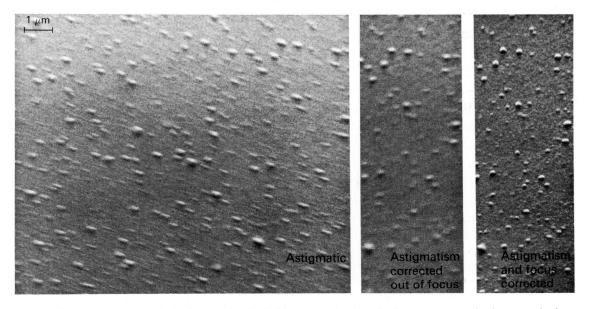

Figure 5.29. Astigmatism results in the scanning probe being non-circular. The image appears to be drawn out in the direction of the greatest dimension of the probe. Comparison of this micrograph with Figure 5.28 shows the difference between poor focus of a correctly compensated beam and an astigmatic image.

rotates through 90° as focus is passed through, and on a three-dimensional specimen the change in direction can be seen in different planes of the specimen.

Mechanical effects

Image features drawn out in one direction can also be the result of mechanical instabilities of the microscope or specimen. In the SEM specimens mounted on the stubs with certain types of adhesive or on sticky tape can drift during recording of a micrograph. Often these mechanical movements of the specimen are not detected during setting up of the exposure, and are first seen in the micrograph. The effect of specimen drift may be subtle. Because the object field is only being sampled a portion at a time the effect may be one of distortion, dependent on the direction of drift in relation to the direction of scan. If the specimen is found to be unstable there is usually little alternative to repreparing it. Examples of the effect of mechanical instability are shown in Figure 5.30.

Contamination

Improvements in column design and in vacuum systems over the past few years have combined to make contamination much less of a problem than it used to be. However, certain types of specimen, careless 'housekeeping', and older designs of microscopes all give rise to the characteristic appearance of contamination deposits.

Contamination builds up over the whole of the illuminated field of view, and becomes noticeable if the magnification is reduced and the previously illuminated area forms part of the now larger field of view. Contamination effects are shown up most prominently on flat, featureless surfaces, e.g. the polished glass surface in Figure 5.31, for which the contrast has been increased in order to show up fine surface detail. Metallic surfaces which have been polished and any specimens with slight traces of oily deposit or finger grease are also prone to show scanning frames where the contamination built up during the final focusing operation at higher magnification. Locally increased electron bombardment during microanalysis of a small area can also result in the formation of contamination patches (see Figure 5.13(*b*)).

The methods for minimising the effects of contamination are based on reduction of the amount of adsorbed hydrocarbon available for polymerisation and reduction of the time of exposure of the specimen to the electron beam. The column and specimen chamber of the microscope should be maintained as clean as possible, and components and specimens should not be handled with bare hands. It is advisable to focus the microscope and take the micrograph as rapidly as possible after the field of view has been brought under the electron beam. Fortunately in the newer microscopes where the viewing screen is several times larger than the final recorded image, there is no longer a need to focus at a higher magnification and then come back to the chosen value.

5 μm

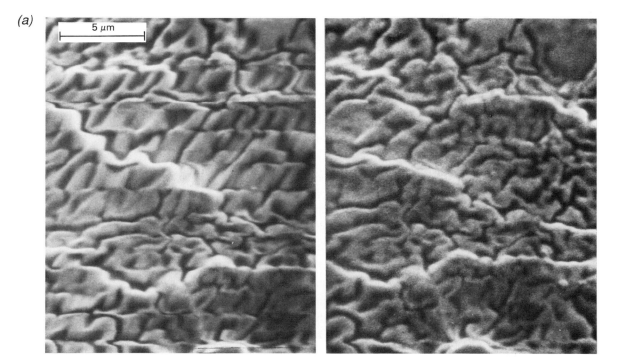

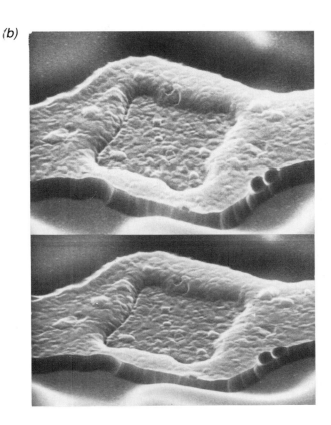

Figure 5.30. Mechanical instability is often the result of the heating effect of the beam, and will differ in effect between the rapid frame-scan of setting up and focusing and the slow scan of image recording. (*a*) The specimen, a soiled human hair, was mounted by gluing the two ends of a short length of hair to the stub surface, and then metal-coating it. Examining the middle of the fibre resulted in instability which combined a slow drift with sudden jerky displacements. The pattern of instabilities varied with the conditions of scan; none was visible during setting up for the exposure. The left-hand micrograph, recorded in a 40 second frame-scan, shows bands of distorted detail due to a slow drift separated by rapid image displacements. The right-hand micrograph was recorded immediately afterwards in a 20 second scan and corresponds more closely with the image seen during setting up. (Beam energy was 10 keV.) (*b*) A more insidious example, which can result from the use of unsuitable mounting techniques, is illustrated by two micrographs of part of a microcircuit taken under identical conditions of magnification and tilt angle. They show an image compression in the lower picture because of a slow specimen drift up the frame during the recording scan.

5.2.7 More instrumental effects

The above sections have shown how scanning electron micrographs can be explained (interpreted) purely on the basis of predictable interactions between a specimen and the scanning electron probe beam. This final section is to warn that in practice emissive mode micrographs will usually be modified by the instrument itself, and the same specimen may look differently in two different models of microscope because of this.

Four, or possibly five, different sources of secondary electron signal have been identified for the SEM, and at least three different backscattered electron signals. These are shown diagrammatically in Figure 5.32.

Secondary electrons SE1 and backscattered electrons BS1 are the informative emissions responsible for the image detail we have considered and explained above. SE1 contains the finest topographical detail resolvable by the probe size. BS1 contains highly resolved atomic number information.

BS2 originates from an area wider than the incident probe, depending on such factors as the primary beam energy, the atomic number of the specimen and its thickness, as seen earlier from Monte Carlo simulations. The secondary emission SE2 will vary with BS2 and will be independent of the SE1 signal, i.e. it serves merely to dilute the contrast of the SE1 image.

SE3 is a secondary electron signal from the pole-piece of the final lens caused by electrons BS1 + BS2 backscattered in that direction. It is again a signal unrelated to the detail under the probe, and serves to dilute the SE1 topographical contrast.

SE4 is a current of secondary electrons coming from the bore of the final lens, principally from the final aperture, whilst SE5 are secondary electrons from parts of the specimen stub, stage and chamber exposed to bombardment by stray high-energy electrons BS3 scattered by the specimen. A fraction of BS3 may enter the E–T detector.

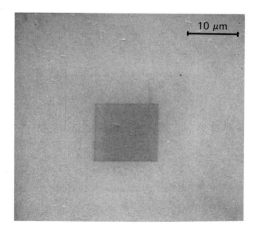

Figure 5.31. The visible effects of contamination in the SEM. The effects of specimen contamination are most readily observed when the magnification is reduced and a previously viewed field can be compared with surrounding areas which have not been electron bombarded.

The relative contributions of the contrast-diminishing emissions depend on the geometry of the individual electron microscope – positioning of the E–T detector, working distance, nature and tilt of the specimen, and the beam energy. Hence, it may be expected to vary between types of microscope and between different conditions of use of the same instrument. It is unfortunate that for the conventional SEM there is no means of isolating the SE1 high-contrast, high-resolution topographical signal from the Z-dependent BS1 and the broader-scale dilutions of SE2–5 and BS2,3.

The electron optics of the in-lens SEMs and the analytical TEM (Figure 3.46), in which the SE1,2 signals spiral up the lens bore and are extracted sideways to a scintillator/photomultiplier detector, eliminates some of the featureless background. If the SE2 signal may be assumed proportional to the backscattered emission BS1+2, its effect can be simulated by taking a proportion of the signal to an efficient backscattered electron detector and subtracting it from the through-lens SE current, varying the proportions to maximise the contrast of the topographical SE image.

For a discussion of some of the factors involving SE and BS1 and 2 near the limit of resolution in scanning microscopy the reader is referred to Joy (1991).

5.2.8 Specimen preparation artefacts

There are well-defined types of artefact in scanning microscopy caused by the use of imperfect techniques for specimen preparation. In some cases, e.g. on an insulating

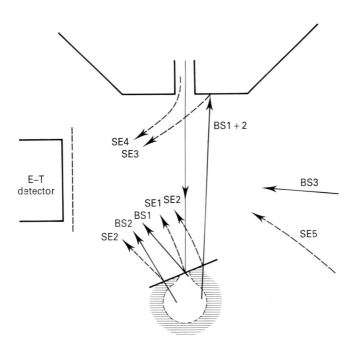

Figure 5.32. Sources of emissive mode signals in the SEM (schematic). For explanation see text. (N.B. The scale at the specimen has been magnified to enable the origins of the emissions to be distinguished.)

specimen charging under examination, the false result is obvious and the remedy is to re-prepare the specimen. Artefacts can occur in the preparation of replicas for SEM and in the examination of beam-sensitive materials. Blisters and cracks in polymeric or organic specimens in the SEM are artefacts which may pass for original structures on first acquaintance but which can be recognised as false on more critical examination. In cold-stage (cryo-) SEM there is the risk of frost formation on the specimen surface during transfer into the microscope. A number of these effects are illustrated in Figure 5.33.

5.3 Recording and printing electron micrographs

Traditionally, transmission electron micrographs have been recorded on photographic plates or sheet film exposed within the vacuum of the microscope. After wet processing, which requires chemicals and a darkroom, the negatives are capable of being enlarged photographically, e.g. to A4 size for a report, or up to 16"×20" (40.6×50.8 cm) for a display print, with high quality on glossy, matt- or pearl-surfaced bromide paper. The results can have archival permanence.

Micrographs from scanning electron microscopes are traditionally recorded by photographing a single slow scan on a high-definition cathode-ray tube. After wet processing in a darkroom the negatives can be printed to any desired size on to photographic paper, but above 2 or 3 times the dimensions of the recording tube the sharpness begins to fall off because of the granular nature of the fluorescent image.

Thus from both TEM and SEM, micrographs of lasting high quality can be pre-

Figure 5.33. (*Opposite*) Common (recognised) artefacts in scanning electron microscopy. (*a*) Specimen charging. Bright and dark lines parallel to the scanning direction (usually the base of the print), distortion of outlines, and glowing bright areas of specimen are symptomatic of specimen charging. Fine particles are particularly difficult to coat with a stable conducting film; the micrograph shows granules of icing sugar. (*b*) Scanning raster effects. In scanning instruments it is possible to have interaction between the scanning raster and specimen detail of regular periodicity aligned parallel or nearly so with the raster. In this figure moiré fringes occur between scan lines and the rulings of a cross-ruled grating. The effect can be overcome by rotating the specimen, by altering the number of scan lines per frame, or by rotating the scan frame on the specimen. (*c*) Handling artefact. Found on a gold-coated microscope slide, with a particle dispersion, this cluster of particles was identified as part of a fingerprint. Great care should be taken not to finger specimens which are to be examined; surfaces which have been inadvertently or unavoidably handled should be washed in aqueous and organic solvents to clean them before starting normal specimen preparation. (*d*) Beam damage in polymers. This flat-bottomed pit was formed in the surface of an acrylic specimen (Perspex) during examination with a reduced-area scan, even though the electron energy was reduced to 10 keV. Specimen tilt angle was 45°. (*e*) Comparison of BSE and SE high-resolution cryo-SEM images of a Pt-shadowed, carbon-coated freeze-fractured yeast cell (Walther *et al.*, 1995) shows the presence of condensed water vapour and hexagonal ice crystals from the transfer into the SEM. These surface artefacts are seen only by SE imaging. (By courtesy of P. Walther, E. Wehrli, R. Hermann & M. Müller.)

(a) Specimen charging
10 μm
10 kV

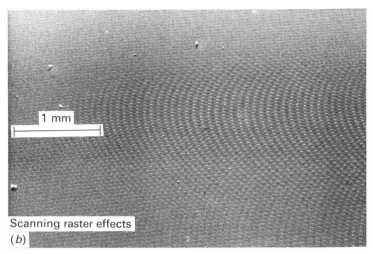

1 mm
Scanning raster effects
(b)

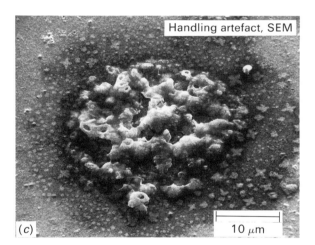

Handling artefact, SEM
10 μm
(c)

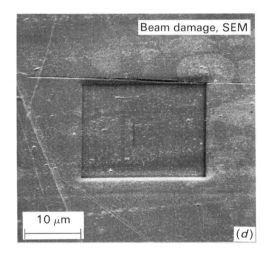

Beam damage, SEM
10 μm
(d)

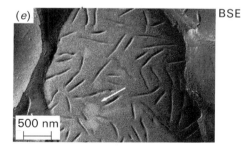

(e) BSE
500 nm

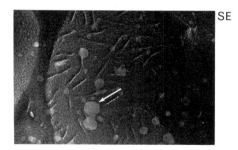

SE

pared, but they require a photographic darkroom and time measured in hours for the processing. There is an exception: on the SEM, 'instant' photographic prints nominally 3¼×4¼ or 4×5 inches (73×95 or 90×115 mm) can be produced on a one-off basis within minutes of exposure, with a photographic negative if desired.

Setting up and taking the micrographs in the first place is far more stimulating than producing the prints afterwards, and making multiple copies, although readily possible, is a labour-intensive chore.

The alternative, which should seriously be considered when purchasing a new microscope or upgrading an older one, is digital imaging and recording on magnetic or optical media. In this technology the image is scanned (by a TV camera on the TEM, or as a natural course in the SEM) and the signal along each scanned line digitised into picture elements (pixels). Completed scans may be stored in a computer memory or frame store as an image of x pixels$\times y$ pixels, frequently 512×512 or 1024×1024.

Electron micrographs are naturally monochromatic, so each pixel can be considered as having a brightness which can be represented by a grey tone. A binary image has two tones, either black or white. A 2-bit image has 2×2 levels of grey; a 3-bit image 2×2×2, and so on. A 4 bit image will have 16 tones, and is the practical limit of reproduction of a printed image. The human eye can distinguish about 30 shades of grey, photographic prints about the same, and negatives 35–50. Figure 5.34 illustrates some of these grey scales in practice. The individual grey levels in an 8-bit image could not be distinguished or reproduced in one colour, but the actual intensity values can be used in comparisons or in quantitative image analysis. The number of bits in each pixel determines the computing requirement for image transmission and storage.

Visually the reproduction of a digital image is considered acceptable so long as individual pixels cannot be distinguished. Tell-tale characteristics of over-enlarged pixels are jagged profiles of lines or edges running obliquely to the scan direction. The unaided eye can just see detail at the 0.1–0.2 mm level, i.e. 50 pixels per cm or 625 pixels on a 100×125 mm micrograph, so a 512×512 pixel resolution would be the practical minimum if the final micrograph is to be printed on to 4×5 material; 1024×1024 would be undetectable at that size, and for A4-size micrographs a resolution of 2048×2048 would be desirable but 1024×1024 adequate (Figure 5.35). Top photographic quality represents perhaps 4096×4096 pixels, or 16 megabytes per

Figure 5.34. (*Opposite*) Tonal rendering of digitised images. A scanning micrograph of a lady's watch was recorded with increasing numbers of grey shades, from 2 (**either** black **or** white, *1-bit*) through 4, 8, 16, 32 (*5-bit*) to 256 (*8-bit*). Photographic negatives were enlarged on to glossy photographic bromide paper. On the original prints there are only slight differences detectable between the 16 and 32 grey-level images, the latter being indistinguishable from the full 256-level image. 1, 2, 3 and 5-bit images are compared here. (By courtesy of A. Brooker, JEOL (UK) Ltd.)

image. Acquiring, writing, storing and printing such an image would make that resolution impractical, and 1024×1024 is a more normal high-resolution image, taking between 10 and 60 s to write to a magnetic or optical disc, compared to several seconds only for direct photographic recording in the TEM and between 40 s and 2 min for recording from a high-resolution c.r.t. on an SEM.

Storage may be on magnetic disc, one image only on a 3½″ or 5¼″ floppy disc but between 10 and 100 on a hard disc or 400–1000 on an optical disc, which has the advantage of being permanent and easily accessed. The optical disc with storage capacity for hundreds of high-resolution images is a well-tried medium, and well-tested routines are available for indexing images and associated data. Erasable 'floptical discs' use a magnetic medium but optical reading, and may have a similar size but much greater storage capacity than ordinary floppy discs.

An important feature which may be a vital one is that, once recorded, the digital image can be recalled at will for comparison or analysis and a relatively stable hard

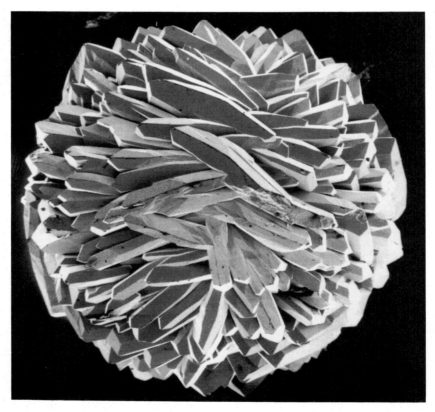

Figure 5.35. Effect of pixel resolution on micrograph quality. The complete hydroxyapatite crystal was recorded at 1024×1024 pixel resolution. Strips from 200 mm square prints show the same region of the crystal recorded at resolutions of (*left to right*) 128×128, 256×256, 512×512 and 1024×1024 pixels. (By courtesy of A. Brooker, JEOL (UK) Ltd.)

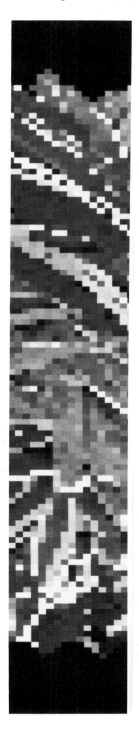

copy can be printed on a video printer in less than a minute without the need for chemicals or a darkroom.

Making hard copy of the digitally recorded images can be done in several ways. If the size, flexibility and quality of photographic printing is important, the image can be put back on to the recording tube, and photographed in the traditional way. If the size and semi-permanence are acceptable, prints may be made using video printers on either thermal or dry-silver imaging paper. Permanent prints on glossy paper may be made at much higher cost by the dye-sublimation process in a video printer. Alternatively, some laser printing procedures are able to provide images of acceptable quality on plain paper, and direct integration of micrographs into printed reports is then possible using desktop-publishing programmes. There is no comparison between the visual impact of a glossy photographic print and a laser print on document paper, however. The quality of glossy dye-sublimation prints at full A4 size has narrowed the gap between photographic and electronic printing recently, however.

Apart from the considerations of darkroom-free recording and reproduction of micrographs, there are other advantages of having an already digitised image, which is suitable for modification, processing and analysis given appropriate software programmes. Moreover, it is possible to link the electron microscope into a computer network with central image storage and processing; even to discuss the images by telephone with someone on the opposite side of the world while the specimen is still in the microscope!

5.4 Stereoscopic microscopy

An electron microscope compresses information from what may be a three-dimensional object into a two-dimensional image. With a little more trouble the micrograph can in suitable cases be given an apparent third dimension. This may reveal important new information and allow new measurements to be made. The principles of 3-D imaging will be described, together with an account of how it is applied to electron microscopy.

Stereoscopy, or the visualisation of objects in three dimensions, is based on the fact that the observer's two eyes have different viewpoints and his brain processes the information from these two sensors to give a mental impression of length, breadth and depth. In stereoscopic photography a camera with two separated lenses (or the same camera used in two different positions) is used to record a scene from two similar but separated viewpoints; when the two photographs are put before the observer so that each eye sees the appropriate image he obtains the impression of seeing the scene in depth (Ferwerda, 1982). Moreover he can make use of this impression to make measurements of the scene in three dimensions, as we shall see later. The pair of photographs which combine to give a three-dimensional impression are termed a *stereoscopic pair*, or *stereo pair* for short.

In an electron microscope the final image shows the specimen as it would be seen from a viewpoint on the axis of the lens system – from the objective lens in the TEM and from the final probe-forming lens in the SEM. The same region of the specimen can be viewed from a different angle if the specimen plane is tilted a few degrees about an axis perpendicular to the lens axis. Thus a stereoscopic pair of micrographs is usually obtained in the transmission microscope by tilting the specimen grid a few degrees on either side of the normal square-on position and recording the same field of view at each of the tilted positions. An alternative procedure involving tilting the beam instead of the specimen has also been reported (Fan & Ellisman, 1994). In the scanning microscope, too, a 'stereo' pair may be obtained by recording micrographs at two different angles of specimen tilt. Also, in the case of an already-tilted specimen, the change in direction of view may be obtained by rotating the specimen stub through a few degrees (Figure 5.36).

In either type of electron microscope, tilting will result in image shift and a loss of focus if the axis about which the specimen is tilted does not actually pass through the field of view. The original field can be brought back by using the x, y shift controls and focus should be restored by use of z (axial) shift if there is one. Refocusing the lens should be a last resort (unless the change is very small) since in either type of microscope it results in a change of magnification; the stereoscopic effect will then be impaired since the brain (or any measuring technique) will be trying to match two images of different sizes. Stereoscopic microscopy is made much easier if the microscope is fitted with a eucentric goniometer specimen stage, in which tilting takes place about an axis through the actual image field. This is not essential, however, and satisfactory stereo pairs can be obtained with a simpler tilting stage provided the specimen field to be examined is not too far from the centre of the grid or stub.

When the micrographs from the two viewpoints are correctly examined as left and right eye views an impression of the specimen will be seen in depth, and additional qualitative and quantitative information becomes available. In transmission microscopy the internal detail in thin sections or foils can be stretched out along the z-axis and hence clarified; in particular the greater specimen thickness in the HVEM can result in a confusing superimposition of detail, all in focus, which can only be separated by stereoscopic viewing. The pseudo-relief in shadowcast specimens can be made more 'solid' by the stereoscopic effect, when correctly done. In scanning microscopy the already solid appearance of specimens can be added to, sometimes with unexpected results, by making stereoscopic pairs of micrographs.

5.4.1 *Viewing the micrographs*
The normal, upright, human observer sees three-dimensionally with a horizontal separation between his two eyes. In stereoscopic photography the two separate pictures are usually recorded through two lenses on the same horizontal line so that the two images can be positioned side by side for viewing. The viewing arrangement has

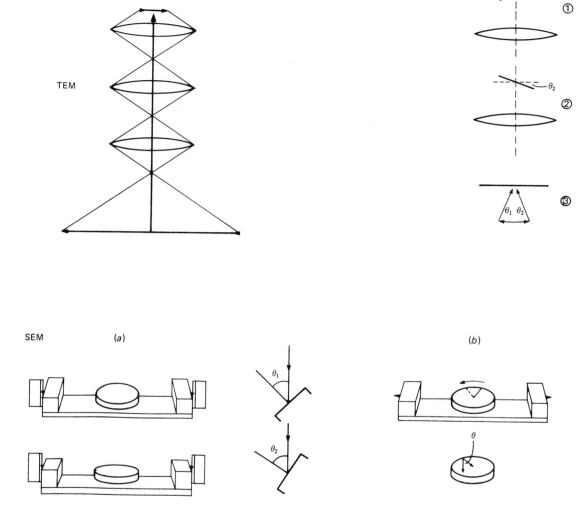

Figure 5.36. Diagrammatic illustrations of stereoscopic microscopy with TEM and SEM.
TEM: The final image is a view of the specimen seen looking up the optical axis of the microscope. To produce different angles of view the specimen is tilted about an axis which is at right angles to the optical axis. In ① the specimen plane is tilted θ_1 about an axis perpendicular to the paper. In ② the tilt is θ_2 in the opposite direction. The two micrographs obtained will show the specimen as seen from two directions separated by $(\theta_1 + \theta_2)$ ③.

SEM:(*a*) Stereo pairs by tilting. A stereoscopic pair of micrographs may be obtained by taking pictures at two different tilt angles θ_1, θ_2. The prints must be rotated 90° to give a vertical tilt axis for viewing, and this may make interpretation more difficult. (*b*) Stereo pairs by rotation. At other than normal incidence, rotation of a stub also results in a change of viewing angle. Since the tilt axis is vertical on the micrograph the stereo pair may be viewed as recorded, side by side.

to be such that the observer's left eye sees only the left-hand image and vice versa for the right eye. Simple viewing devices have two eyepieces separated by about 6.35 cm, the average interocular separation, and pairs of prints or transparencies with the same separation are placed at a suitable distance in front of the eyepieces so that the observer sees the images at infinity, fused into one 'solid' impression. Figure 5.37 shows typical viewing devices in common use, whilst Figure 5.38 is a stereoscopic pair of photographs mounted with the correct separation. With practice, many observers can dispense with the viewing device and see such pictures in 3-D by relaxing their eye muscles so that their eyes are looking straight ahead. Each eye then sees only the appropriate photograph and the brain fuses them together. If the observer

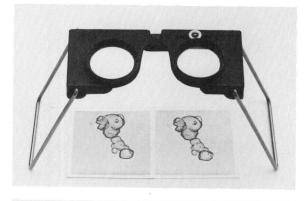

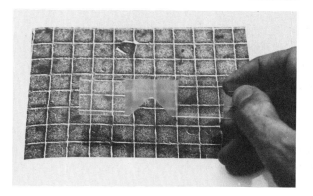

Figure 5.37. Simple viewers for stereoscopic pairs. 'Spectacles on stilts' (*upper*) enable a pair of prints up to 6.3 cm wide to be seen, enlarged, in 3-D. The separation between the two halves of the viewer is variable to accommodate differences in interocular separation between observers. The moulded plastic transparency viewer (*centre*) accepts pairs of transparencies in 4.25×10.1 cm cardboard or metal frames. The 'Nesh' viewer (Neubauer & Schnitger, 1970) (*lower*) consists of a pair of opposed plastic prisms moulded together in the form of a lorgnette. Stereo pairs of prints, which may be of any size, are mounted above one another, left eye view uppermost. When viewed at the correct distance the two images are each displaced vertically by one-half of their height and are seen superimposed in 3-D.

then covers up one of the prints, he can assess how much his appreciation of the subject was assisted by the information about the third dimension.

(N.B. If the left and right eye views are inadvertently interchanged a *pseudoscopic* image results, and the specimen is seen inside out. In the case of a transparent subject this may not matter, but a solid object appears unreal, almost ghostly. The remedy is to exchange left and right pictures, and a solid 3-D impression will be obtained when they are correctly separated and aligned.)

In stereoscopic microscopy the left and right eye viewpoints result from tilting the specimen about a fixed axis. In order to see the stereoscopic effect the micrographs must be placed side by side with the direction of the tilt axis vertically between them. It is in satisfying this last condition that an important difference between light and electron optics becomes apparent. This is the rotation of an image or a scanned frame which occurs in a microscope column because of the spiral paths of the electrons through the electromagnetic lenses. Unless the stereo pair of micrographs is correctly orientated there will be a reduction in the stereoscopic effect, or even none at all if the tilt axis in the image plane happens to lie along the line joining similar features in the two prints. The severity of the resultant problem differs between TEM and SEM, and will be dealt with separately for the two types of instrument. (N.B. On many electron microscopes the tilt axis is parallel to one of the stage shift directions (usually x) and its direction can be determined by operating this shift control and seeing which way the image moves.)

Figure 5.38. Stereoscopic pair of photographs of primroses, mounted for viewing. This pair can be seen in 3-D by use of the simple 'spectacle' device shown in the previous figure or, with practice, without any optical aid, as described in the text. The viewing device is preferable, however, since it gives an enlarged image. After the 3-D impression has been studied the reader should cover or close one eye and compare the 2-D effect. Much more may now be read into this than before the 3-D experiment.

Stereo viewing for transmission microscopy

As the image proceeds down the microscope column each lens introduces an angular rotation which depends on the magnification of that lens. Thus the direction of the tilt axis in the specimen plane (often running from front to back of a microscope with simple tilting stage) may have no simple relationship to the corresponding direction in the image plane.

In order to be able to use stereoscopy in the TEM, therefore, *either* the stereo pairs must be viewed through a stereo viewer and rotated about their centres until the orientation of maximum 3-D effect is reached *or* the microscope should be calibrated for each magnification so that the necessary angular re-orientation of the micrographs is known. Once calibrated, the stereoscopic effect can be obtained simply by applying appropriate rotation to the micrographs, making due allowance for any inversions of the image which may have occurred between the fluorescent screen and the print or transparency which is being viewed. Fortunately, one of the results of computer-control of multi-lens TEM columns is that manufacturers have been able to balance-out rotations by the various lenses over large parts of the magnification range, and the direction of the tilt axis can be aranged to be parallel to one edge of the micrograph, and the prints can be viewed side by side. Because micrographs are seldom printed as small as 6 cm wide the simple apparatus used for stereo photography is often supplemented by a mirror stereoscope designed for prints up to about 25 cm square (e.g. Figure 5.45 below). The Nesh viewer provides an ingenious and convenient way of viewing stereo pairs of any size with 'landscape' format. A final practical detail which ought to be mentioned concerns the direction of shadow-casting in micrographs for stereo pairs. In order to have consistency between the story told by the shadows and that told by the stereoscopic effect it is helpful if the specimen grid is orientated in the microscope so that the direction of shadowcasting is parallel to the tilt axis of the specimen stage. It will then be certain that the shadow-casting will come from the top or bottom of the print, according to whether black or white shadows are required, when the prints are oriented for maximum stereo effect. Stereo pairs of transmission micrographs of several types of subject are shown in Figure 5.39.

Stereo viewing for scanning microscopy

Since there are no imaging lenses in the SEM the problem of image rotation on change of magnification does not occur. Instead, it is necessary to ensure that the axes of the scanned frame are aligned with the tilt axis of the specimen stage. This is usually designed to be the case when the specimen is situated at a particular working distance from the final lens; the tilt axis is then parallel to the horizontal (x-axis) edges of the micrograph. At longer or shorter working distances the orientation of the scanned frame is rotated about the beam direction and ceases to be parallel to the

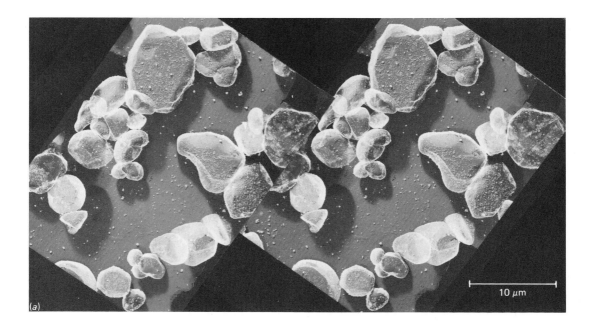

Figure 5.39. Stereoscopic pairs of transmission micrographs. (*a*) Shadowcast replicas of surface-treated phosphor particles. ZnS particles were shadowcast and carbon-coated before being dissolved out of the extraction replicas. The stereo pair, taken with a specimen tilt of 9° between the two micrographs, enables the distribution of coating material over the particles to be readily studied. (N.B. The awkward shape is a typical result of the displacement of the specimen tilt axis in the final image plane.) (*b*) Shadowcast two-stage replica of a diamond-ground glass surface. The depth of visible surface damage was measured with a stereometer between the two levels marked with arrows. A more direct method not involving a viewing device is illustrated in Figure 5.44. (*c*) A million volt transmission micrograph of a thin section of developed photographic emulsion showing the filamentary nature of the silver deposit. Specimen tilted 20° between micrographs.

specimen stage tilt axis or at right angles to the stub rotation axis. Since the orientation of scan on the display cathode-ray tubes is fixed the effect is one of rotation of the tilt axis away from the borders of the micrograph. This can be corrected for viewing a stereo pair in 3-D by re-orientating the prints as with transmission micrographs but to a very much smaller degree. When the tilt and rotation axes are accurately parallel to the *x*- and *y*-axes of the micrograph, respectively, it is evident that for effective stereoscopic viewing the micrographs must be turned on their sides if the two views were obtained by tilting, but can be viewed side by side if stub rotation was used. Of these the latter is the more satisfactory since the brain is happier with the combined perspective and lighting effects, as in the normal micrograph, than with the same thing turned on its side. The appearance of the two presentations is shown in Figure 5.40. These problems need not occur if the SEM has a scan rotation unit, which can be adjusted until the tilt axis is aligned with the *y*-axis of the micrograph. The stereo pair produced by tilting is then viewed side by side. If the SEM has a split-screen presentation and a scan rotation unit, stereo pairs can be made for immediate viewing on a single sheet of 4″×5″ Polaroid®-type material, which is close to the ideal size for a simple stereo viewer.

Stereo pairs on 35 mm negative film can be conveniently prepared for hand viewing or projection if the two images are recorded on separate frames of film with reversed contrast, so that they develop as positives. The two frames are cut out and mounted in standard 4.25×10.1 cm stereoscopic mounts, with 23×24 mm or 24×30 mm windows.

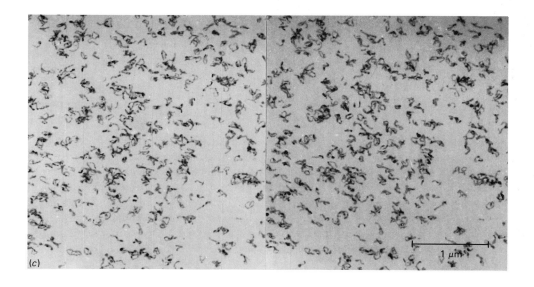

(c)

Anaglyphs

A long-established technique for viewing stereoscopic pairs of photographs without a special viewer is to print or project the two images superimposed on one another but differentiated by colouring one image red on white and the other blue or green on white. The resulting apparently overlapping images are separated by viewing

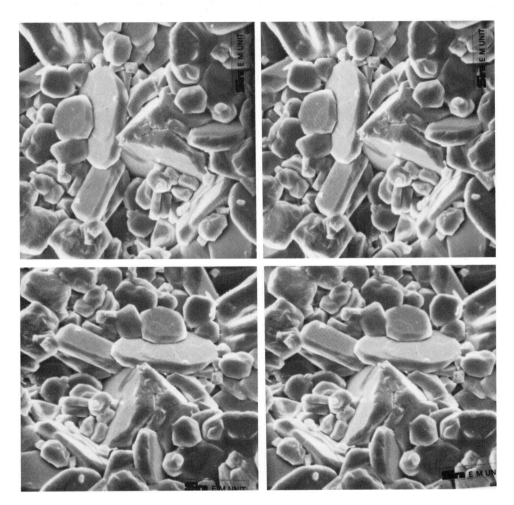

Figure 5.40. Stereoscopic pairs of scanning micrographs. The same small area of an etched alumina surface is seen in two stereoscopic pairs of micrographs. The upper pair was obtained by altering the stub tilt angle from 25° to 30° between the two pictures, whilst the lower pair was obtained by rotating the stub 10° about the stub axis, at a constant tilt angle of 30°. In order to align the prints with the effective tilt axis vertically up the page the upper pair have been turned on their sides. To produce a given amount of stereoscopic effect the angular movement of the specimen must be greater if the movement is a rotation than if it is a change of tilt. The relationship [sin(rotation angle at constant tilt) $\times$ sin(angle of tilt) = sin(change of tilt angle)] can be useful in this context. (Magnification $\times$ 3000.)

them through coloured 'spectacles' containing one red and one blue or green filter. The filters and the coloured inks or dyes must be carefully matched so that each filter transmits its own colour of image but none of the other. A coloured image will not be seen through a filter of its own colour but will appear as black on colour through the other filter. Each eye thus only sees the image appropriate to it. For example, as illustrated in Figure 5.41, if the left eye view is projected in blue and the right eye view in red, use of spectacles with a red filter in the left eyepiece and blue in the right gives the left eye a black image on red and the right a black image on blue. The full stereoscopic image is therefore seen in black and white. This type of presentation is called an *anaglyph*.

Unlike the type of stereoscopic presentation already described, any difficulties are found in the preparation of the anaglyph itself; viewing is straightforward as prints or projected images of any size and there are no adjustments needed to suit the eyesight of individual viewers.

Anaglyph images can be printed photographically through Wratten filters on to colour film or paper. Alternatively, positive images may be projected, superimposed, through the appropriate filters on to a smooth white screen, and photographed. The important point in either case is to achieve accurate vertical alignment of the two images, since the eye has no tolerance for divergences in this direction. This technique can only be applied to stereoscopic photographs in colour under specially contrived conditions, but this is no hardship in electron micrography. However, unless the microscopist has an enthusiastic photographic department it is unlikely that the difficulties of preparating anaglyphs will be any easier to work with than the viewing problems of the more frequently used double print. The latter, moreover, has the advantage that there is a good quality monoscopic image in black and white to be seen even if the stereoscopic effect is not used.

Anaglyphs with digital imaging: Setting up a stereoscopic pair for viewing as an anaglyph is made particularly easy on a microscope equipped for digital imaging with a

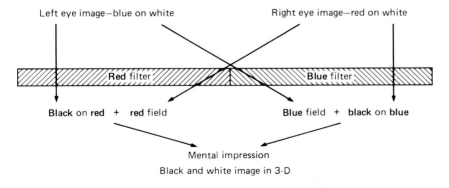

Figure 5.41. Principle of anaglyph presentation of stereoscopic pairs.

frame store and colour monitor. Some manufacturers have a dedicated software programme to make the procedure even simpler.

Once the tilt axis has been identified and aligned with the vertical edges of the
micrograph (monitor) the specimen is manipulated into the first tilted position and
the image stored. This can be recalled and displayed in blue or green (for example)
on the monitor. The specimen is tilted to the second position and translated until the
live image is in exact vertical alignment with the stored blue image. Displaying the
live image in red shows up the parallax between different planes in the image. If the
image is now viewed through the red/blue spectacles it will be seen in 3-D and can
be optimised by translating the specimen so that the red and blue images of the most
distant features in the field are well separated on the screen.

The anaglyph can be recorded for display elsewhere by photographing the
monitor on colour film or by using a colour video printer. The two views can also be
recorded separately in black and white for reconstruction in a conventional stereoscopic viewer.

Auto-stereograms

Readers may have encountered *Nimslo* prints or 3-D postcards in which a scene can
be viewed in apparent 3-D without the aid of any viewing device or filters. These
stereo photographs are known as *lenticular stereograms* and are associated with the
Frenchman, Maurice Bonnet (Marraud & Bonnet, 1980). The same principle of
operation can also be applied to stereoscopic micrographs, although a great deal
more work is necessary compared to the stereo pair for viewing through an optical
aid. A series of micrographs taken in the normal way are printed and viewed using a
special lenticular screen.

The image in a lenticular stereogram is formed and viewed through a moulded
plastic screen consisting of narrow cylindrical lens elements aligned parallel to the
vertical edge of the picture (Figure 5.42(*a*)). Considering the action of a pair of these
lens elements (Figure 5.42(*b*)) it can be seen that pencils of light are focused to
narrow strips by the cylindrical lenses, and that the position of each strip along the
image plane varies with the angle of incidence of the pencil on the lenticular screen.
So different viewpoints have different positions along the image plane behind the
lens element, and, in reverse, looking at the lens from different angles (e.g. from left
and right eye viewpoints) different image strips will be seen by the two eyes. There
is consequently a mechanism for sorting out left and right eye views without the need
for any further optical aids, polarised light, etc.

For a good lenticular stereogram about 20 different viewpoints are used, and a

Figure 5.42. (*Opposite*) Lenticular stereogram. (*a*) Moulded ribbed screen used in contact with
photographic film or print for recording and viewing an auto-stereogram. (*b*) Pencils from different
directions I, II are focused by the cylindrical lenses into strips in the image plane.

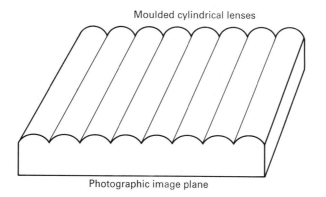

Moulded cylindrical lenses

Photographic image plane

(a)

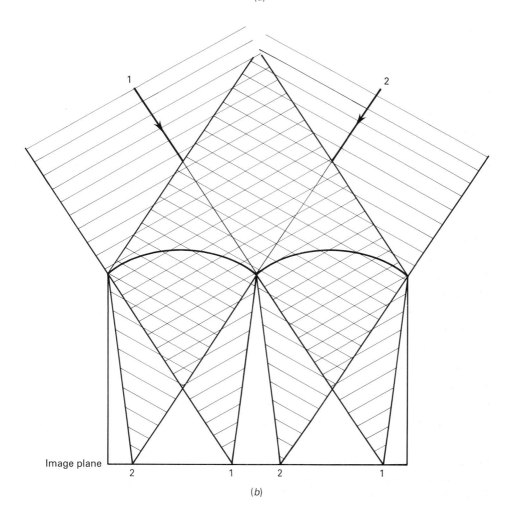

Image plane

(b)

series of micrographs would be taken with equal increments of tilt between the normal range of angles, e.g. 21 micrographs at half-degree increments from −5 to +5 degrees. Each of these micrographs will be projected on to the lenticular screen at a slightly different angle, so that the strip images lie side by side. Assembling the images into a stereogram is the job for a specialist, but large display micrographs, usually in the form of backlit transparencies, may be seen at microscopy exhibitions.

Making measurements from stereo micrographs

Although stereo micrographs can be invaluable merely by enabling the relative heights of different parts of a specimen to be seen unambiguously, they also provide the starting point for quantitative measurements in the third dimension, when stereoscopy becomes stereometry. Measurement of depth relies on either direct or indirect determination of the *parallax* or degree of lateral movement of one particular feature on the micrograph relative to another as the viewing direction is altered. To help us to understand this method let us consider the simple arrangement shown in Figure 5.43. Figure 5.43(*a*) is a plan view showing a row of three sticks A, B, C pegged out in a straight line with equal spacing AB and BC. The line BN is normal to the line of sticks and two viewpoints P_1, P_2 lie $\theta°$ on either side of BN. From P_1, P_2 the appearance of the row of sticks is identical (shown in the frames at P_1, P_2). Now consider the case where stick B is moved forward to B′ by a distance h (Figure 5.43(*b*)) in the direction BN. From P_1 it appears as though B has moved towards C; from P_2 it appears that B has moved towards A. (Readers with stereoscopes can check that B is in fact still centrally placed!) The amount of sideways movement of B between the two views is called the *parallax* resulting from an angular movement 2θ of the observer's position. It can readily be seen (Figure 5.43(*c*)) that the movement of B seen from P_1 is B′X$=h\sin\theta$ and similarly from P_2 the movement of B is B′Y$=h\sin\theta$. The total parallax between P_1 and P_2 is B′X + B′Y$=2h\sin\theta$ so that, if the lateral movement between P_1 and P_2 is measured, e.g. from two photographs, the depth differential h can be readily determined. When the two photographs are micrographs the calculated value of h is M times the distance on the actual specimen, where M is the magnification of the micrographs on which the parallax is measured. Thus we can say in general,

Vertical height separation $h=X/(2M\sin\theta)$

where X is the x parallax and θ is one-half the angle of tilt between the two views. Figure 5.44 shows how this formula can be used to determine the height separation between two points on stereoscopic pairs of micrographs.

The parallax may be measured on the prints by means of a ruler. However, if the height difference between two extended planes is required it may not be possible to find detail on the micrographs from which to measure the parallax. In this case it is converient to use a commercial stereometer as used for aerial surveying, such as the one illustrated in Figure 5.45. The prints in their correct orientation are placed on the stereometer table and observed in 3-D through the stereoscope. In the instrument illustrated an attachment enables the image of a bright spot of light to be 'floated' in the field of view and, by operating a micrometer, settled in any plane of interest. The micrometer is read and the spot settled in a second plane. The change in micrometer reading is the parallax difference between the two levels, and the usual equation can be used to calculate the height difference on the original specimen. This method used on the micrographs in Figure 5.44 enabled the depth of the large pit in the ground glass surface to be determined as 12.6 μm.

Stereoscopists have found that the brain will fuse up to about 5 mm of parallax at the normal viewing distance of 25 cm. If the depth of specimen detail h is known (equalling the section thickness, for example) then the maximum permissible

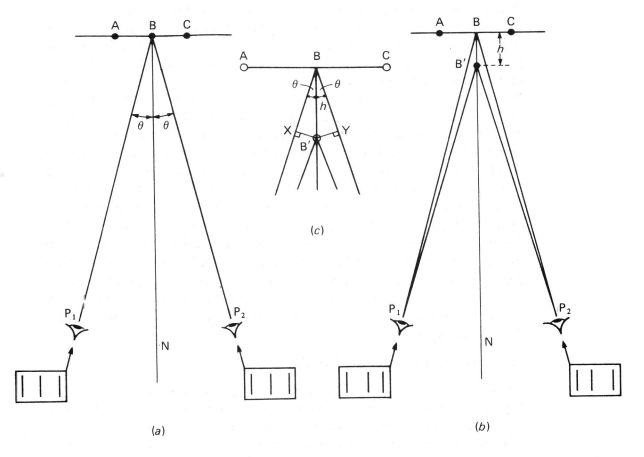

Figure 5.43. Diagrams showing the origin of parallax in stereoscopic observations and its use for the measurement of the third dimension. For explanation see text.

product $M\sin\theta$ can be calculated from the parallax equation above, and the conditions of taking the stereoscopic pair of micrographs chosen to give appreciable but not excessive stereoscopic effect (Hudson & Makin, 1970).

Making quantitative measurements on scanning micrographs of tilted specimens is more complicated (Lane, 1969; Boyde, 1973), because the magnification is affected

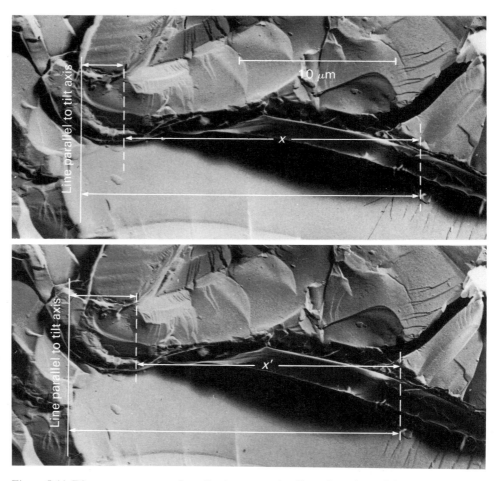

Figure 5.44. Direct measurement of parallax in stereo pairs. Central portions of the stereo micrographs in Figure 5.39(*b*) are shown further enlarged to a total magnification of 4100 diameters. In order to measure the height difference between the levels marked with arrows, surface features at the two levels are selected which appear on both micrographs. Arbitrary lines parallel to the tilt axis (*y*-axis) are drawn on the micrographs. These facilitate the measurement of the *x*-separation, x, x', between the features on the two prints. The difference $x - x'$ is the parallax between the features corresponding to a total tilt angle of 9°, magnified by ×4100. The measured value on the prints is 8.0 mm. Putting these values into the relationship $h = x$ parallax$/2M\sin\theta$ the value of h is calculated to be 12.45 μm. This is in good agreement with the value of 12.6 μm determined experimentally using the stereoscopic effect and a mirror stereometer of the type shown in Figure 5.45.

by working distance and tilt angle, but computer programmes are available to take care of the mathematics. A system for analysing stereo pairs using a digitising tablet and microcomputer is described by Roberts & Page (1981).

Dynamic stereo-microscopy

This is the name given to a technique whereby the specimen in a scanning microscope can be seen in 3-D in 'real-time' by the simultaneous production of images from two viewpoints. Instead of tilting or rotating the specimen the axis of the scanning beam is tilted between two directions, giving left and right eye viewpoints of the specimen. Scanning at TV rate (50 or 60 Hz), alternate frames are scanned from the two opposite directions; there is a choice of methods for viewing the 3-D image. Monochrome images can be displayed on separate monitors for each eye, or on a single monitor viewed through special liquid-crystal shutter eyeglasses synchronised with the image signals. Alternatively, they may be seen as anaglyphs in red and blue or green on a colour monitor.

Figure 5.45. Mirror stereometer for examining stereoscopic pairs of micrographs and making measurements of parallax. (1) Optional light sources and micrometer for measurement of parallax. (2) Binoculars with adjustable separation to accommodate different observers. (3) Light-box for illuminating transparent images. (4) Fluorescent strip-light for illuminating prints. (By courtesy of Cartographic Engineering Ltd.)

In the original commercial embodiment of the technique (Chatfield *et al.*, 1974) the additional beam-deflection coils were fitted between the final lens and the specimen. In a more recent alternative system the deflection uses the alignment coils of the electron gun. When digital imaging and frame averaging are employed the 'real-time' anaglyph image is of a high quality, and the effects of change of magnification and specimen manipulation (tilt, shift, etc) can be seen immediately in full 3-D by viewing through the appropriate red/blue 'spectacles'.

This type of stereo-microscopy is valuable when the topography of an unknown specimen is being evaluated. It becomes indispensable when *in-situ* experiments are being carried out, such as when a specimen is advanced towards a cutting edge in order to cut thin shavings from it. The video signals may of course be recorded on tape for later replay and detailed study.

Since the amount of parallax is a measure of depth in the field of view, dynamic stereo-microscopy provides a very convenient way of making measurements in the third dimension, and augments the on-screen cursor generally provided for 2-D measurements from the screen image.

Factors inhibiting the wider use of stereo-microscopy

The technique of stereo-microscopy can be seen to offer genuine advantages when the structure of specimens is to be inferred from electron micrographs. Although the technique is by no means new, its use is still somewhat rare; it may be helpful to consider the reasons for this. Some of these are concerned with obtaining stereo pairs and others with viewing them. One of the former is undoubtedly the fact that many microscopes are not equipped with eucentric goniometer specimen stages. This means that recording the identical area on a stereo pair may still be a chancy operation if the specimen has to be chased with the shift controls as tilting or rotation are carried out: it is very easy to lose sight of the chosen features after taking the first micrograph, or even to forget what the field looked like. The SEM or TEM user with a Polaroid camera, video printer or frame store has an advantage here since he can have a visible record of the first view. It is possible to draw outlines of salient features with wax pencil on the visual display; indeed, this may be essential when a split-screen record is being attempted. However, the luckiest operator is the one with a software programme for stereoscopic anaglyphs, who can colour and manipulate two images until he gets them right.

The prospective purchaser of a new SEM, whose major interest is in three-dimensional topographical information, would be well advised to investigate the commercial instruments offering dynamic stereo-microscopy as a working technique.

The TEM user whose specimen is wholly or partly crystalline faces a difficulty because the appearance of his specimen will alter with the angle of viewing owing to diffraction as well as spatial changes. He must therefore try to ensure that there are equivalent diffraction conditions in both micrographs of a stereo pair.

When viewing and making measurements from transmission micrographs it is a nuisance to have to calibrate for image rotation in order to find the direction of the specimen stage tilt axis on the final micrograph. As far as the author is aware there is no simple commercial device available to enable the stereo pair to be rotated under a stereo viewer in order to find the orientation showing maximum 3-D effect. However, on modern multi-lens microscopes this problem has usually been designed out by the manufacturer.

When the microscopist has succeeded in taking and orientating his stereo pair he is faced with a different problem of showing them to others, and incorporating them in reports and publications. A surprisingly large number of people are unable to see stereo pairs in 3-D under any conditions. Some authorities put the proportion as high as 50%. The remainder have a wide range of interocular distances and eyesight defects to be accommodated in viewing. The best quality hand viewers for stereo pairs mounted as transparencies are provided with separate adjustments for inter-ocular separation and focus; handing such a viewer around a group of people is a daunting experience. Paradoxically the simpler and cheaper viewers without any adjustments appear to be more generally acceptable. Properly mounted trans-parencies can be very effectively shown to an audience by projection, but the most common system using polarising filters and spectacles to separate left and right eye images require a double projector, a metallised ('silver') screen and a darkened room.

Probably the simplest method of viewing in 3-D is to use small prints under a viewer as already illustrated in Figure 5.37. Even this has to cater for differences in interocular separation, however; moreover the prints must be less than 6.35 cm wide and may have to be produced by reduction printing. Larger stereo pairs, preferably of landscape format, can be placed above one another for viewing through a prismatic Nesh viewer. Finally, the stereo pairs can be prepared as anaglyphs, as already described. This type of presentation is probably the most expensive to produce, but at least there is no problem of interocular separation to be considered in viewing the result. Using digital imaging and a colour video-printer, anaglyphs are much easier and therefore more likely to be used in practice.

In spite of the numerous deterrents mentioned above it is encouraging that stereo pairs are seen in publications from time to time, especially from authors such as Boyde (e.g. Boyde & Ross, 1975) and Steere (e.g. Steere, 1981). The pairs are usually presented as two appropriately spaced small prints. A very interesting presentation, enabling side by side comparison between mentally reconstructed images, is the array of complementary stereo pairs shown in an earlier section on freeze-etching (Chapter 4).

5.5 Image processing and analysis

In the early days of electron microscopy it was sufficient to have micrographs which provided information about the appearance and size of microstructures in a specimen at a resolution which was previously unattainable. Since then, as the understanding of the imaging process itself has improved, it has been realised that microscopy is able to provide detailed quantitative and analytical information about the specimen. Using micrographs to make comparisons between laboratories or between processes has necessitated extracting as much information as possible from the electron distribution in the image.

It may be possible to arrange the examination so that the analysis takes place while the specimen is in the microscope (i.e. in *real-time*); otherwise it is necessary to record the image, reproduce it and if necessary manipulate it in order to emphasize a particular type of data (*image processing*) before carrying out *image analysis* to quantify the information.

5.5.1 *Quantitative electron microscopy*

Quantitative image analysis is a procedure which is applied to a microscopic image to extract numerical data from it about the size, shape and distribution of particular features which may be distinguished in the image by their brightness or their shape. Some of the basic measurements required from image analysis are shown in Figure 5.46. Frequently occurring examples requiring analysis are size distribution of particles in pigments, emulsions and cultures, and phase identification and quantification in polished sections of metals or minerals.

The most basic way of analysing an image is to take a ruler or a graticule to the print or projected transparency and make measurements over what is judged to be a representative area of the image. This may be adequate for some purposes – establishing the largest and smallest dimensions, for example – but to make a large number of measurements for a statistical analysis requires time and patience. With a human operator there is, moreover, a possibility of measurement criteria varying throughout such a series of measurements.

The next step upwards is a semi-automatic device which still requires the operator to work on a print but which stores and sorts the measurements and takes care of the statistics. A position-sensitive digitising tablet will continuously output the coordinates of a stylus moved across its surface. If a micrograph is placed on such a tablet and the tablet is interfaced to a simple computer, dimensional measurements can be recorded as the stylus is moved over relevant features on the micrograph. A small computer with magnetic 'floppy discs' for data storage and for the operating programme is able to perform many tasks of image analysis in this way.

The ultimate level in image analysis is the completely automatic image analysing computer with a manual override or editing facility for use on 'difficult' images. The image can be presented to the instrument as a print or transparency scanned and digitised by a TV camera, or a stored or 'real-time' image from a microscope. The computer digitises the image, assigning a brightness value or 'grey level' to each of

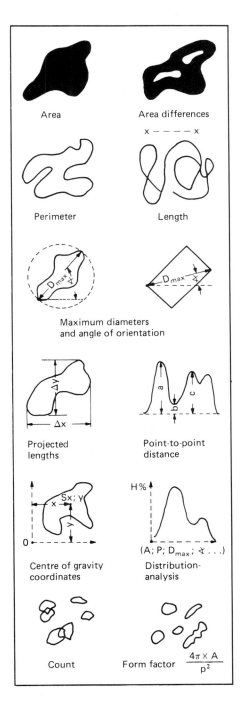

Figure 5.46. Some of the quantities measured in image analysis (By courtesy of Kontron Elektronik GmbH.)

the pixels or picture elements, and stores the data in its memory, whence it can be recalled for processing and analysis. Once an image has been digitised and stored it can be manipulated in a variety of ways until the desired results are obtained and printed out. Numerous computer programmes are available to carry out all the basic measurements noted in the Figure 5.46.

5.5.2 Image processing

One of the advantages of digitised electronic images is that they can be modified to give them improved characteristics for analysis. Poor contrast and noise can be treated to give more uniform and more distinguishable features so that automatic image analysis has a better chance of being carried out correctly. Filtration processes work on small groups of picture elements (pixels), e.g. 3×3 or 5×5, modifying the central pixel intensity according to those around it, and then moving on in steps through the whole image. Boundaries can be sharpened or noise reduced by appropriate filtration.

The presence of repetitive structure in a noisy micrograph can be shown and clarified from Fourier transformation of the micrograph. This can be carried out on a transparency using a laser diffractometer on an optical bench (Horne & Markham, 1973) or by computation from a digitised image. Deleting all but the regular features and reconstructing the image via an Inverse Fourier Transformation produces a clarified image of the repeated structure.

One of the advantages of human observers is their use of experience to decide whether features in a micrograph are overlapping or touching, making measurements accordingly, by estimation. Automatic image processing can simulate this decision-making by making use of the procedures of erosion and dilation. In the former all boundaries are retracted step by step, which has ultimately the effect of separating overlapping features. Expanding these individually by the same amount as the erosion returns the sizes to their original values, but with knowledge of any overlapping situations.

One of the essential operations before quantitative measurements can be made on micrographs is to define the features to be measured. The process of segmentation or thresholding requires a decision on the range of image intensities or densities to be counted as features for measurement.

One common potential source of error is a varying background density because of instrumental deficiencies. Uneven illumination, hot spots, or a tapering specimen across the field can result in errors in analysis unless a 'shading correction' is applied to the image to even out the background.

5.5.3 Determination of mass thickness by image analysis

When an electron beam passes through an amorphous object in an electron microscope the intensity I of the beam in the image plane is related to the intensity I_0 incident on the specimen by an equation of the form $I = I_0 e^{-C\rho t}$, where C is a constant dependent on the conditions of the observation, beam energy, and objective aperture size, but effectively independent of the specimen; ρ is the density of the specimen; and t is the thickness of the specimen. The product ρt has the dimensions mass/unit area and is known as the *mass thickness* of the specimen. In a non-homogeneous specimen the localised mass thickness variations give rise to the density variations of the micrograph. Thus, if the value of C is determined with a standard specimen of known mass thickness the mass thickness of any other specimen can be determined from measurements made under similar conditions.

Current densities in the image can be measured directly with a Faraday cup placed at different parts of the image field. Direct measurements of current density have the advantage of speed but the disadvantage that no permanent record is obtained and the measurement can only be repeated by setting up the experiment again and relocating the same area of specimen in the microscope. Fortunately, the electronic image contrast can be deduced indirectly from the photographic record or a digitised electronic image. The density of photographic blackening of most emulsions used in electron micrography is linearly related to the electron dose up to a density of at least unity, and sometimes higher. So, a microdensitometer can be used to determine the ratio I/I_0.

By a further extension of this principle Bahr & Zeitler, (1965a) measured the mass of individual particles on a micrograph, even those with irregular outlines. Differences between the light transmitted by a field containing a particle or particles and the same area of plain background gave the additional scattering due to the particle itself, and proportional to the mass of the particle. The system was calibrated by using objects of known masses, photographed under identical conditions in the microscope. Figure 5.47 illustrates the range of biological materials which can be 'weighed' in this manner. At the upper and lower ends of the scale, values are directly comparable with those from other techniques but in the middle of the scale, 10^{-13} to 10^{-16} g, there would seem to be no method of comparable accuracy for the determination of the mass of single objects.

On an analytical TEM operated in the STEM mode the probe can be placed on features in the specimen and the electron scattering measured with an annular dark-field detector. Such measurements have been reported by a number of workers, e.g. Wall & Hainfield (1986).

One of the requirements of mass thickness determination by this method is that the specimen be amorphous. Diffracting crystals introduce a very strong directional dependence of electron scattering which bears no relationship to the intensity–mass thickness proportionality being used here. Usable linearity has been reported for crystals away fron their Bragg reflecting conditions, however, and films of very small crystallite size (e.g. a few nanometres) may show a linear transmission –

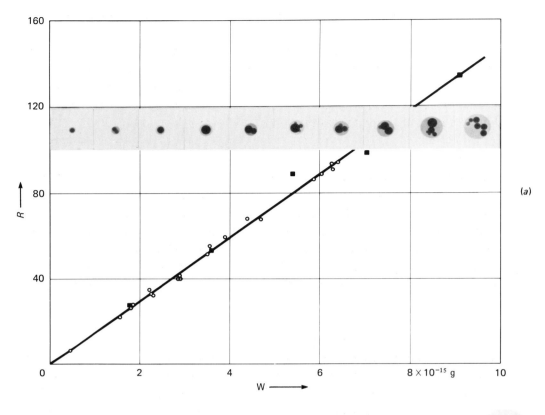

(a)

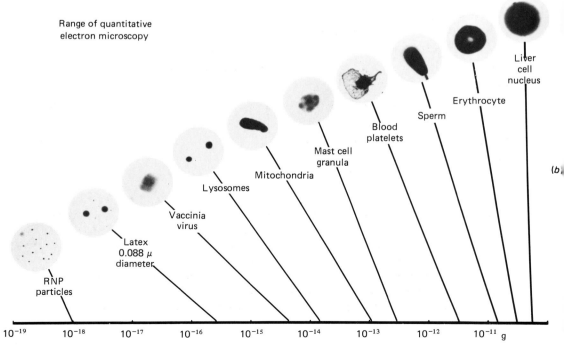

(b)

thickness relationship. An example of profile plotting for such a film is given in Chapter 8.

The linearity of electron response of photographic materials at the low density end of their scale can be used for quantifying beam–specimen interactions for a variety of purposes. For example the influence of various operating parameters (e.g. electron energy or objective aperture diameter) on image contrast may be measured conveniently from micrographs of suitable test objects.

5.5.4 Stereology

Stereology is the name given to the study of three-dimensional structures from two-dimensional sections or projections of them. It has a strong statistical basis, and derives its data from the interactions between a geometrical sampling grid and an image field. The techniques and their application are beyond the scope of this book. A review of the literature and discussion of the techniques of stereology and morphometry is given in Weibel (1969). Articles on the technique are published from time to time in *Journal of Microscopy* and in the twice-yearly *Acta Stereologica*.

5.6 Suggested further reading

A reader wishing to go thoroughly into the manner of formation of TEM and SEM images will find much of interest in Reimer (1985, 1993a,b). Various facets of quantitative electron microscopy are dealt with in the symposium proceedings edited by Bahr & Zeitler (1965b). Good reviews of the principles and practice of image analysis are given in Joyce Loebl (1989), and Russ (1990). Image analysis and mathematical morphology is the subject of Serra (1984, 1988). Williams (1977) and Weakley (1981) are both concerned with making measurements from micrographs in biology.

The general topic of electron microscopy at molecular dimensions is discussed in Baumeister & Vogell (1980).

A book which suggests potential sources of artefacts in specimen preparation and use of the microscope, particularly in biology, is that edited by Crang & Klomparens (1988). Instrumental deficiencies and their correction are described by Chapman (1986).

Figure 5.47. (*Opposite*) Quantitative mass determination by transmission microscopy. (*a*) Calibration curve of apparatus for determining particle mass from density measurements on transmission micrographs. Photometer reading *R* is plotted against the calculated masses of Formvar films (squares) and polystyrene latex particles (circles). Typical test objects are shown in the inset micrographs. Measurements on the same photometer from micrographs made under similar operational conditions in the microscope enables the masses of unknown samples to be determined. The range of such objects is illustrated in (*b*). Mass measurements can be made by this method on specimens of irregular shape and unknown composition. (By courtesy of G. F. Bahr & E. Zeitler.)

Stereo electron microscopy has relatively little literature, but there is a section contributed by Boyde in Robards & Wilson (1993). The principles and discussion of stereoscopic photography and the mounting and viewing of stereo pairs are well described in Ferwerda (1982). A simple but informative book of stereoscopic photography (which comes complete with a viewer) is Burder and Whitehouse (1992).

The range of modes available in scanning electron microscopy are described and illustrated in Newbury *et al.* (1986), Lyman *et al.* (1990), and Goldstein *et al.* (1992).

Analysis in the electron microscope **6**

Although the moving force behind the development of electron microscopes was the potential of shorter wavelength illumination to reveal finer detail than the light microscope, the richness of interactions between specimen and the illumination makes the electron microscope capable of determining much more than morphological structure. In addition to the analysis of sizes and shapes already mentioned (Image Analysis, Chapter 5) it is possible to obtain information on the elemental composition of specimens, and in the TEM about any crystalline compounds which are present. All this on a micrometre scale at worst, and at best down to sub-nanometre dimensions. On some specimens surface and bulk compositions can be distinguished and the presence and distribution of impurities can be shown.

This chapter will describe the types of analysis possible in electron microscopes, using the techniques:

> Electron diffraction
> X-ray microanalysis
> Electron energy loss spectroscopy
> Auger electron spectroscopy
> Cathodoluminescence

6.1 Electron diffraction

6.1.1 *Electron diffraction in the TEM*

In the TEM the strong interaction between crystalline specimens and the electron beam can be used to derive crystallographic information in a number of different ways. Not all are possible on all microscopes.

High-resolution diffraction

It is possible to use many commercial TEM columns as conventional diffraction cameras (Chapter 2, page 52–3), with camera length typically about 300–400 mm and very well-resolved patterns because the illuminating spot can be made very small. A special specimen stage, usually called a high-resolution diffraction accessory, is placed at the bottom of the column near the final projector lens, which is not energised. The other lenses in the column are used to provide a very narrow parallel beam of electrons at the specimen. No lenses are used between the specimen and the screen, and the camera length L can be measured accurately. The geometry and physics of the interaction are exactly as described in Chapter 2 and Figures 2.17 and 2.18. The resolution of diffraction, i.e. the ability to distinguish between diffraction spots of close separation, is given by $d/\Delta d$ and equal to $L/\Delta R \times \lambda/d$. This can be as high as 10^6 in the lower stage.

The high-resolution diffraction accessory requires an airlock for introduction of the specimen, a stage for tilting it in the beam, and possibly a charge-neutralising device so that crystalline insulators can be examined by reflection diffraction without charging-up under examination. The additional cost of the diffraction attachment, as well as the fact that the specimen being studied by electron diffraction cannot simultaneously be examined by transmission microscopy, makes this type of study less popular than the so-called *selected-area diffraction* (s.a.d.) technique in which a small area of specimen can be selected from a high-resolution transmission image and its electron diffraction pattern produced on the screen of the microscope by operation of a switch or push button. This procedure can, in principle, be carried out in all microscopes with three or more imaging lenses without the need for any instrumental modification or accessory stage.

Selected-area diffraction

The technique relies upon the fact that the electron beam after diffraction by the specimen forms a miniature diffraction pattern (rings or spots) in the back focal plane of the objective lens (Figure 6.1(a)). If the first projector lens (or the diffrac-tion lens in a four-lens imaging system) is now weakened so that it is focused in this plane instead of the image plane of the objective lens (i.e. the first intermediate image plane), the pattern is projected – enlarged – on to the screen of the micro-scope. If the lens currents for this mode of imaging are preset the microscope can be switched alternately between imaging the specimen and projecting its diffrac-tion pattern (Figure 6.1(b)). Moreover, if a diaphragm containing an aperture of diameter D is inserted into the plane of the first intermediate image it will pass only those electrons coming from a part of the object of diameter D/M, where M is the magnification of the objective lens. With this area-selecting aperture in place, switching to the diffraction mode results in a diffraction pattern being formed only by those electrons originating from the chosen small area of object. If the magnification of the objective lens is, say, $\times 40$ a diaphragm 100 μm in diameter will isolate electrons from a region of the specimen 2.5 μm across, and a form of micro-identification is therefore available. With suitable care in operating the microscope selected-area diffraction patterns can be formed with electrons from chosen regions as small as 1 μm across. Below this size there is a probability that, because of the spherical aberration of the objective lens, electrons in the diffraction pattern came from parts of the specimen outside the area which had been selected, and the tech-nique may therefore give misleading results (Riecke, 1961). Figure 6.2 shows how a small selection aperture can be used to isolate the diffraction pattern from part of a larger field of view.

When the microscope has a third condenser lens C3 and a Riecke–Ruska condenser/objective there are further possibilities for small-area selection for diffraction. If C3 is used to project the illumination to a focus at the front focal plane of the Riecke–Ruska lens, the illumination becomes parallel at the specimen (ideal for diffraction). The diameter of the beam can be determined by the C2 aperture, and diameters smaller than 1 µm at the specimen can be selected safely.

If C3 is used to project parallel illumination into the Riecke–Ruska lens, this can be used to focus the illumination into a small spot at the specimen, and it can be

Figure 6.1. (*a*) Ray diagram showing the formation of the diffraction pattern and intermediate image by the objective lens. Electrons diffracted in the same directions OA, O′A′ come together to form a spot P in the back focal plane of the lens. Similarly the diffraction spot Q is formed by electrons diffracted at an equal angle on the opposite side of the undeviated beam. (*b*) In a three-lens microscope the intermediate lens (or Projector 1) is normally focused on the intermediate image formed by the objective lens. When P1 is weakened to focus on the back focal plane of the objective lens the final image is an enlarged diffraction pattern. The selected-area diaphragm ensures that only electrons coming from a chosen region in the specimen contribute to the diffraction pattern.

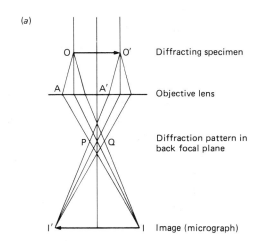

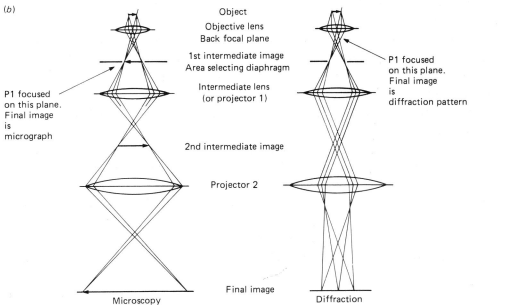

positioned on sub-micrometre detail in the specimen. Switching to diffraction then results in a *microdiffraction* pattern being formed, as illustrated in Figure 6.3.

One of the less obvious advantages of carrying out microscopy at 1 MV arises in connection with the use of selected-area diffraction. Not only are the diffraction patterns clearer than at 100 kV because of improved penetration through the specimen, but the precision with which the smallest areas can be selected is some 50 times improved; thus it should be possible to form selected-area diffraction patterns reliably from regions of specimen as small as 20 nm across in the HVEM.

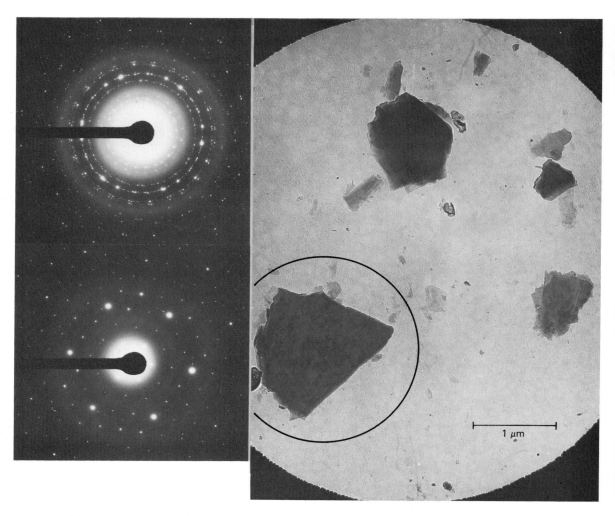

Figure 6.2. Selected-area diffraction (s.a.d.). The micrograph shows a field of crystalline particles outlined by a large selection aperture (6 μm at the specimen). The corresponding 'spotty ring' type of diffraction pattern (*upper left*) indicates that the crystals are oriented in a number of different directions. Using a smaller selecting aperture (2.3 μm at the specimen) positioned as indicated, the pattern is formed from the chosen crystal alone, giving the single-crystal pattern seen at lower left.

Low-angle diffraction

Since the s.a.d. pattern is projected by a lens system its size can be varied to suit particular requirements, e.g. studying very large or very small lattice spacings. This is equivalent to varying a camera length L, and effective lengths between 100 mm and several metres can be obtained under normal s.a.d. operation. Very long camera lengths, up to hundreds of metres, can be obtained using a procedure for *low-angle diffraction*. The objective lens is switched off and the diffraction pattern is formed in

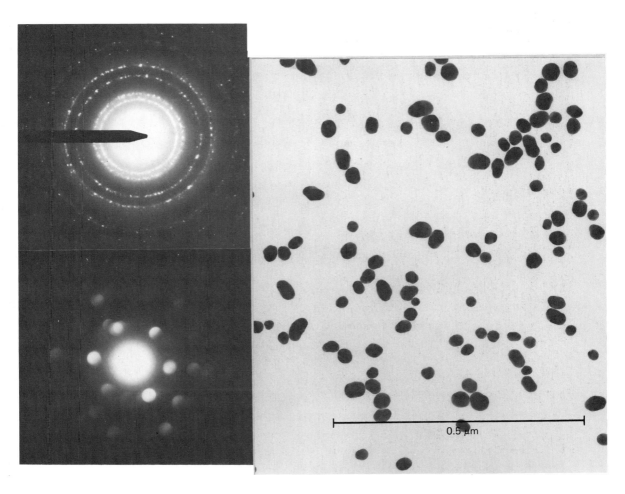

Figure 6.3. Selected-area microdiffraction. Similar to Figure 6.2 except that the field of view is less than 1 μm across and hence too small for conventional s.a.d. of individual particles. A single particle has been selected for diffraction analysis by focusing the illumination into a spot covering the particle. The diffraction spots are discs rather than points because the electron beam is no longer parallel at the specimen, but a cone; each diffracted beam is therefore also a cone, which becomes a disc in the film plane. Discs can overlap and obscure others; hence analysis of smaller crystals from microdiffraction patterns is less precise than conventional s.a.d.

the back focal plane of the diffraction lens, and projected by the following lenses in the column. Using this technique diffraction patterns of measurable dimensions may be obtained from materials with d-spacings of tens of nanometres, such as polymers and macromolecules. Under normal s.a.d. operation the diffracted beams from such specimens would be so close to the undeflected beam as to be unmeasurable. For example, if $d=10$ nm and 100 keV electrons ($\lambda=3.7$ pm) are used, $R=0.37$ mm when $L=1000$ mm, and $R=37$ mm when $L=100$ m. (N.B. It is important to appreciate that the scale of a diffraction pattern depends not only on L but also on λ (diffraction camera equation, page 52.) Patterns can be enlarged or diminished by decreasing or increasing the electron gun voltage, respectively.

Making measurements from diffraction patterns

Conventionally, a diffraction pattern will be recorded on a photographic film and is best analysed directly from the processed film, since there is a risk of introducing dimensional distortion in copying it. The exposure of a pattern should not be heavier than is needed to record the positions of the rings or spots, as it is more difficult to judge the centre of a large and very dense silver deposit. The centre of a ring pattern always receives a heavier electron density than the outer regions; if it is desired to study both regions, separate recordings should be made to give the best density to each. Patterns can be measured with a low-powered magnifier and a transparent scale over a light-box, on a travelling microscope, or with a scanning microdensitometer. Great care is needed if the precision of measurement is to be better than 5%, i.e. within 0.5 mm on a 10 mm diameter ring.

In the electron diffraction camera and its equivalent, the high-resolution diffraction stage of the TEM, the relationship $\lambda L=Rd$ can be used for precise determinations of lattice spacings once L has been measured and λ calculated. If the gun voltage is not known accurately it is better to determine the product λL (the camera constant) experimentally by calibrating with a standard substance such as thallium chloride or aluminium.

Diffraction measurements in the s.a.d. mode of the TEM are less direct because of the lens system between the specimen and the projected diffraction pattern. The camera constant must be determined for as many of the range of camera lengths and gun voltages as are going to be used; care must be taken that the calibration grid and the unknown specimen occupy exactly the same plane in the microscope. This is no problem with a eucentric goniometer stage, but can be a potential source of error otherwise. To achieve the highest possible precision with the s.a.d. technique it is usual to evaporate the standard specimen on to the surface of the unknown one, so that the two patterns are produced simultaneously. There is a further complication in so far as the projected pattern may be subject to radial distortion by the lenses, and it is recommended to determine λL separately for various zones of the pattern, since it may differ between inner and outer regions.

Lattice spacings may be measured to 1% by very careful working, but 5% is a more practical aim which may still be sufficiently precise to enable materials to be identified by electron diffraction, using the crystallographic data provided by the JCPDS Powder Diffraction File (JCPDS, 1994). Selected-area diffraction can also be very useful for the purely qualitative assessment of the degree of crystallinity of a specimen, as for example in the study of thin films formed by vacuum- or chemical deposition techniques. It is very widely used in metallurgy to identify the nature and orientation of crystals in thin metal foils. To take full advantage of the s.a.d. technique for crystal identification a double-tilt goniometer specimen holder is essential. This will enable the specimen grid to be tilted in any plane and reflections to be obtained from lattice planes which may be a long way from their Bragg diffracting angles in the normal specimen orientation. Goniometer stages and special tilting specimen cartridges for top-entry stages allow tilt angles of 45° or even 60° to be obtained under observation in the microscope.

Computer-aided analysis: The time taken in film processing and measurement can be saved if the microscope has a TV display and facilities for image storage and analysis, since not only can the calculation of *d*-spacings but the identification from JCPDS files can be computer-aided. It can also be carried out in real time, whilst the specimen is still in the microscope.

Tilted-beam diffraction patterns: If the illuminating beam is tilted about a point in the specimen plane the whole diffraction pattern is displaced across the optic axis by the same angle. The beam tilt control can be calibrated in terms of *d*-spacing in such a way that diffraction patterns can be analysed by measuring the tilt needed to bring selected diffraction rings or spots onto the optic axis of the microscope. This procedure is particularly relevant in the analytical TEM with well-developed beam tilt and scanning facilities.

Convergent beam electron diffraction (CBED)

This is a technique based on electron diffraction which can be used for characterising appropriately presented crystalline materials. The technique was first reported by Kossell and Möllenstedt in 1938, but was not seriously exploited until the STEM attachment on TEMs made convergent beam formation a routine procedure. Much of the pioneering work was due to Steeds and co-workers at Bristol University, but CBED is now widely used in materials science studies.

In essence the technique involves forming a very small, highly convergent focused electron spot a few tens of nanometres across on an ideally parallel-sided crystalline specimen and projecting the resultant transmitted electron diffraction pattern on to the final screen. Because of its convergent nature the beam employs a range of angles of incidence simultaneously, and the pattern consists of broad spots each containing a great deal of fine structure. By suitable choice of condenser-aperture diameter, electron wavelength and crystal orientation (a double-tilt goniometer specimen stage is considered essential) convergent beam diffraction patterns of a kaleidoscopic complexity can be obtained which, at the least, can be used to 'fingerprint' the particular type of crystal producing it. When interpreted quantitatively by an expert they give detailed information about the crystal structure of the material. Patterns made with only elastically scattered electrons in an energy-filtering microscope show remarkable clarity and fine structure (Deininger & Mayer, 1992) (Figure 6.4).

For examples of CBED patterns and their significance and limitations the reader is referred to publications by Steeds and others (e.g. Steeds, 1979; Tomokiyo, 1992). A useful bibliography was compiled by Sung & Williams (1991).

6.1.2 Electron diffraction in the SEM

Several phenomena resulting from the interaction between the electron beam and a crystalline specimen can be obtained in the SEM. The first effect, the formation of *electron channelling patterns*, was reported by Coates (1967) and has since been studied and explained by numerous workers (e.g. Joy *et al.*, 1982). The effect is found when extensive single crystals, e.g. 'slices' of silicon for microcircuit manufacture, are examined in the emissive, backscattered or specimen current modes in the SEM at low magnification with a defocused electron probe, i.e. under near-parallel illumination with a large angular swing of the probe between the extremes of the frame. At certain angles of incidence of the beam on the crystal the Bragg condition

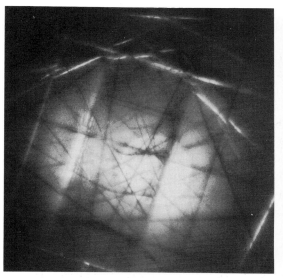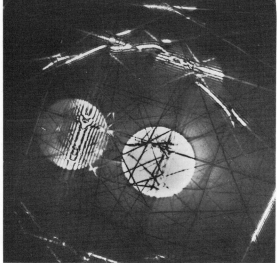

Figure 6.4. Convergent-beam electron diffraction pattern from a 270 nm thick silicon (220) sample; 100 nm probe, 100 kV. (*a*) Normal image, (*b*) zero-loss energy-filtered image. (By courtesy of J. Mayer, Max Planck Institute for Metals Research, Stuttgart, Germany.)

$2d\sin\theta=n\lambda$ is fulfilled and anomalous backscattering occurs from the crystal. The micrograph is crossed by an array of bright and dark lines and bands resembling the patterns (Kikuchi patterns) seen in transmission electron diffraction patterns from large perfect crystals. During the course of one wide-angle scan the beam may satisfy the Bragg condition for a number of different lattice planes in the crystal and for $n=-3, -2, -1, 1, 2, 3$, etc., for each set of planes. For a given d-spacing the angle for electron channelling to take place will depend on the gun voltage. Thus, for the (111) reflection from silicon θ will be 0.63 degrees for 30 keV electrons, 0.79 for 20 keV, 1.11 for 10 keV and 1.58 for 5 keV electrons. At a magnification of ×20 and a working distance of 20 mm from the pivot point the angular swing of the beam is 7° on either side of the axis. Figure 6.5 gives a schematic diagram of the electron channelling effect.

Selected-area channelling patterns
The above technique requires an extended specimen of many square millimetres, but no special modifications are necessary to the SEM. Van Essen & Schulson (1969) showed that if the scanning geometry was modified so that the pivot point of the

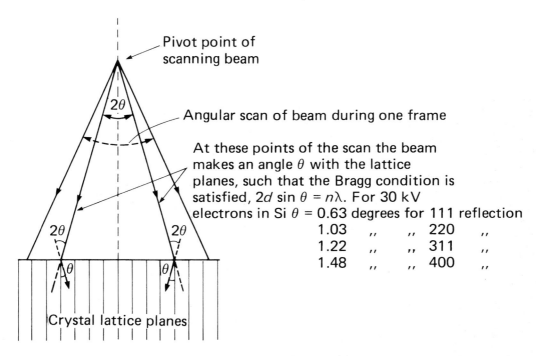

Figure 6.5. Schematic diagram shows angle of incidence of the scanning beam on the specimen varying more than $\pm\theta$ during each line-scan. At angles satisfying the Bragg equation an abrupt change in electron backscattering coefficient occurs. For a given set of lattice planes there will be sets of parallel lines in the BSE image corresponding to $n = \pm 1, \pm 2, \pm 3$, etc. in the Bragg equation.

beam was lowered on to the specimen surface the beam would rock through a range of angles of incidence at a selected small area on the specimen, which could be as small as 5–10 μm across in a dedicated SEM or down to 1 μm or smaller in a STEM attachment on a TEM. The technique of *selected-area channelling pattern* (SACP; Figure 6.6 (*a,b,c*)) is used to study the orientation of individual crystallites in poly-crystalline materials. The technique requires the specimen to be smooth and free from surface damage such as the disturbed layer (the Beilby layer) resulting from metallographic fine polishing (Davidson & Booker, 1970).

The visibility of the bands in channelling patterns is much improved if the video signal is put through a derivative-processing unit. However, this has the effect of subduing any lines parallel to the scan direction, and derivative-processing is dis-couraged by regular users of the technique. If it is used it is recommended to record three pictures of each pattern with (i) no derivative-processing, (ii) derivative, (iii) derivative plus 90° of scan rotation, to ensure that nothing is inadvertently lost from the pattern. (If the opposite edges of bands in a channelling pattern are of different tone, i.e. one black, one white, then the pattern was derivative-processed, and lines in one direction will be more difficult to see).

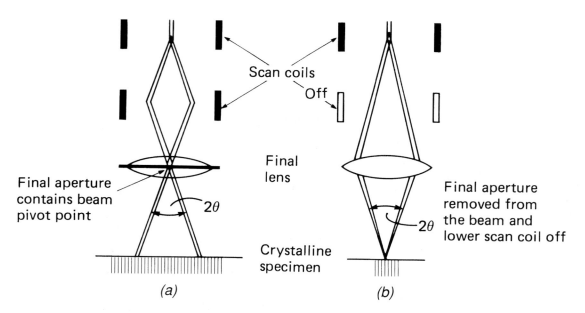

(a)   *(b)*

Figure 6.6. Electron-optical arrangement for (*a*) general area channelling, and (*b*) selected-area channelling patterns. In (*a*) an essentially parallel beam of electrons scans over several square millimetres of surface. In (*b*) the electron beam is focused at a point on the specimen surface and rocks through the angle 2θ about this point. (*c*) Selected-area channelling patterns from 1 μm areas of silicon ((100) pole) and tungsten ((110) pole).

Electron backscatter patterns

Two further diffraction techniques are possible in the SEM, but require additional facilities in the specimen chamber. These are *electron backscatter patterns* (EBP) and *x-ray Kossel line patterns* (XKLP or XRKP). In the former (Harland *et al.*, 1981; Dingley & Randle, 1992) an electron beam incident on a crystalline surface at a steep angle (60°–70° of specimen tilt) is scattered in preferred directions relative to the crystallographic planes in the surface and a photographic film or plate positioned facing the specimen will record a pattern of lines by direct impact of the scattered electrons. The pattern contains information from a shallow surface layer and can be used to study the orientation of surface deposits and to identify small particles.

X-ray Kossel patterns

X-ray Kossel patterns (Dingley, 1981) are produced when the characteristic x-radiation produced in the specimen by the impact of the primary electron beam interacts with the crystal planes in the specimen to give sharply defined diffracted beams of x-radiation. The Kossel patterns may be used quantitatively to measure lattice spacings with a precision as high as one part in 25 000. Usable x-radiation can be produced from light-element specimens by placing powdered metal on their surface and focusing the electron beam on individual particles (e.g. copper particles on a diamond surface).

(c)

Silicon
(100) pole

Tungsten
(110) pole

6.1.3 Electron diffraction in STEM
Micro-microdiffraction

In a STEM modification of a TEM the scanning electron probe is travelling parallel to the optical axis of the microscope when it strikes the specimen (see Figure 3.47). Electrons diffracted by a crystalline specimen will be focused into a diffraction pattern in the back focal plane of the objective lens, just as in the TEM mode. The pattern will remain stationary, even though the beam is scanning across the specimen, and if the lower imaging lenses of the microscope are operated in the diffraction mode an enlarged image of the diffraction pattern will be projected onto the final screen and may be recorded on film or via a TV camera. This is sometimes known as $\mu\mu$-diffraction, and a wide range of camera lengths will be available as in TEM diffraction. In the STEM imaging mode the same electrons would be collected together with the undeviated beam by the STEM detector, to form the bright-field image.

6.2 X-ray microanalysis on electron microscopes

The technique of electron probe x-ray microanalysis owes its beginning to Castaing (1951), whilst Duncumb introduced the scanning microprobe analyser in which imaging and analysis could be combined (Cosslett & Duncumb, 1956). It was Duncumb, too, who developed EMMA, the first dedicated analytical TEM, in which two wavelength-dispersive spectrometers were fitted to the column of a 100 kV high-resolution TEM (see Figure 3.15) (Cooke & Duncumb, 1969). Ironically, this development followed one year after the report by Fitzgerald *et al.* (1968) of the coupling of a solid-state energy-dispersive x-ray spectrometer to an electron beam instrument. All subsequent x-ray analysis in the TEM has used the energy-dispersive spectrometer (EDS or EDX) because of its lack of focusing restrictions, the ability to measure all x-ray peaks simultaneously, and the very much easier task of fitting it to a TEM column.

On the scanning electron microscope either a wavelength-dispersive spectrometer (WDS or WDX) or a solid-state device or, in many cases both, can be fitted. Elemental analysis is the most common facility added to an electron microscope.

In the following sections we shall see what can be achieved, and what are the possibilities and limitations. The physics background to x-ray generation was explained in Chapter 2, and the principles of the two types of x-ray spectrometer are described and compared in Appendix 3.

6.2.1 X-ray spectrometers on electron microscopes
Geometrical (hardware) requirements

There are different problems in fitting x-ray detectors to SEMs and TEMs, caused by the nature and accessibility of the specimen area in the two types of instrument.

The SEM has a capacious specimen chamber with attachment ports to which spectrometers of energy- or wavelength-dispersive type can be fitted. The requirements of the Everhart–Thornley electron detector and the x-ray spectrometers are similar in so far as they are fitted on the side of the chamber towards which the specimen may be tilted; they both need a direct line of sight to the source of their respective emissions.

If a focusing crystal spectrometer is fitted with its plane vertical, the specimen height for analysis will be critically fixed by the focusing requirement of the crystal (see Appendix 3, page 414). A reflected-light microscope (perhaps with a video camera to make viewing the image more convenient) may be fitted permanently to the chamber so that when the specimen has been adjusted to visual focus it is also at the spectrometer focus. A crystal spectrometer in the horizontal orientation gives more latitude in specimen height, but will not receive any x-radiation from an untilted specimen. The solution is to tilt (incline) the plane of the horizontal spectrometer to give an x-ray take-off angle of 30° or so, as shown in Figure 6.7.

The lack of focusing requirement for an EDX detector means that its positioning is not critical, but it must face the specimen and be as close as possible. The detector mount is usually made retractable, with a sliding vacuum seal, so that it can be pulled back from the specimen area when it is not actually being used. The level of the crystal will be above the specimen, to give a positive take-off for x-radiation from an untilted specimen. If the Si(Li) crystal is several centimetres from the specimen it

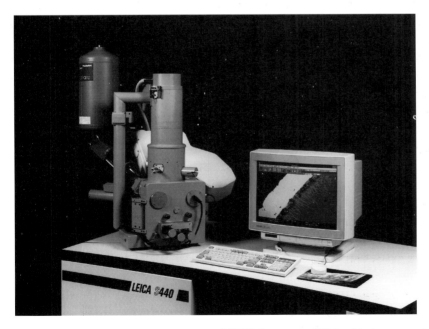

Figure 6.7. Inclined WDX spectrometer and EDX detector on SEM, with the same x-ray take-off angles. (By courtesy of LEO Electron Microscopy Ltd.)

intercepts only a small proportion of the emitted x-radiation. For example, a 10 mm^2 detector 5 cm from the emitting specimen subtends at most a solid angle of 0.004 steradian, and has a collection efficiency of only about 0.06%; this is still much higher than a crystal spectrometer.

The detector and its pre-amplifier stage are strongly cooled to minimise noise in the output signal. This usually means that EDX detectors are instantly recognisable because of the prominent 7.5 or 10 litre dewars storing several days' supply of liquid nitrogen coolant. Construction of the detectors has now improved so that they may be allowed to run dry and warm up if they are not required for use. This is a welcome relaxation for weekends and holidays. Commercial solid-state detectors are now available with flat-sided and even 'cup-sized' dewars. 'Dry' detectors are also available, in which thermoelectric or some other form of cooling is used.

Mounting an EDX detector on the TEM is complicated by the location of the specimen within the objective lens. In earlier microscopes it was not possible to position a Si(Li) detector within sight of the specimen grid. In more recent microscopes either parts of the specimen holder and/or lens are cut away to give a direct line-of-sight between the specimen and detector, or the lens is designed with specimen plane accessible to a detector introduced from the side, either level with the specimen or slightly above it. The detecting crystal can be brought very close to the specimen in some side-entry designs. The combination of a closer approach and larger (e.g. 30 mm^2) crystals enables solid angles of detection of 0.18–0.21 steradians to be achieved in some designs of TEM. The maximum efficiency of detection is necessary in instruments which are to be used for analysis with the smallest electron probes. The VG HB603 FESTEM, for example, which can analyse with sub-nanometre-sized focused beams, can be fitted with two windowless detectors together subtending a total of 0.5 steradian solid angle at the specimen (Lyman *et al.*, 1994).

Both wavelength- and energy-dispersive types of x-ray analyser require a control system and a means of outputting the results.

In WDX the control system is required to vary the orientations of the analysing crystal and detector, and to measure the rate of detection of x-ray photons at all settings. On earlier spectrometers the output was a trace of count-rate against angle 2θ on a chart recorder. Nowadays a graphical output is obtained on a video display monitor, and the peaks are identified automatically from a computer databank.

In EDX the control box is required to provide the necessary bias voltage to the Si(Li) or Ge crystal and the safety circuit which prevents the voltage from being applied to the crystal if it is not adequately cooled. Amplified current pulses from the crystal are analysed for amplitude and sorted into appropriate channels of a display of x-ray counts versus energy. A computer, varying in size and speed between a PC AT or XT and a work station, is usually employed. On computer-controlled microscopes the same computer can be used for EDX, saving the cost of duplicating the facilities.

Si(Li) or Ge

Energy-dispersive spectrometers are widely available with a lithium-drifted silicon detecting crystal, usually 10 mm^2 on a routine SEM and 30 or even 60 mm^2 on TEM and high-resolution SEM. This has reached a highly developed state and earlier drawbacks which restricted spectrum acquisition to a few thousand counts per second have been developed out. Compared to the WDX, the energy-dispersed spectrum consists of very broad peaks with full-width half-maximum (FWHM) of 130–135 eV at best, measured at 1000 cps at the Mn Kα peak of 5.9 keV. At lower energies the peaks narrow to a FWHM of 75 eV at F Kα of 0.677 keV. Higher counting rates and larger-area crystals normally result in higher FWHMs, i.e. poorer resolution, but for analysing spectra the number of counts in a spectrum is more important than the resolution, if a choice has to be made.

The Si(Li) detector is ideally used in the ranges 0–10 or 0–20 keV. All detectable elements (Be to U) have characteristic emissions in the range 0–10; if there are ambiguities due to overlapping peaks these can usually be resolved by measuring in the 0–20 keV range. Above 20 keV, needed to excite the K lines above $Z=44$ (Ru), the higher voltages of the TEM (e.g. $\geqslant 50$ kV) are necessary to excite the lines efficiently. The detection efficiency of Si(Li) falls off in this range because high-energy x-ray photons pass right through the crystal without interaction.

One of the artefacts of semiconductor detectors, the silicon escape peak, occurs at all beam energies above 1.838 keV, the Si K$_{ab}$ energy. Never more than a small percentage of the height of its 'parent' peak, the small escape peak 1.74 keV lower in energy can be mistakenly interpreted as a characteristic peak of another element.

After a number of years of development the manufacturing processes for detectors made of high-purity germanium (Ge or HPGe) have resulted in a detecting crystal for electron microscopes with competitive properties for EDX.

Advantages of Ge: Resolution is improved over Si(Li) to 115 eV at Mn Kα (1000 cps), reducing to 65 eV at F Kα. At 10 000 cps the FWHM broadens out to the same value as the best Si(Li) 1000 cps value, i.e. 133 eV. The narrower peaks enable some previously unresolved by EDX to be shown more precisely, and help with automatic quantitative analysis (Figure 6.8).

The higher density of germanium increases the absorption of higher energy x-radiation, and gives increased detection efficiency for use on TEMs, particularly those operating at intermediate and high voltages.

Disadvantages: Because the K absorption energy of Ge is 11.103 keV there will be no escape peaks associated with x-ray lines below this energy, but they will be possible from $Z=34$ upwards. The Se Kα escape peak at 1.322keV falls very close to the same element's Lα peak at 1.379 keV! Because of the higher energy of Ge Kα photons

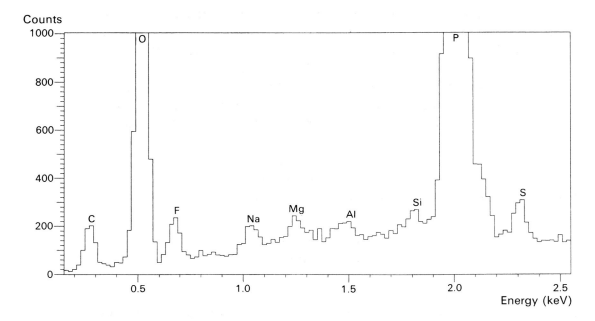

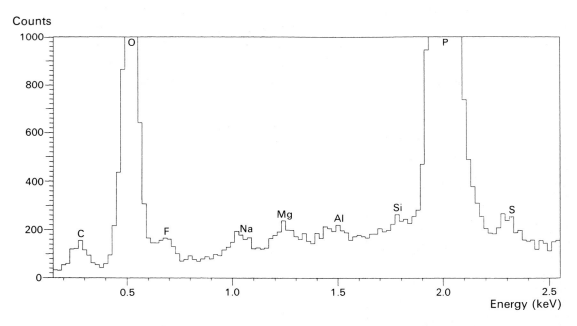

Figure 6.8. Energy-dispersive spectra from calcium phosphate mineral obtained from Ge (*upper*) and Si(Li) (*lower*) detectors. Both detectors fitted with Super Atmospheric Thin Windows extending analysis to light elements. Note the much sharper detection of all minor peaks when using Ge, which would assist in quantitative analysis. (By courtesy of Oxford Instruments Microanalysis Group.)

there is a higher probability of escape from the detector than in the case of Si(Li), and escape peaks from a Ge detector will be more prominent. There will also be a discontinuity in the sensitivity of the detector at the absorption edge 11.103 keV.

It is evident that Ge has an advantage over Si(Li) if higher resolution at higher count-rates are necessary, or the K lines of heavier elements are to be studied in the TEM. The possible implications of higher escape peaks should be borne in mind with germanium, although they will be taken into account by the analytical software with the EDX system.

6.2.2 Practical use of x-ray spectrometers on electron microscopes

There are two main questions which are asked in connection with specimen composition:

(i)   What is this feature or this field of view made of?
(ii) How are individual elements distributed throughout the field of view?

(i) Analysis of specific features

This is the most common use of x-ray microanalysis in electron microscopy, and can be carried out in all instruments to which an x-ray spectrometer can be fitted. In a static TEM image the illumination is focused down to fit the feature (Figure 6.9); in a sequentially scanned image there is a choice between restricting the scan to a selected area or of stopping the spot on a selected point. [As far as the writer is aware, the size of spot on the image tube (or the circle used in some displays) is never scaled

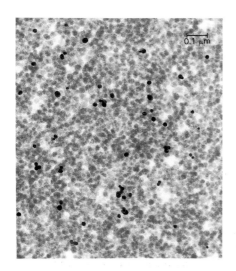
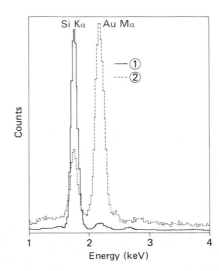

Figure 6.9. Analysis of specific features. General area analysis of a thin section of gold on silica (gold 6% by weight) shows only a very small gold peak (Curve 1). Concentrating the electron beam on a single metal particle clearly shows this to be gold (Curve 2).

to the actual area being analysed on the specimen – a discrepancy which could lead to misinterpretation of the analysis.]

The analysis will provide information on the elements present and can be qualitative or quantitative. For quantitative analysis the geometry of bulk specimens in relation to the spectrometer must be precisely known, ideally by having a flat, polished surface, so that corrections to the measured spectrum for penetration, absorption and fluorescence effects (the so-called ZAF correction) in the specimen can be calculated accurately. In thin transmission specimens the corrections are usually negligible but there can be side-effects due to the excitation of x-radiation from features outside the field of view, by backscattered and stray electrons.

(ii) Elemental distribution

This is primarily a job for a scanning instrument and provides information on the spatial distribution of individual elements within a field of view on the specimen. The information is provided as either a single line-scan or as a distribution 'map'. The technique can be operated using either a WD or an ED spectrometer. In either case the spectrometer is operated so that only x-radiation of a single chosen energy or wavelength is being recorded. This is readily done on a crystal spectrometer; on an EDX system it involves setting a single-element window so that an output pulse is obtained whenever a photon is received with an energy within a selected channel or narrow band of channels situated on the strongest clear peak of the element. The individual pulses are used in distribution mapping; in line-scans they are processed by a ratemeter circuit to give an output voltage proportional to the rate of receipt of pulses.

In a single-element line-scan the electron probe is scanned slowly in a straight line across a chosen region of the specimen. The ratemeter output from the spectrometer is applied to a chart recorder or to the Y-modulated input of the microscope's recording c.r.t. The Y-deflection of the pen or c.r.t. trace indicates the variation of concentration of the chosen element along the path of the probe. By making a double exposure on the recording camera it is possible to superimpose a trace showing the variation of concentration of a chosen element along a line shown on the secondary or backscattered electron image.

An elemental distribution map is produced by scanning a whole field with the spectrometer tuned to a chosen spectral line or energy. Pulses from the detector are used to brighten the c.r.t. spot momentarily, producing bright spots at the appropriate points along the scan as each x-ray photon is received. As the scan proceeds an increase in the concentration of the element being analysed results in a more concentrated rash of speckle on the recorded image. At low count rates several complete frame-scans may have to be recorded in order to put sufficient dots on the record for variations in concentration to be recognised. Unfortunately, there is always a background of dots on the record even where the selected element is totally absent. This comes from the Bremsstrahlung continuum which will contain some radiation of the chosen wavelength. Because of this background it is difficult to distinguish low concentrations of a few percent in a distribution map, particularly with an ED spectrometer. On the other hand, when larger concentrations are involved, a comparison of distribution maps for several different elements can enable much to be learned about local variations in the composition of the specimen. Colour displays enable the distributions of several elements to be seen on a single map.

A 'background-subtracted' distribution map showing more clearly the true variation of elemental concentration over a specimen can be obtained if the local count-rate is digitised and stored at the appropriate address in a computer memory until the whole field has been covered. Before reading out the stored data a number of counts equivalent to the contribution by the continuum is subtracted from each of the picture points. The clarified distribution can be typed out as a set of numbers, plotted on a yes/no basis for presence of the chosen element, or transformed into a set of grey levels or colours representing elemental concentration and displayed or plotted. Examples of line-scans and a distribution map are given in Figure 6.10; other examples were given earlier in Figure 3.25.

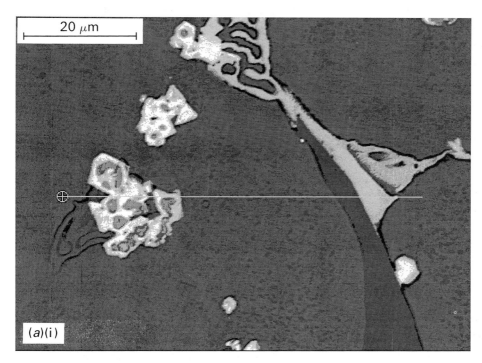

Figure 6.10. Spatial distribution of elements. (*a*) Distribution of tantalum, cobalt and chromium across a polished section of a cobalt-based superalloy. (i) The line of the analysis is marked (*above*) on a backscattered electron image, and (ii) concentrations of individual elements along the line are shown (*overpage*) on separate graphs. The number in the heading of each line-scan is the counts full-scale for that graph. (By courtesy of Oxford Instruments Microanalysis Group.)

Points which must be taken into account in interpreting x-ray spectra and distribution maps from rough-surfaced specimens are:

absorption of x-radiation from the specimen by material (which may not be evident on the electron micrograph) between the emitting area and the detector;

absorption of x-radiation from the specimen and re-emission at a longer wavelength (fluorescence);

emission of characteristic radiation by material outside the picture area (and not necessarily part of the specimen) which is bombarded by electrons backscattered from the specimen.

Figure 6.10 (*cont.*) (*a*) (ii)

Ta Mα, 300

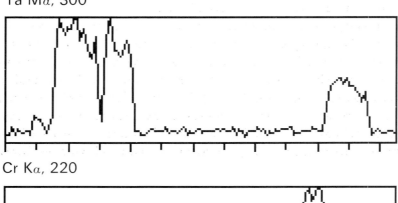

Cr Kα, 220

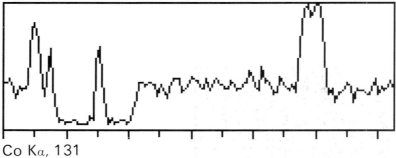

Co Kα, 131

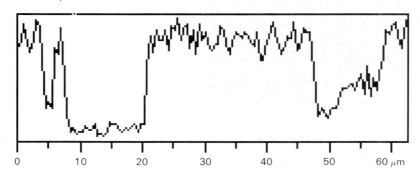

0   10   20   30   40   50   60 μm

An example of the variations which can occur in x-ray spectra for the reasons outlined above is given in Figure 6.11. This shows spectra accumulated by an EDX detector in an SEM when an electron probe was stationed at points on the three faces of the crystal of sodium chloride shown in the micrograph. Although the specimen, which was carbon-coated, contained only Na and Cl two of the analyses show varying proportions of aluminium, silver and iron under unfavourable conditions of detection.

A rough-surfaced specimen can modify the detection of its own emission in a dramatic and misleading way. Some warning of the existence of possible absorptions due to surface roughness can be obtained if the Everhart–Thornley detector and the x-ray detector (both of which have a narrow acceptance angle) are situated close together in the specimen chamber. Using the electron detector in its BSE mode will reveal shadowing effects due to specimen topography. An example of this is shown in Figure 6.12. The two detectors will not be coincident, however, and the analogy between the two radiations should not be carried too far.

A further important practical point concerns the effects of electron scattering within specimens which are being analysed. As explained in Chapter 2 the combined effects of spreading and penetration lead to x-radiation being generated in a volume within the specimen whose effective size is much greater than might be anticipated from the diameter of the original electron probe (Figure 6.13(a)). The actual size of the emitting volume varies with beam energy and the Z number of the specimen. Higher-energy electrons become more effective ionisers after they have been slowed down somewhat by collisions with the specimen, and the x-radiation is generated in a volume which is droplet shaped.

(b)

Figure 6.10 (*cont.*) (*b*) Split-field micrograph with backscattered electron and Cu Kα x-ray distribution images of part of a TEM specimen grid.

Figure 6.13(*b*) is a graph of the spread of aluminium, copper and gold x-ray emissions at different electron energies. It can be seen that if high spatial resolution is required in an analysis (e.g. in an elemental distribution map), the electron energy should be reduced to a value close to the absorption energy of the particular x-ray series which is being studied. Even so, the resolution for x-ray analysis of bulk specimens is never as good as that of the same scanning microscope in the electron emissive mode. Figure 6.13(*c*) shows the effect of electron energy and beam diameter on x-ray spatial resolution for an iron specimen ($Z=26$). A limiting resolution of about 0.25 μm is obtained, i.e. nearly two orders of magnitude larger than the secondary electron resolution.

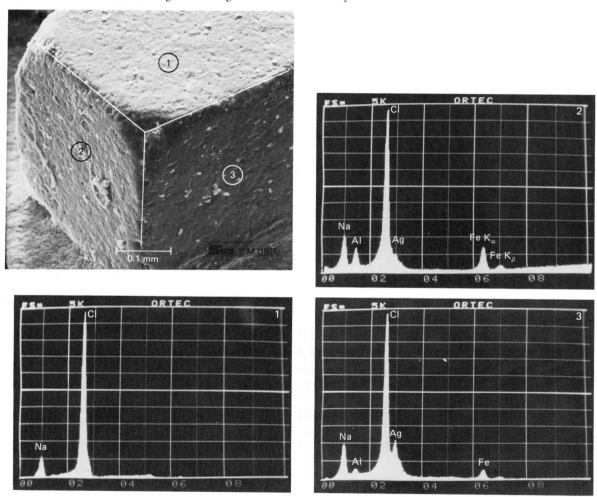

Figure 6.11. Absorption and fluorescence effects in x-ray spectra. X-ray spectra obtained in an SEM from three faces of a cube of sodium chloride under a static electron probe (30 kV, specimen tilted towards the electron collector). Absorption of x-radiation from faces ① and ② results in a lower peak height relative to the background radiation from the aluminium stub, the silver paste holding the cube to the stub, and from the iron pole-piece of the final lens. (The latter peak should be removable by more effective collimation of the detector.)

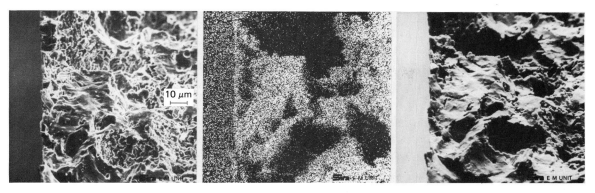

Figure 6.12. Scanning micrographs of the fracture surface of a magnesium–aluminium–silicon ceramic in electron emissive (*left*), Al Kα x-ray distribution (*centre*) and backscattered electron (*right*) modes. It can be seen that the elemental distribution map is strongly biased by surface roughness effects, which are not obvious in the secondary electron image alone.

Figure 6.13. Spatial resolution of x-ray analysis with focused electron beam. (*a*) The spreading of an electron beam within a target material results in the emission of x-radiation from a much larger volume than merely a cylinder under the electron spot, except when the target is thin compared to the electron range. (*b*) Curves relating the effective x-ray emitting area in a specimen to the electron beam energy, for elements of low, medium and high atomic numbers. (*c*) Effect of electron energy and probe diameter on x-ray spatial resolution in bulk iron. ((*b*) by courtesy of Kevex Corporation, after Beaman & Isasi, 1972; (*c*) by courtesy of Edax International.)

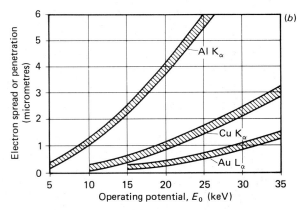

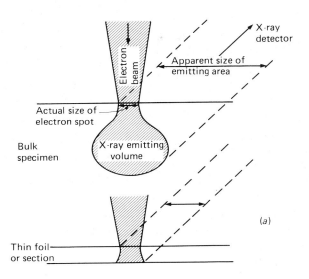

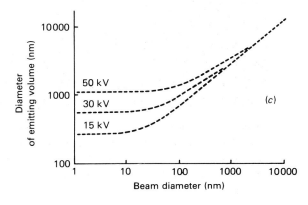

Analysis of thin specimens

Electron beam spread in thin specimens, e.g. ultramicrotomed sections, is less than in bulk material, since the bulbous bottom of the droplet shape is chopped off, leaving the narrower neck (Figure 6.13(a)). This result is useful in the analysis of thin foils and biological thin sections, when x-ray spatial resolutions of less than the specimen thickness can be obtained – even smaller if the specimen has been frozen and dried.

The calculated broadening (Newbury & Myklebust, 1979) of a beam of 100 keV electrons due to passage through specimens of increasing thickness is given in Table 6.1. Values given are the broadening of the beam, in nanometres, at the stated film thicknesses.

The lower limit of detectability of materials by electron-stimulated x-radiation in electron microscopes will depend on whether they are analysed in bulk or in thin sections. However, on the conservative estimate of 1% concentration being detectable the minimum mass would lie between 10^{-14} and 10^{-20} g.

Trace-element detection (XRF) in the SEM

One of the factors limiting the sensitivity of detection by EDX of small quantities of matter is the presence in all spectra of a background of *Bremsstrahlung*. This is inevitable when the characteristic x-radiation is excited by electron bombardment, but is absent when x-ray emission is stimulated by absorption of higher-energy x-radiation, i.e. by the process of x-ray fluorescence. Using this principle the detection sensitivity is increased by more than two orders of magnitude, at the expense of a loss of spatial resolution.

A thin metallic foil placed above the sample in an SEM will emit its characteristic x-radiation under electron bombardment. This radiation will in turn excite background-free fluorescent radiation from the specimen immediately adjacent to it. Figure 6.14(a) shows the construction of a foil-holder designed to be mounted on the end of an EDX detector. When the holder is wound into position above the specimen, the precise localisation of the analysed area of specimen will be unknown (within about 300 μm), but the sensitivity of detection of small concentrations is increased up to 1000-fold. The analysis can be optimised for a specific element if the material of the foil is chosen with a characteristic emission peak just above an absorption edge of the chosen element. This is illustrated (Figure 6.14(b)) by two spectra from a standard glass sample, irradiated by x-radiation from Mo and Ge targets. To achieve enhanced detection of a wide range of elements it is necessary to use several different target materials. Rh, Mo, Ge can be used for higher Z-, and Ti, Al for light elements, with correspondingly lower electron energies for excitation (e.g. 15 keV). These last foils will be thinner, too, and it is advisable to scan the electron spot to reduce the risk of melting them. The targets have to be thick enough to absorb all the electron beam, but thin enough to transmit a high proportion of K radiation from the target and the high-energy continuum.

Commercial attachments are available for carrying out trace element detection on SEMs. It is most important to check that the EDX detector is screened from the exciting x-radiation whilst receiving a maximum signal from the specimen. (Pozsgai 1991).

Quantitative analysis in the electron microscope

General principles: We have seen how it is possible on electron microscopes to determine (within limits imposed by the type of spectrometer used) which elements are present in a selected small volume of a specimen. In many cases this knowledge may be sufficient for the purpose of the study. However, it may be desired to know how much of each element is present – either roughly, as in a semi-quantitative analysis, or within a percent or so in a quantitative analysis.

For a quantitative result, x-ray intensities are measured for each line in the emission spectrum, and the composition of the sample is deduced from a knowledge of relative emission efficiencies under the conditions in which the spectrum is acquired.

Bulk specimens: In bulk specimens the x-radiation may be generated well below the surface of the specimen, and may have to pass through a considerable thickness of specimen before being emitted in the direction of the detector. The x-ray signal will be diminished by absorption and added to by fluorescence, so the intensities of the measured spectrum must be adjusted (corrected) to give the x-ray spectrum actually generated within the specimen. The length of path through the specimen depends on the x-ray 'take-off angle', and this must be known accurately. Geometrical effects are most easily calculated if the surface of the specimen is flat and polished; this is a prerequisite for the most accurate quantitative analysis, as in the dedicated electron probe x-ray microanalyser to be described later. The 'ZAF-correction' adjusts the actual spectrum for atomic number, absorption and fluorescence effects; it is applied iteratively, since before the corrections can be calculated the specimen composition must be known at least approximately.

When the intensities have been determined for each element in the sample these are related to amounts of material either by use of standard reference specimens of known composition, which are put under the same beam under the same conditions, or by use of calculated sensitivity factors.

Thin specimens: The absorption and fluorescence corrections for transmission specimens are very much reduced or negligible, because of the thin-ness of the specimen. On the other hand, perfectly parallel-sided specimens of known thickness are rare, and higher peak intensity could be due either to a higher concentration of the element or a thicker region of the specimen.

Accordingly in materials science TEM specimens, for which specimen composition in terms of weight or atomic percentage of each element is usually required, the elements are considered in pairs.

Table 6.1. *Electron beam spreading in thin foils*

Element	Film thickness (nm)				
	10	50	100	300	500
Carbon	0.22	1.9	4.1	16	33
Aluminium	0.41	3.0	7.6	30	66.4
Copper	0.78	5.8	17.5	97	244
Gold	1.71	15.0	52.2	599	1725

(*a*)

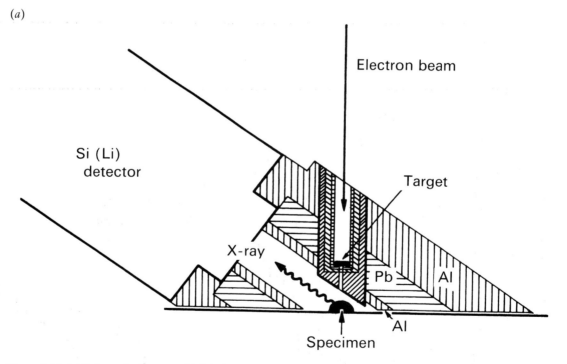

Figure 6.14. (*a*) Schematic diagram of foil-holder for XRF analysis in the SEM. The unit is designed to be mounted on the end of an EDX detector, and screens the detecting crystal from all x-radiation except from the SEM specimen beneath the foil. (*b*) Comparative spectra from NIST standard reference glass 612 using (*i*) Mo K radiation from a 100 μm foil at 39 keV and (*ii*) Ge K radiation from a 70 μm foil at 25 keV. Numbers on some peaks are the parts per million concentration of that element in the specimen. ((*a*) reproduced, with permission, from Pozsgai, 1982; *b*(*i*) from Pozsgai, 1985; and *b*(*ii*) from Pozsgai, 1993.)

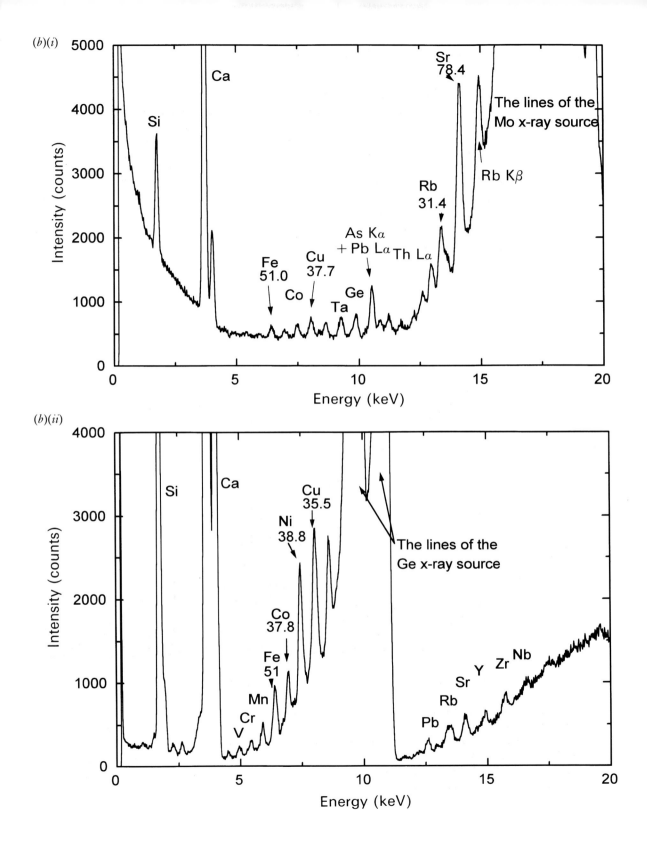

For elements A, B for which the peak intensities are I_A, I_B, respectively and their concentrations are C_A, C_B, the ratio of concentrations $C_A/C_B = k_{AB} \times I_A/I_B$, where k_{AB} is a constant for that pair of elements at the voltage of the analysis.

Similarly for the pair of elements A, C,

$$C_A/C_C = k_{AC} \times I_A/I_C$$

If A, B, C are the only components of the specimen, then $C_A + C_B + C_C = 1$.

If the elements are known to be in the form of compounds, e.g. oxide, carbonate, etc., for which the other elements are not measured, appropriate adjustment are made to the numbers. The k-factors are unique to a given TEM/EDX system operated at a particular beam energy. They can be determined by the analysis of standard specimens of known elemental proportions. They may also be calculated from theory (Cliff & Lorimer, 1975).

In biological analysis the specimen is normally composed of elements C, H, O, N which are not readily determined, and what is required is the amount of another element, e.g. Ca, per unit mass of the whole, for the analysed volume of the specimen. Use is made of the fact that the intensity W of the x-ray continuum in the spectrum is proportional to the total mass M_t of the analysed volume, whilst the peak intensity I_x of the chosen element x (e.g. Ca) is proportional to the mass M_x which is present in the analysed volume. The mass fraction C_x of the element is given by

$$C_x = k \times I_x/W,$$

where k depends on experimental factors, including gun voltage.

The subject of quantitative analysis of biological materials is a complex one. Factors which have to be considered in any particular analysis include the nature and concentration of standards, the stability of the specimen during analysis and, in hydrated specimens, the degree of hydration of the specimen (Hall, 1979).

In any TEM there is the risk that the x-ray spectrum contains a contribution from the microscope itself. X-radiation generated by the impact of stray electrons on components of the column and specimen stage can be minimised by using appropriate apertures in the condenser system, using a beryllium specimen holder and keeping the column clean. A well-designed collimator over the end of the detector should protect the latter from x-rays and electrons from sources outside the specimen area. A very real problem is due to the specimen itself, which causes electron backscatter onto surrounding metalwork. The often-recommended 'hole test' cannot simulate this source of extraneous x-ray signal.

The reader can find out more about all methods of quantitative analysis in electron beam instruments from such publications as Reed (1993), Scott *et al.*, (1994), and Hall (1979).

Problems specific to biological microanalysis

The elemental analysis of biological material presents a number of difficulties. What is required is a knowledge of composition whilst the subject is living. The analysis can only provide a statement of what is in the specimen after living processes have ceased and the specimen has been prepared for the electron microscope. For morphological examination this involves the addition of stabilising chemicals, e.g. glutaraldehyde or osmium tetroxide, and the removal of water or its substitution by another substance, e.g. an embedding resin. In addition, contrast in biological tissues may be so poor that heavy-metal stains have to be added in order that the different components can be identified.

It can be seen that cryo-preparation provides a ray of hope for biological microanalysis. Not only is there the possibility of freezing soluble salts *in situ*, but no additional chemicals are involved, unless a cryoprotectant such as glycerol is used to restrict the size of ice crystals. As already discussed in Chapter 4 ice crystal formation should be minimised, as dissolved salts can be swept along by the advancing ice crystal formation. The fastest possible cooling rates must be achieved to avoid redistribution of dissolved salts. There is the advantage that fixation by quick freezing can be almost instantaneous compared to the protracted process of chemical fixation. It is possible, therefore, to study changes in composition caused by physiological processes which can literally be 'frozen' at different stages of the process.

Having obtained the frozen specimen with, ideally, vitreous ice, at −170 °C or thereabouts, it is necessary to prepare it in a suitable form for analysis. The best procedure is to cut frozen sections and examine them in a cryo-specimen holder in a suitably equipped TEM.

Alternatively the cryo-sections may be freeze-dried or cryo-substituted, hardened and sectioned. In either case there is a possibility of losing or redistributing any dissolved salts, but the treatment has a much stronger chance of success than a non-cryogenic route.

Readers desiring more detailed information should consult Echlin (1992) and/or several books on biological x-ray microanalysis in the *Suggested further reading* list on page 301.

Automatic control of microanalysis

X-ray microanalysis is a field in which a number of routine operations can be controlled by microprocessors. The modern wavelength-dispersive spectrometer can be operated by a computer. The energy-dispersive spectrometer is also computer-based, and accumulation, manipulation and analysis of spectra can be performed automatically. Specimen stages on the SEM and dedicated electron probe microanalyser can be operated electrically via stepping motors. Hence, routine quantitative analysis of ideal specimens on these instruments can be left to the computer to carry out. Examinations of a research nature, however, and analysis on the TEM, require the observation and curiosity of a trained operator and automation is suitably restricted to setting up the spectrometers and manipulating the data. One potential source of trouble in automated analysis is a slow drift of the specimen or the beam so that the area or point analysed is continuously changing. Even this can be compensated for by having the computer interrupt the analysis periodically and acquire a morphological image which is compared with that at the start of the analysis. If the image has moved (drifted) because of specimen or instrumental instability a correcting shift is applied to the scan coils before the analysis is resumed (Vale & Statham, 1986).

6.2.3 *The dedicated electron probe x-ray microanalyser (EPMA)*

This is an analytical SEM which has been optimised for the rapid non-destructive quantitative elemental analysis of solid materials. As a 'dedicated' specialist instrument it has more comprehensive facilities for carrying out continuous, often automated, elemental analysis than would be associated with a general attachment to an electron microscope. It does not require a spatial resolution better than a micrometre and is much used in metallurgy and geology for analysing second phases, grain boundary precipitates, etc. It is only non-destructive in so far as the material analysed is not destroyed, but material may be lost in preparing the flat, polished, surface which is needed for an accurate analysis. Also, since the analysis has a resolution in depth of a few micrometres it will be a bulk analysis only if the composition of the surface is representative of the bulk. The ultimate detection limit can be lower than 1 part in 10^4, but depends very much on the element(s) concerned and the type of spectrometer used.

The commercial electron probe x-ray microanalyser is now an instrument consisting of a three-lens column (Figure 6.15) which puts rather more current into a larger focused spot than the scanning microscope. Constancy of beam current is essential over a long period, and detection and feedback circuits ensure constant current within 1% over 12 hours. The specimen is placed at the focus of two or more (up to twelve in one design) crystal spectrometers, which allow intensity measurement to be made of at least two wavelengths simultaneously from elements between boron ($Z=5$) and uranium ($Z=92$). An energy-dispersive x-ray (EDX) detector may also be fitted, with the same x-ray take-off angle as the wavelength-dispersive spectrometers. This can give a simultaneous qualitative or quantitative analysis of elements from sodium ($Z=11$) upwards, which may be adequate for the purpose in hand, without requiring the lower detection limits and light-element capabilities of the crystal spectrometers. Secondary and backscattered electron imaging of the analysed field is also provided, the A + B operation of a split backscattered detector (Chapter 5, page 223) being particularly useful in showing phase differences which can then be analysed quantitatively. Unusually for an electron beam instrument the

EPMA is also fitted with a fixed focus metallurgical light microscope. This has a very shallow depth of field and it is arranged that when the specimen is in sharp focus for the light microscope it is also at the focus of the crystal spectrometer. For convenience of viewing, the image from the light microscope may be captured by a TV camera and displayed on a video monitor.

The specimen stage, which can accommodate several mounted specimens and blocks of calibration standards, can be positioned electrically by remote control, so that analyses can be performed on a number of predetermined areas of the specimens. Using a computer (which may be part of the EDX or microscope-operating system) and appropriate interfaces, the spectrometers and counting equipment may be programmed so that the specimen stage will move to a predetermined point which

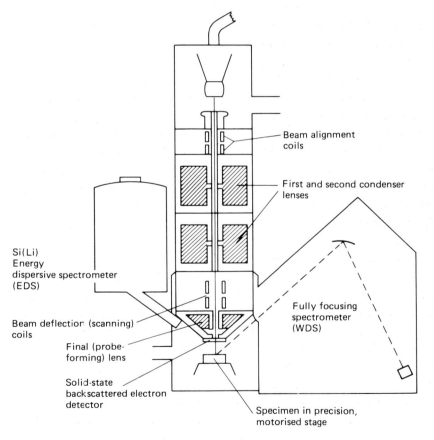

Figure 6.15. Schematic diagram showing the essential parts of a dedicated electron probe microanalyser (EPMA). Although only one wavelength-dispersive spectrometer is shown, in vertical orientation, multiple spectrometers or monochromators may be fitted around the column. Also fitted, but not shown, will be a wide-aperture light microscope whose plane of focus is also the focus of the wavelength-dispersive spectrometer(s). The microanalyser will have an electron gun operating to at least 30 kV, with optional 40 or 50 kV.

will be analysed before moving to the next point and repeating the procedure. Hours of continuous unattended analysis can be carried out, so that the operator is presented with a printed summary of the analyses. The output may be either a series of point analyses, or elemental distribution maps. Several elements may be superimposed on a single map if each one is printed in a different colour.

The maximum sensitivity of a crystal spectrometer is only attained when the source of x-radiation lies at the focus of the spectrometer. Ideally, therefore, only flat, polished specimens should be used. These are necessary in any case if the ZAF corrections are to be accurate. Quantitative analyses accurate to 1% or less may be made (Love & Scott, 1981). Specimens for EPMA are typically 25 mm in diameter and 10 mm thick, prepared by mounting powders, pieces of mineral or blocks of metal in a suitable resin and then polishing an end-face flat with polishing papers or diamond paste. Insulating specimens will be thinly coated with carbon to prevent specimen charging during analysis, which would cause the analysing beam to drift away from the chosen spot.

The analysis may be made more tolerant of specimen roughness by using an inclined crystal spectrometer (Appendix A3) on the EPMA column, but this is not a popular arrangement as a single inclined spectrometer takes up as much room around the column as several vertical spectrometers, and may have a poorer detection sensitivity because it uses smaller crystals.

6.3 Electron energy loss spectroscopy (ELS or EELS)

When an electron beam traverses a specimen in the TEM, STEM or FESTEM for every characteristic K-shell x-ray photon generated by inelastic collision there will be an electron which has lost the characteristic energy $E \, K_{ab}$. There will be, similarly, characteristic energy losses corresponding to L, M absorption edges of the specimen. If the energy losses suffered by the beam in passing through a localised region of the specimen are measured it should be possible to infer the specimen composition. This is the principle behind electron energy loss spectroscopy, a technique confined to relatively thin transmission specimens and most usefully applied to light elements, e.g sodium and lower atomic number elements, which are more difficult to analyse by other means. 'Relatively thin' means thin enough for electrons to have suffered only one inelastic collision in passing through them; ideally less than 30–50 nm for biological material at 100 kV and thinner for denser specimens. The efficiency of collection of transmitted electrons can be much higher than that of x-radiation but the signal-to-background ratio is much poorer.

The electron energy distribution in an image field is also used in the technique of Electron Spectroscopic Imaging (ESI). In this case the energy spectrometer is used as an energy filter, and an image can be formed from electrons which have lost only a chosen amount of energy.

Spectrometers for EELS may be accessories fitted at the bottom of the column of a conventional TEM, or after the specimen in a FESTEM. Alternatively, they may be built into the column of an energy-filtering TEM as described in Chapter 3. All of these separate electrons of different energies by passing them through a magnetic field. Slower moving electrons (those which have suffered most energy loss) are deflected more than those which have suffered zero energy loss. Electrons of a specific energy may be selected by means of an appropriately positioned slit and pass to the detection system (Figure 6.16 (*upper*)). The spectrometer is 'tuned' to pass different energies by varying the magnetic field. Imaging electrons are selected for

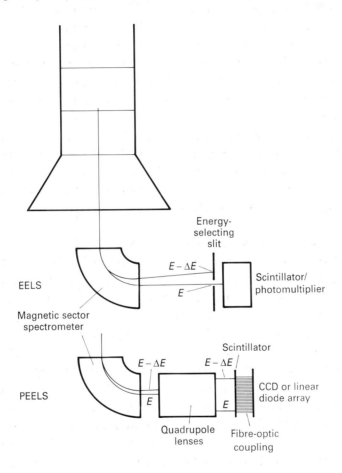

Figure 6.16. EELS spectrometers as accessories to TEM. (*Upper*) Serial EELS. Magnetic sector disperses electrons from a chosen area of the image and a narrow slit transmits electrons of a narrow range of energies only, to be detected and counted by a scintillator/photomultiplier combination. The energy range is covered by progressively varying the magnetic field of the spectrometer. (*Lower*) Parallel EELS, PEELS. The initial dispersion by the magnetic sector is about 2 μm per electron volt energy change. This is increased by a further lens system before falling on a row, or rows, of detectors whose outputs form up to 1000 channels of a PEELS spectrum. The individual detecting elements may physically be 20–25 μm wide.

energy analysis by moving a chosen feature in the TEM image over the spectrometer entrance slit, or in the STEM by stopping the probe over the selected area.

An energy spectrometer produces a spectrum of the energy distribution in the transmitted electrons relative to the primary beam energy. A typical distribution follows the pattern of Figure 6.17. The majority of the electrons in a spectrum are found in the initial 'zero-loss' peak and in plasmon loss peaks between about 5 and 30 eV, involving interactions with valence or conduction electrons. Beyond this a smoothly falling background has superimposed on it the 'ionisation edges' of atoms whose x-ray absorption energies are reached. These are the peaks used for analysis. Those in the range up to about 5 keV are used for analysis; for light elements the range is up to 2 keV. Ahn & Krivanek (1983) have prepared an 'atlas' of typical peaks.

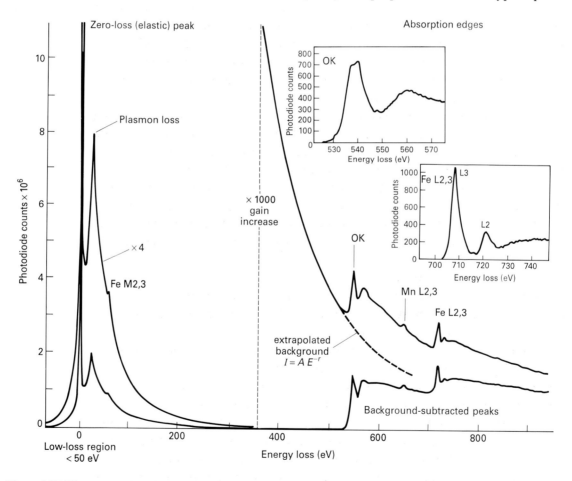

Figure 6.17. Electron energy loss spectrum from a mineral. Characteristic peaks for zero-loss (elastically scattered) electrons and plasmon interactions are followed by a falling curve on which are superimposed absorption edges of individual elements. Extracted edges for oxygen and iron, after subtraction of the background, are shown as insets. (Original spectra by courtesy of C. Colliex.)

The spectra may be analysed quantitatively (Leapman, 1992). Estimation of a peak intensity (i.e. the area included under a peak of a chosen eV width) depends on correctly predicting the shape of the falling background in the absence of the absorption edge. Empirically this has been found to take the form $I = AE^{-r}$, where I is the intensity at energy loss E, and A, r are constants for a particular spectrometer; other, less empirical, models are being pursued. This is the stage at which quantitative analysis is subject to the greatest uncertainties. A computer performs a curve-fitting routine and, as with all spectral computations, statistical fluctuations (noise) have a strong influence on the final result. Long counting times and smooth curves are thus necessary for the highest precision of analysis.

Elemental distribution mapping is possible using electron spectroscopic imaging. This involves recording energy-filtered micrographs at energies above and below the chosen absorption edge and subtracting one from the other. This can be done photographically, but it can be done more quickly and precisely if the background is modelled from digitised images stored in memory. An illustration of the complete process to map the Fe distribution in a liver section 30 nm thick is given in Figure 6.18.

Each absorption edge has its own characteristic fine structure. Information on chemical bonding, molecular structure and dielectric constant may be obtained from a detailed study of energy loss curves. This type of information is contained in peak displacements of a few eV (see also ESCA, Chapter 7); whether this can be resolved on a particular microscope depends on the design of the spectrometer and on the energy spread of the primary electron beam. This is wider (1.5 eV) from a thermionic tungsten filament than from a cold field emitter (0.2–0.4 eV). An energy spectrometer of high resolution is therefore a natural component for the FESTEM, in which elemental analysis by EELS is possible down to a spatial resolution of about 1 nm.

6.3.1 Parallel EELS (PEELS)

The magnetic sector energy analyser used in EELS is tuned-in for one energy loss value at a time, as in the crystal spectrometer in WDX. Hence, to scan a complete energy spectrum requires a time $N \times D$, where N is the number of channels and D the dwell time in each. If, instead of scanning the electron energy spectrum over a slit with a single detector behind it, a row of N individual detectors is used, the whole spectrum can be measured simultaneously, and a much faster analysis is possible. This requires the manufacture of a row of identical detectors with minimal deadspace between them, and is achieved by the same solid-state technology which produces CCD chips for video cameras. The resulting technique is Parallel EELS or PEELS (Figure 6.16 (*lower*)), which has become an accepted variant of EELS (Wilson, 1993). Additional energy dispersion is provided by multiple quadrupole lenses, to stretch out the energy spectrum over the detector array.

The PEELS approach represents not only a saving in time but a more efficient utilisation of the available signal, from below 0.05% to more than 10%, with a 1024 channel detector array; specimen damage from electron bombardment and contamination are correspondingly reduced. It is necessary to make allowance for any lack of uniformity of output from one detector element to another, and computer routines have been developed to check on this effect. Elemental distribution maps of trace elements can be obtained from parallel EELS (Balossier *et al.*, 1991).

An electron suffering a second (or third, etc.) inelastic collision in passing through a specimen detracts from the absorption-edge signal and adds to the uninformative background. Specimens for EELS must therefore be thin enough for single scatter-

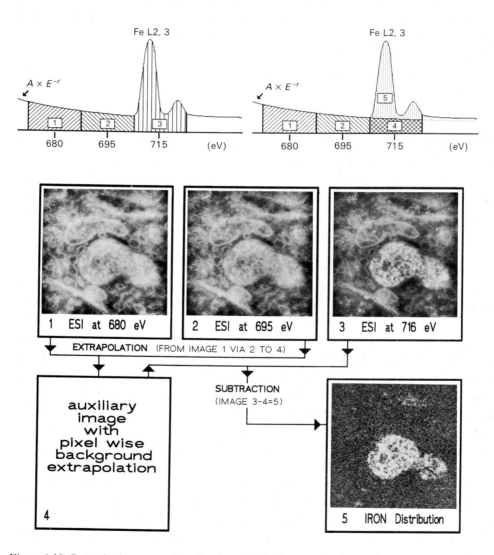

Figure 6.18. Stages in the preparation of an iron distribution map in ultra-thin liver section by ESI (By courtesy of LEO Electron Microscopy Ltd.)

ing only to occur. This thickness may be up to about 50 nm for biological tissues, but quite a lot less for a materials specimen. The criterion for single scattering only to have occurred is sometimes taken to be that the ratio of the intensities of the zero loss to the first plasmon peak should be greater than 5.

This situation is improved at higher beam energies, and more distinct spectra of energy loss peaks are obtainable on Intermediate Voltage TEMs.

Spectrum imaging

A useful technique which permits detailed and varied analysis of a selected field of a specimen to be carried out away from the microscope involves recording a full EELS spectrum from each pixel of a digitally scanned area. The procedure can require a very large amount of computer storage, e.g. up to a GB (10^9 bytes), but enables different analytical procedures to be tried and compared using the same set of data (Jeanguillaume & Colliex, 1989).

Effect of electron energy

The mean free path for plasmon excitation and for inner shell ionisation are increased in higher-voltage operation. A factor of three between 100 kV and 1 MV has been found, and significant improvement would be anticipated for intermediate voltages. The visibility of absorption edges (jump factor) is improved in higher-voltage microscopes, and at 1 MV detectability has been extended out to the germanium K-edge at 11 500 eV (Jouffrey *et al.*, 1980).

6.4 Auger electron spectroscopy

The emission of Auger electrons in addition to x-radiation provides a further means of elemental analysis, as mentioned in Chapter 2. The Auger electron is a result of a secondary process (illustrated in Figure 2.14) following the initial ejection of an electron from an inner shell by incident electron bombardment. The K vacancy is filled by an L electron but the K radiation resulting from this is absorbed in ejecting a further electron from the L shell; this is the Auger electron and has an energy characteristic of the atom from which it came. The two vacancies in the L shell are filled by electrons from higher bands, with emission of L and possibly M x-radiation. The energy of the Auger electron can be measured by means of a spectrometer such as a cylindrical mirror analyser; elemental analysis down to $Z=2$ is possible from the energy spectrum.

Auger electrons are excited from the same droplet-shaped volumes as the x-radiation we considered earlier, but only those which originate within one mean free path below the surface, i.e. are emitted without energy loss, are useful for analysis. Auger electron spectroscopy therefore only analyses the outermost few nanometres of a surface, and the analysis must be carried out in ultra-high vacuum to minimise the thickness of gas layer adsorbed on the surface.

The general efficiency of Auger electron emission is poor, and high current densities are necessary to obtain a sufficient signal-to-noise ratio from all but the light elements. This is the region where conventional x-ray analytical processes are difficult to operate because of absorption effects, and hence Auger electron spectroscopy can be used to supplement or even supplant more conventional microanalytical methods. If the intensity of the Auger emission is made large enough by increasing the beam current (but at the same time worsening the spatial resolution compared to secondary electron emission images unless a field emission source can be used) it is possible to produce scanning Auger images (Scanning Auger Microscopy, SAM) showing the distribution of particular elements, in a similar manner to elemental x-ray distribution maps already described. Insulating specimens present difficulties since charge-dispersing surface coatings are impracticable with such a surface-specific analytical technique.

This is not a technique which can be easily added to a conventional electron microscope, although Bleeker & Kruit (1991) showed the feasibility of a technique for extracting Auger electrons from a commercial TEM. Most reports of analysis using the technique originate from dedicated surface science equipment based on a bakeable stainless steel ultra-high vacuum system and a field emission electron gun.

6.5 Cathodoluminescence (CL) imaging

Much can be found out about the band structure of materials, phosphors, geological and even biological samples by observing and measuring light emission under electron bombardment. This can be studied using a special stage under a light microscope, retaining the natural colours of the luminescence emission, and also in electron microscopes with higher spatial resolution. CL imaging and analysis can be carried out in analytical TEMs (e.g. Yamamoto *et al.*, 1984), but is more often carried out in SEMs, where there is more space to mount efficient detection systems around the specimen.

Since the excited volume and light intensity are small, it is necessary to have an efficient light collection system for CL studies. An ellipsoidal mirror over and around the specimen, with a hole for admission of the electron beam, reflects light to a photomultiplier, which converts light photons into electrical pulses which may be used for imaging on a display tube.

CL emissions can vary both in intensity and in wavelength (colour). The simplest detection system collects all the light and gives an image of the panchromatic brightness variation over the specimen. More detailed information is obtained if the CL emission is analysed by passing it through a monochromator or colour filters before detection. A still more useful system analyses the CL emission into red, green and blue (RGB) components which are detected separately. The amplified signals are recombined on a colour display, so that the operator can examine the colour variation of CL as though he had a direct view of the specimen in the specimen chamber (Saparin & Obyden, 1988).

Special detectors can be fitted to extend the detection of CL into the infrared. A total range of 0.16–2 μm can be analysed and imaged with a commercial grating monochromator.

CL emissions and their colour can be helpful in showing the presence and location of different phases or impurities in many types of crystalline materials, and the technique is widely used in one or other of the forms outlined above (Figure 6.19). Specimen cooling can considerably increase the intensity and sharpness of emission peaks, and liquid-nitrogen- or helium-cooled specimen stages are available.

6.6 Suggested further reading

Useful introductory articles on EELS are written by Joy (1979), Ferrier (1977) and Budd & Goodhew (1988). An atlas of EELS peaks has been compiled by Ahn & Krivanek (1983).

The progress of AES on electron microscopes can be followed through the publications of Venables and co-workers, e.g. Venables *et al.* (1994). Another description of XRF in the SEM is by van Riessen & Terry (1982). A review of EPMA is given by Mackenzie (1993).

The biologist wishing to use x-ray microanalysis will find much useful information in Meek & Elder (1977), Roomans & Shelburne (1983), Gupta & Roomans (1993), and in Sigee *et al.* (1993); also other analytical techniques in Hawkes & Valdrè (1990).

Interpretation of electron diffraction patterns will be helped by several books, e.g. Andrews, Dyson & Keown (1971), Alderson & Halliday in Kay (1965), Misell & Brown (1987). For a detailed discussion of electron diffraction processes see the two volumes edited by Cowley (1992, 1993). Microdiffraction is discussed in detail in Spence & Zuo (1992), whilst four books of examples of convergent beam diffraction

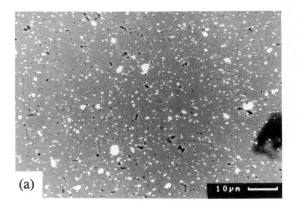 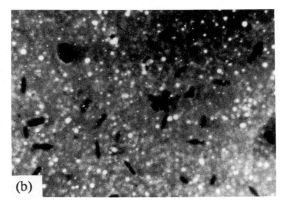

Figure 6 19. Comparison of (*a*) backscattered electron and (*b*) panchromatic CL images of an alumina + 4% zirconia ceramic. The CL image picks out those particles of the zirconia additive which are monoclinic, and also reveals the presence of a dark impurity shown by EDX to be magnesium silicate. (From Page, 1993)

patterns are Mansfield (1984), Tanaka & Terauchi (1985) and Tanaka *et al.* (1988), (1994).

A book on the production and interpretation of electron backscatter patterns is Dingley *et al.* (1994).

A number of analytical techniques in electron microscopy are described by Loretto (1993) and Joy *et al.* (1986).

The *Proceedings of the 13th Pfefferkorn Conference* (1994) are devoted to the topic of cathodoluminescence. The subject is also dealt with in Yakobi & Holt (1990).

Specialised EM- and other microscopical and analytical techniques

7.1 Dark-field electron microscopy

In a transmission microscope the bright-field image is formed by electrons which have traversed the specimen without significant change in direction. Darker areas show where electrons have been deflected out of the beam and intercepted by the objective aperture, by atomic scattering if the specimen is amorphous or by Bragg reflections from a crystal.

It is often more informative to form the image using the deflected electrons rather than the transmitted beam, since one is then using only electrons which have actually interacted with the specimen. Examination modes based on this principle are known as *dark-field* modes, since the final screen is dark unless there is a specimen present to deflect electrons into the image (see, for example, Figure 7.1). The principle is the same as that used in dark-ground light microscopy and is similarly useful

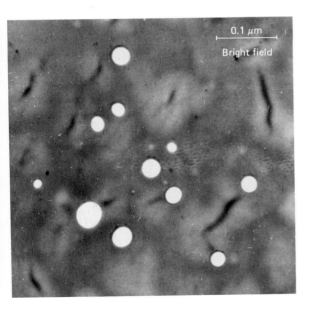
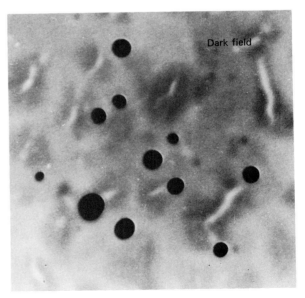

Figure 7.1. A *holey carbon film* imaged in bright- and dark-field. Since one image is formed with electrons missing from the other one they are negatives of one another. The effect is most marked in the holes; when there is no scattering the dark-field image is completely black.

in showing up low-contrast detail more strongly. There are several ways in which dark-field observations can be made.

7.1.1 Displaced objective aperture

In normal bright-field microscopy the objective or contrast diaphragm is placed symmetrically about the axis of the objective lens and scattered electrons are absorbed by its periphery. If the diaphragm is displaced sideways it will absorb the direct beam and transmit electrons scattered or diffracted in its direction, so that the final image is formed only from these electrons (Figure 7.2(*a*)). If the specimen is crystalline the diaphragm may be moved so as to transmit only the electrons from a specific diffraction ring or spot so that areas of specimen with a particular crystallographic orientation will be highlighted in the image. If a very precise selection of scattering angle is desired a small-diameter diaphragm is used. Since the diaphragm is normally situated close to the back focal plane of the objective lens its position relative to the diffraction pattern is easily seen by switching to *diffraction* operation of the microscope. This mode of operation is very easy but has the disadvantage that off-axis rays are used for imaging and image quality is adversely affected by both spherical and chromatic aberrations in the lenses. Fine detail can appear streaked, particularly if the chosen diffracted beams are far distant from the central spot.

7.1.2 Tilted-beam illumination

In this mode, alternatively called '*high-resolution dark-field*', to distinguish it from the method above, the objective aperture is kept centred on the optical axis and the direct beam is moved off-axis by tilting the illumination (Figure 7.2(*b*)). The process of tilting the beam can be followed in the diffraction mode so that electrons in a particular diffraction spot can be made to pass down the axis of the microscope. Because the imaging is then symmetrical about the microscope axis a high-quality dark-field image results from this method.

The general effect of dark-field operation using a selected diffraction spot or direction is similar whether the method used is displacement of the aperture or tilting the illumination, but fine structure is better imaged by the latter method, as can be seen in the micrographs of Figure 7.2(*c*).

Both of the above techniques are suitable for crystalline specimens in which the diffracted electrons are strongly concentrated in the direction of the objective aperture. In amorphous specimens, however, the objective aperture only transmits a small fraction of the illumination scattered by the specimen and image brightness is therefore low, necessitating long exposures to obtain a useful image density in the micrograph.

Conical scan dark-field

When the object of dark-field microscopy is to reveal the distribution and size of a polycrystalline second phase in a composite material the simple techniques of tilted-beam or displaced-aperture can give misleading results because of the small proportion of the diffracting crystallites which actually appear in the final image. The magnitude of the discrepancy can be appreciated by comparing the diffraction pattern in Figure 7.2(c) with the indicated size of the displaced objective diaphragm.

An elegant and effective solution to this situation is provided where the beam tilt coils can be programmed to give the desired illumination tilt $\theta°$ from the optic axis and continuously rotate the plane of tilt around the axis, i.e. to illuminate the specimen with a hollow cone of angle 2θ. All the appropriately orientated crystals in the field of view will now diffract along the optic axis and contribute to the high-resolution dark-field image. The result of such an operation is shown in Figure 7.3. An alternative solution using the STEM mode is described later.

7.1.3 Strioscopy

The technique of hollow cone illumination is also used in this mode of dark-field microscopy. The final condenser lens is critically focused in the plane of the specimen and contains a special annular diaphragm so that the illuminating spot is in fact the apex of a hollow cone of electrons travelling at an angle to the optic axis. In the

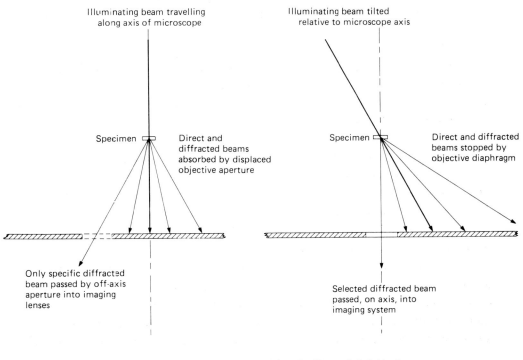

(a) Dark field microscopy by displaced
objective aperture

(b) Tilted-beam dark field microscopy

Figure 7.2. Dark-field microscopy with the TEM. (*a*),(*b*) Schematic diagrams of two common methods of dark-field imaging.

(c)

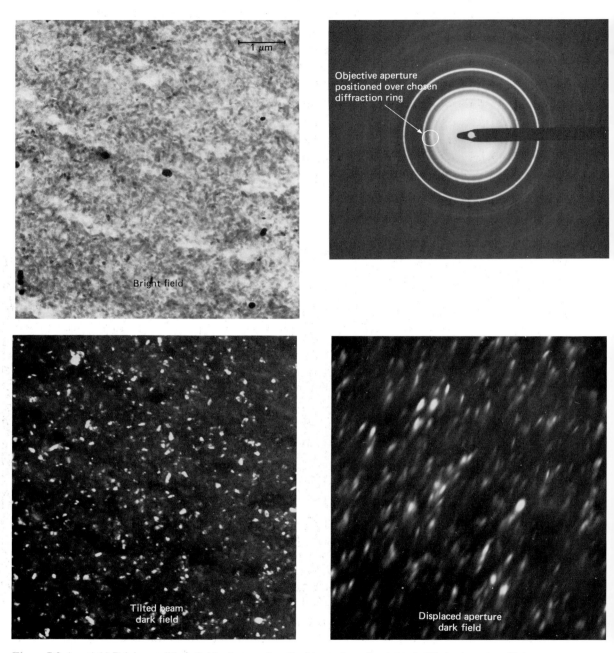

Figure 7.2 *(cont.)* (c) Bright- and dark-field micrographs of a thin section of resin-embedded microcrystalline alumina, and a diffraction pattern indicating the origin of the dark-field images. The crystallite size can be measured precisely from the tilted-beam dark-field image. Note that the two dark-field images are formed by different crystallites. Only a small proportion of the specimen contributes to either image by these dark-field techniques.

absence of a specimen the conical illumination forms a ring in the back focal plane of the objective lens. An objective aperture is used with diameter just smaller than the ring, so that no direct electron illumination passes into the final imaging system (Figure 7.4). In the presence of a specimen, however, the interior of the hollow cone is filled with scattered electrons, which pass through the aperture to form a high-resolution image of quite acceptable brightness. Strioscopy provides a useful method for forming high-contrast images of suitable amorphous specimens. The principal requirement of the technique is that specimens be thin, with support films and extraneous matter reduced to an absolute minimum. The fact that focused illumination is used means that the specimen must be capable of withstanding this degree of beam intensity, and also that the magnification used must be sufficiently high to allow the visible field to be filled by the spot, otherwise the illumination will be uneven over the field.

Figure 7.3. Conical-scan dark-field imaging. The right-hand micrograph is a bright-field image of a catalyst with very finely divided Pd on a single-crystal ceramic support. The diffraction pattern (*centre*) shows faint rings due to the Pd and bright reflections from the support. The left-hand micrograph is a dark-field micrograph with conically-scanned illumination tilted to image with the electrons in the Pd ring indicated on the diffraction pattern. The size and distribution of the Pd can be seen in good contrast. A conventional displaced-aperture dark-field image would have shown only a fraction of these particles. (Micrographs from Thompson, 1981; by courtesy of Philips Electron Optics.)

If focused illumination is liable to cause damage to a particular specimen the intensity of focused illumination can be reduced to a certain extent on most microscopes by decreasing the emission from the electron gun or using a smaller aperture in the first condenser lens, C1. There is a limit to the reduction which can be achieved by this means, however. A greater flexibility of illumination can be obtained if a third condenser lens is used to produce the actual focused spot on the specimen. C2 can then be used as usual for the intensity control. One commercial microscope, the AEI 801S (no longer made) had a mini-lens built in as C3 (Figure 7.4(*c*) to make this control possible. The technique should be an appropriate one for the modern multi-lens TEMs.

Strioscopy has been found to be a particularly useful technique in high-voltage transmission microscopy, in which the high transparency of specimens can result in low-contrast images. If two neighbouring regions of a specimen scatter 5% and 10% respectively of the incident beam the contrast between them in bright-field will be only $(95-90)/95 = 0.05$ and not easy to discern, whereas in dark-field it will be $(10-5)/10 = 0.5$, which is clearly visible. However, as stated before, it is important not to dilute the contrast with scattered electrons from a support material, which contribute no useful information about the specimen. This applies to all dark-field techniques, in fact, as does the desirability of keeping the specimen thin. Scattered

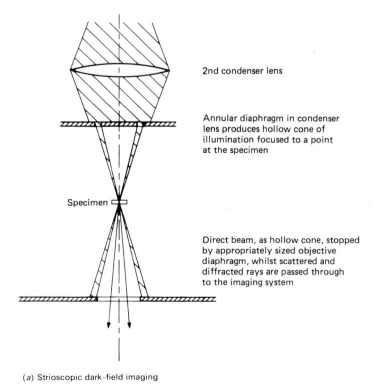

Figure 7.4. (*a*) Schematic diagram (*right*) showing the principle of strioscopic dark-field imaging, and (*opposite top*) a strioscopic dark-field micrograph (at 100 kV) of unshadowed and unstained filaments of DNA extracted from T4 bacteriophage by Kleinschmidt's method. (By courtesy of G. Dupouy, L. Enjalbert and L. Lapchine.) (*b*) Commercial microscope with third condenser lens to permit strioscopic imaging with easier control of image brightness. (By courtesy of Kratos Ltd.)

2nd condenser lens

Annular diaphragm in condenser lens produces hollow cone of illumination focused to a point at the specimen

Specimen

Direct beam, as hollow cone, stopped by appropriately sized objective diaphragm, whilst scattered and diffracted rays are passed through to the imaging system

(*a*) Strioscopic dark-field imaging

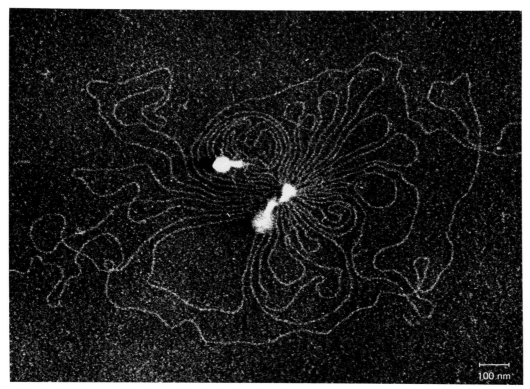

100 nm

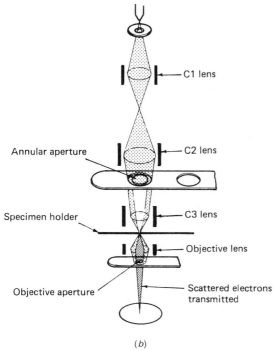

(*b*)

electrons which have suffered energy loss will degrade the resolution of the image because of chromatic aberration and should be kept to a minimum.

Another procedure for suppressing the direct beam has been used experimentally by some workers, but has not been generally adopted because it either presents great practical difficulties or introduces troublesome asymmetries into the imaging system. This procedure involves mounting an electron-opaque beam stop within the objective aperture, and has been achieved by various means (e.g. Shechtman & Brandon, 1970). The successful development of the strioscopic method has reduced the need for alternative techniques, however.

As briefly mentioned earlier, in Chapter 5, focusing a microscope in dark-field is difficult because of the absence of Fresnel fringe effects. The easiest solution is to focus the objective in bright-field and then switch to dark-field. Alternatively a through-focal series at different focus settings may be recorded in the expectation that one of the exposures will be close to true focus.

7.1.4 Dark-field STEM imaging

There are two main methods for forming dark-field STEM images; commercial microscopes use one or other or both of them. The most direct analogue of TEM dark-field is to use the objective diaphragm to intercept all but the chosen region of the diffraction pattern. These electrons go through to the STEM detector and provide the signal for the c.r.t. image. This method suffers from the usual drawback that only a proportion of the crystals orientated to give the chosen reflection actually appear in the dark-field image. The other method uses a separate annular dark-field detector concentric with the bright-field disc detector. The annular detector collects electrons in all orientations around the optic axis, but is less selective than the objective diaphragm since it also accepts electrons scattered through wider angles than the chosen ring. On the other hand, it provides an excellent bright image of all scattering centres in the field of view. The effective inner diameter of the annulus can be varied with the camera length control, so that discrimination between scattering through high and low angles can be provided. This type of STEM detector was illustrated in principle earlier in Figure 3.48.

7.2 Spot scan

The analytical TEM in the STEM mode emulates the SEM with only an electron collector after the specimen. If the transmission detector is removed and the objective and projector lenses are activated as for high-resolution TEM imaging, the result is a scanned image in which the resolution is that of the TEM but the electron irradiation of the specimen is reduced to a single scan by a small spot. The technique is particularly beneficial if the spot is a small, intense, coherent one from a field emission gun (Busing & Otten, 1991). Setting up the focus, stigmation, etc., is carried out on a reduced raster, possibly with the scan deflected to a different area, if there is significant risk of beam damage. The end result should look the same as a normal TEM image.

7.3 Microradiography

An electron beam focused to a small spot on a metallic foil gives a small source of x-radiation, which may be used in several useful techniques. Its use for background-free trace-element analysis was described in Chapter 6. Another possibility is x-ray projection microscopy or microradiography, a technique originally exploited by Cosslett & Nixon (1960).

Because of the penetrating power of hard x-radiation, if a thin specimen is placed close to the microscopic x-ray source an enlarged image of the internal structure of the specimen will be obtained on an x-ray detector placed further away from the source, and parallel to the specimen plane. If d and D are the source–specimen and source–detector distances a magnification D/d times will result. The x-ray source can be made as small as a few micrometres, and this will be the limiting factor in resolution of detail.

If the x-ray image is projected on to a CCD chip, a digitised image can be obtained and quantitative observations made (Erre *et al.*, 1993). Displacing the x-ray spot slightly will enable a stereoscopic pair of microradiographs to be produced, and hence information about the density distribution in depth within the specimen.

Applications of the technique can include examination of the internal structure of biological materials, e.g. insects, and fault-finding in encapsulated microelectronic devices. Contrast can be maximised by choosing an x-ray target material with strong emission at an energy just above an absorption edge of a major constituent of the specimen. X-ray microscopy is reviewed in Duke & Michette (1990).

7.4 Voltage contrast in the SEM

Electrostatic and magnetic fields at the surface of specimens can have a marked influence on the behaviour of low-energy electrons such as secondary electrons. This effect can be optimised in order to image the surface distribution of potential over a specimen such as a microcircuit in which different parts of the circuit are at different potentials. An extension of this technique of voltage contrast is stroboscopic microscopy, used to study the voltage distribution over an operating microcircuit. If the circuit is oscillating with a certain frequency, the electron beam in the SEM is pulsed (blanked) at that same frequency, so that micrographs with voltage contrast show the potential distribution at a chosen phase of the oscillation.

This technique is very useful in microcircuit development, particularly in showing the location of defects in the circuits, in which the conductors may be only a micrometre across.

7.5 EBIC, OBIC and SOM–SEM

These are techniques usually used to study semiconductor materials in microelectronics development. A specimen, e.g. a 'chip' of doped silicon, is mounted on the specimen stage of an SEM and the specimen current amplifier used to provide an imaging signal. The electron beam of the SEM creates *electron–hole pairs* in the semiconductor, which are collected to provide a beam-induced signal. Rastering the beam over the specimen results in an Electron Beam Induced Current (EBIC) image showing regions of changed electronic properties, e.g. internal crystallographic features, in the semiconductor (Holt & Joy, 1989).

If a scanned focused optical beam is used similar information is obtained by Optical Beam Induced Current, OBIC. This has the advantage over EBIC that there is no risk of damaging the semiconductor by use of a high-energy electron probe. If this is carried out in an SEM the technique is described as Scanning Optical Microscope–SEM, SOM–SEM (Maher, 1985; Battistella *et al.*, 1987). The scanning light-spot is produced by cathodoluminescence when the electron beam is scanned over a thin phosphor screen or a (YAG) single crystal. The light emitted is focused on the semiconductor specimen by a light-microscope objective.

In all cases the resolution of the technique depends on both the spot size on the specimen and the interaction volume in the specimen.

7.6 Scanning electron acoustic microscopy (SEAM)

A technique with very specialised applicability, this is a means of obtaining information about internal structures and mechanical flaws within rigid, bulk specimens.

The electron beam in an SEM is pulsed on and off by beam blanking at a frequency chosen by the operator, as well as being scanned in a raster over the specimen. Localised heating by the beam creates thermal shock waves in the specimen, which propagate through it and are picked up by a transducer glued or clamped to the side of the specimen. The signal from this transducer is amplified and used to modulate the c.r.t.. When the frequency of pulsing is chosen appropriately the acoustic image contains effects which, with experience of the technique, can be interpreted in terms of mechanical defects within the sample.

Figure 7.5 is an example of a SEAM image and the type of defect which can be located by this technique.

7.7 Spin-polarised SEM (Spin SEM)

Secondary electrons emitted from a surface in a magnetic field, or which is itself magnetised, will be spin-polarised. If this polarisation can be detected, it will be possible to image magnetic domains such as the stripes on magnetic tapes and discs, without having to remove them from their substrates for Lorentz microscopy by TEM.

A Mott spin-polarisation detector (for description see reference below) can be used to separate electrons with different spins, and has been used in an SEM to provide images showing 20 nm resolution of magnetic detail (Matsuyama & Koike, 1994).

7.8 *In-situ* microscopy

This term describes the use of electron microscopes to study the effects of processes carried out within the microscopes themselves. It applies to scanning and transmission microscopes and can involve the study of a specimen undergoing deformation or thermal treatments, or the formation of a new specimen by vacuum-deposition in the microscope.

Mechanical straining stages exist for TEM and SEM. In heating and cooling stages specimen temperatures can be varied up to $+ 1000\,°C$ and down to $-250\,°C$. Heating is normally by radiation, and care must be taken that the magnetic field due to the heating current in the oven does not interfere with imaging in the TEM. Cooling is by liquid nitrogen or liquid helium.

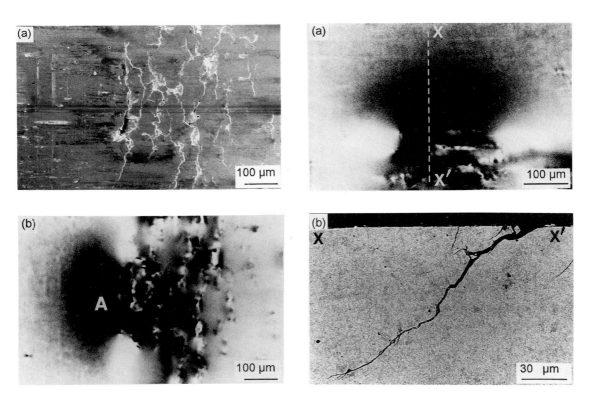

Figure 7.5. Scanning electron acoustic microscopy, SEAM. (*Left*) (*a*) SE image of surface of a damaged gear tooth. A is a region chosen for study by SEAM. (*b*) SEAM image of same area. Region of dark contrast adjacent to surface cracks indicates direction of sub-surface crack. (*Right*) (*a*) As previous micrograph, but scan direction rotated 90°. Note that the dark contrast is unaffected by the rotation. (*b*) Metallographic section cut through plane XX' showing the origin of the SEAM contrast. Note the close correspondence between the horizontal extent of the microcrack and the size of the SEAM dark contrast. (From Shaw, Evans & Page, 1994)

The specimen stage of an SEM can be used as a workshop in which specimens may be prepared by mechanical probing or cutting (Reiss, 1992), or by sputtering away with a focused ion beam (FIB). With biological specimens it is helpful if the specimen to be dissected has been prepared by a process such as OTOTO (p.186) so as to be inherently conductive when fresh features are exposed.

The early stages of formation of vapour-deposited thin films have been studied using evaporation on to a substrate on a TEM specimen grid. To reduce the effects of contamination from the microscope atmosphere it is usual to have additional pumping to the specimen area, and to evaporate under ultra-high vacuum. The crystallographic state of the growing thin film can be studied very sensitively using the Grigson technique (Grigson, 1965) whereby the electron diffraction pattern from the deposit is scanned periodically over a slit in front of an electron collector. The early appearance of diffraction rings or spots on top of the background of scattered electrons can be detected with great sensitivity by this method.

Reactions between a specimen and a controlled gaseous atmosphere can be observed in the TEM (or ideally in the HVEM) if an environmental cell is built into the specimen stage. The specimen on its substrate is placed in a small gas cell sealed top and bottom by thin membranes of collodion or similar plastic. The electron beam passes readily through these windows, and a flowing gas stream passes over the specimen. If the TEM image is photographed with a TV camera, the progress of any reaction may be recorded on video tape for future playback and analysis.

Very fast transient reactions, such as evaporation or crystallisation of a thin metal film under a pulse of laser radiation, may be studied *in situ* (e.g. Bostanjoglo *et al.*, 1991). Praprotnik *et al.*, (1994) operated a liquid-metal ion source inside a megavolt TEM and studied its physical characteristics.

7.9 Electron beam lithography and nano-lithography

Another specialised technique used principally in the microelectronics industry is the use of electron or positive-ion bombardment to expose or to etch photo-resist films in the manufacture of microelectronic circuits and components.

A photo-resist is an organic compound whose properties are modified by exposure to ultraviolet light, making it more or less soluble in a solvent 'developer' (and referred to as a positive or negative resist in these two cases). Polymethyl methacrylate (PMMA) for example, is rendered more or less soluble in organic solvents by exposure to UV light. A uniform thin layer of PMMA exposed to UV light through a patterned mask can subsequently be 'developed' so that the PMMA remains only in regions which were exposed to the UV and is washed away elsewhere. The unprotected substrate can then be selectively etched away to leave a relief pattern. Arrays of lines or interconnections in microcircuits can thus be prepared by the use of suitable masks and etchants for copper, aluminium, gold, silicon and chromium, typical materials in microcircuit construction. The edge-sharpness of circuit elements which can be created in photo-resists using UV light is limited by diffraction of the light, and irradiation with an electron beam can be used as a substitute. By suitably programming the beam positioning and beam blanking in an SEM, patterns can be exposed in suitable resist materials, and processed to give similar effects to light exposures, but with better-defined edges.

Nano-lithography is one aspect of nano-technology which enables structures to be manufactured (or destroyed or modified) on a nanometre scale. Using the very fine electron probes obtainable from field emission electron guns, localised effects can be created. At the simplest, characters can be drawn and permanently fixed in films of surface contamination, and can be 'read' from a high-magnification micrograph of the same area. Letters may be permanently etched in layers of suitable materials by the impact of fine probe beams (Prewett & Mair, 1991). The possibility of 'writing' an encyclopaedia on the head of a pin now exists! More commercially viable will be the advances in microminiaturisation in the communications industry.

7.10 Reflection electron microscopy (REM)
This is a very specialised technique for the examination of solid surfaces in the TEM, which has had two distinct stages of development.

In the 1950s and 1960s the recommended way of examining surfaces was by replication, and procedures for this were developed to a high state of perfection, as already described in Chapter 4. Direct examination of surfaces was only possible in a few TEMs which had suitable beam tilt and deflection and specimen handling facilities. The specimen surface was aligned almost parallel with the axis of the objective lens and the illuminating electron beam was incident on it at a small angle. Solid specimens could not be inserted in the objective lens bore, but had to be mounted outside the lens, with a working distance measured in centimetres rather than millimetres. The specimen was viewed against the illumination, into empty shadows, and because of the extreme tilt the magnification varied greatly in different directions on the micrograph (as in the SEM, but with tilt angles of 80°–85°). Nevertheless, the technique was used commercially to examine the cutting edges of razor blades, and a cine film by Halliday showed a metal turning operation *in situ* (Halliday, 1962). Because of the long working distance the effect of lens aberrations ruled out resolutions better than a few tens of nanometres. In 1964 the commercial introduction of the SEM created a far more satisfactory method for the direct examination of surfaces.

More recently, however, REM has been re-introduced for the examination of subnanometre steps in metal deposits on mica and cleaved crystal surfaces. Small fragments of specimen are mounted on side-entry rod-type specimen holders in the normal position for high-resolution microscopy, but mounted with their surfaces parallel to the optic axis so that the beam grazes the surface of the specimen (Figure 7.6). The successful use of the technique takes advantage of the very much improved vacuum now available in electron microscopes, since clean, gas-free surfaces are necessary.

A very shallow strip of the surface is seen in focus at high resolution, and sub-nanometre surface features are clearly visible in the REM images. Where the sample is a single crystal use can be made of reflection diffraction spots to set up the imaging conditions. A useful resume of the technique is given in Hsu (1992).

7.11 Other microscopical and analytical techniques

The main purpose of this book is to describe the capabilities of the various members of the electron microscope family, so that readers can assess for themselves how they could be served by these instruments. However, something must be said about the relationship between electron microscopy and various other methods of examination or analysis which are at present available. In the same way that it is important to know where scanning and transmission microscopes can give complementary information, it is equally important to know where the electron microscope can add to information from other techniques, and vice versa. Since this is a book on electron microscopy the treatment of the other techniques will be brief and incomplete but references will be given to more detailed literature on them.

The other techniques divide into groups (Table 7.1) according to whether they are alternative forms of microscopy, revealing mainly topographical information, or whether they provide the means for analysing the composition or crystalline nature of surfaces or bulk materials.

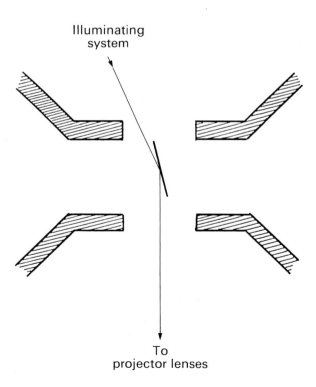

Illuminating system

To projector lenses

Figure 7.6. Schematic diagram of tilted specimen and illumination in high-resolution reflection electron microscopy.

Table 7.1. *Other microscopical and analytical techniques*

1 *Alternative forms of microscopy*	Field ion microscopy, FIM
	Low-energy and photoelectron electron microscopy, LEEM, PEEM
	Scanning probe microscopy (STM, AFM, etc.)
	Scanning acoustic microscopy
2 *Methods for analysis*	Ion probe analysis and SIMS, SIPS
	X-ray and UV photoelectron spectroscopy, XPS, UPS
	Auger electron spectroscopy, AES, SAM
	Laser microprobe mass analysis, LAMMA
	Ion scattering spectroscopy, ISS
3 *Methods for the analysis of crystalline surfaces*	Low-energy electron diffraction, LEED
	Reflection high-energy electron diffraction, RHEED
4 *Methods for bulk analysis*	X-ray diffraction, XRD
	X-ray fluorescence analysis, XRF

7.11.1 *Alternative forms of microscopy*

Field ion microscopy (FIM)

This is a technique which is at present limited to very specialised types of specimen, but which has the ability to show the arrangement of surface atoms in solids and to identify individual atoms. The technique is linked with the principal worker in its development, Erwin Müller, who first developed the field emission microscope (Müller, 1937). In this instrument the 'specimen' was a sharply pointed metallic wire at the centre of a highly evacuated spherical vessel lined internally with a fluorescent powder and a thin electrically conducting film. A potential difference of several kilovolts applied between the wire and the surface of the sphere, the latter being positive, resulted in a very high electric field (10^6 volts per centimetre) at the metal tip. Electrons pulled out of this tip by the strong field travelled along the radial lines of force to the fluorescent screen to form a much enlarged picture of the distribution of field emission over the surface of the tip. The magnification was simply the ratio of the radii of curvatures of the screen and the tip, which could be of the order of 10^6. Careful study of the patterns enabled crystal planes to be identified on the tip but the resolving power was inadequate to show individual atomic sites. Müller discovered that he could obtain a more detailed picture showing individual atomic sites by filling the vessel with a low pressure of gas, e.g. helium or neon, and reversing the polarity of the applied voltage. The image was now formed by positive gas ions and the new instrument was called a *field ion microscope* (FIM) (Figure 7.7) since ionisation of the

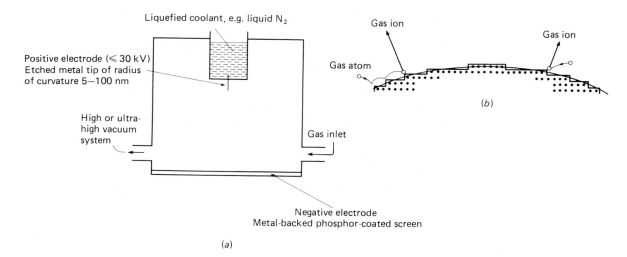

Liquefied coolant, e.g. liquid N$_2$

Positive electrode ($\leqslant$ 30 kV)
Etched metal tip of radius
of curvature 5–100 nm

High or ultra-
high vacuum
system

Gas inlet

Gas ion

Gas ion

Gas atom

(b)

Negative electrode
Metal-backed phosphor-coated screen

(a)

(c)

Figure 7.7. The field ion microscope. (*a*) Schematic diagram showing essential components of a FIM. (*b*) Atomic arrangements at the rounded tip of the emitter (specimen). Crystalline metallic point, polished or etched to a radius of curvature ~100 times the interatomic separation, shows a number of steps in the atomic structure at its surface. Gas atoms bounce over the surface and are preferentially ionised at the edges of steps; the positive ions thus formed are driven along the radial lines of the electric field out to the negative electrodes and a phosphor screen. This results in a field ion micrograph as below. (*c*) Helium ion micrograph from an iridium tip. (By courtesy of T. F. Page and B. Ralph, University of Cambridge.) (*d*) Neon ion micrograph of a 13% Rh–Pt alloy tip (*left*) and a similar tip after oxidation and partial removal of the oxide (*right*). (By courtesy of A. R. McCabe, University of Oxford.)

gas depended on localised electric field effects at the surface of the specimen (Müller & Tsong, 1969).

Specimens for the FIM must have a specialised nature, being sharply pointed metals cooled to a low temperature. The screen image shows the arrangement of projecting atoms in the surface of the specimen. The technique of field evaporation has been developed to remove layers of atoms in a controlled manner so that the atomic structure may be studied in depth. Dislocations in crystal lattices have been identified and studied. A further development by Müller is the atom probe microanalyser (Müller *et al.*, 1968), by means of which a specific atom from a FIM emitter can be analysed for mass-to-charge ratio. This is therefore the most refined analytical technique yet devised. Moreover, both the FIM and IAP (imaging atom probe) are capable of giving a new insight into technological materials and processes (e.g. McCabe & Smith, (1983); Miller & Smith, (1989); Mackenzie, (1992)).

LEEM and PEEM
Two techniques for the study of surface effects on smooth surfaces in ultra-high vacuum are Low-Energy Electron Microscopy (LEEM) and Photoelectron Electron Microscopy (PEEM) which use electrons and ultraviolet light, respectively. A small area of emitting surface is imaged via an electrostatic lens on to a fluorescent screen. Images are characterised by the appearance of lines and features indicating surface steps, pitting, etc. Adsorption of gaseous monolayers can be studied. Reviews of LEEM are provided by Veneklasen (1992), and Bauer (1994).

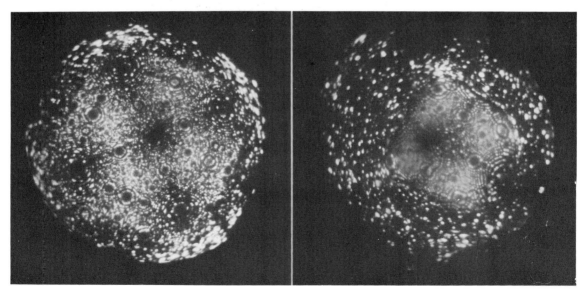

(*d*)

Scanning probe microscopy (SPM, STM, AFM, etc.)

Originating from the Scanning Tunnelling Microscope (Binnig & Rohrer, 1982) there is a family of instruments which are able to image the surfaces of specimens down to an atomic scale with basically very simple equipment, and not requiring elaborate specimen preparation or exposure to high vacuum (unless specifically desired).

The STM consists essentially of a very fine stylus with an atomically sharp tip which is held a nanometre or so away from the specimen surface (Figure 7.8(*a*). Application of a small voltage between the tip and the specimen (which must be a conductor) results in a measurable small 'tunnelling' current passing between the two. In scanning tunnelling microscopy the tip and specimen are traversed relative to one another and either the tip is maintained at a constant height and the variation in tunnelling current is plotted or, more usually, the tip is moved up and down by a servo mechanism to maintain a constant tunnelling current. On a simplistic viewpoint the oscillations of the tip trace out the surface topography of the specimen, going up and down over individual atomic positions. Scanning the tip in a raster over the sample builds up an impression of an area of surface, on an intimate scale. The fine control of the STM tip is made possible by an arrangement of piezo-electric ceramic tubes, which expand and contract in response to changes in the voltage applied to them.

An STM consisting of a tip-holder, power supply, amplifier and chart recorder for outputting the 'micrograph' can be made very cheaply, but when it has been engineered for high stability in air, liquid or vacuum, at high or low temperatures, and results are processed and graphically displayed on an image-analysing computer the cost can approach that of a small SEM. Some instruments have been developed to operate on the specimen stage of an SEM, using the latter to pre-select suitable areas for STM examination. The analytical procedure, Scanning Tunnelling Spectroscopy (STS), results if the scanning tip is held stationary over the specimen and the tip voltage–tunnelling current curve plotted over a range of a few volts positive and negative. If this is carried out at each point of an area-scan the surface electronic structure may be compared with the topographical structure atom by atom (Tromp *et al.*, 1986).

A non-conducting sample will not give a tunnelling current, and the Atomic Force Microscope (AFM) is a variant in which the stylus is mounted on the end of a rigid cantilever and drawn across the specimen under a very light mechanical loading. Vertical displacements of the cantilever are detected very sensitively, e.g. by reflecting a laser beam from the cantilever and detecting the movements of the reflected spot (Figure 7.8(*b*). The AFM approaches the STM in atomic resolution, and can be used on a wide variety of specimen types. If the tip is made magnetic the MFM is formed.

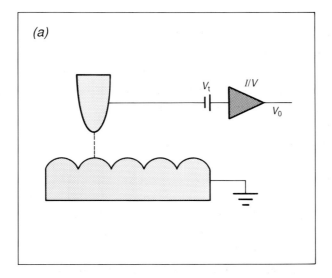

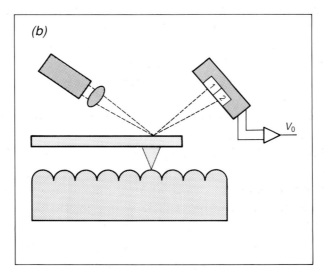

Figure 7.8. (*a*) The STM forms images of electrically conductive samples with sub-angstrom resolution. Using it, both surface topography and electronic parameters are measurable. The electrical current flowing between the tip and the sample depends on the applied voltage V and the tip-to-sample gap. When the gap changes by less than one ångstrom, the tunneling current changes by a factor of ten.

(*b*) The AFM makes high-resolution three-dimensional images of samples which need not be electrically conducting. Probe motions of less than 1 ångstrom and forces as small as a nano–newton are measurable as the tip is scanned across the sample. A probe tip is held on a spring cantilever. A laser beam is reflected from the cantilever into a four-section photodetector. When the probe moves up or down the ratio of light intensity on the photodetectors changes, providing a signal to the system.

Scanning probe microscopy is popular on specimens for which electron microscopy has proved inadequate, e.g. because of poor image contrast or the restrictive need to operate in vacuo. It has the advantage of being quantitative, giving dimensions directly which could only have been obtained indirectly (e.g. by stereoscopic measurements) in the electron microscope. The ability to produce results from specimens in their native state (e.g. wet or at atmospheric pressure) makes the techniques attractive.

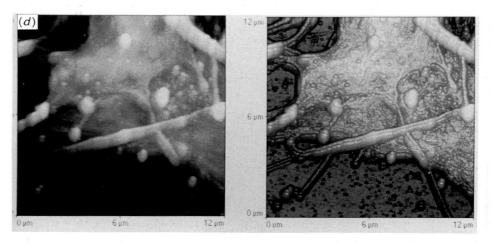

Figure 7.8. (*cont.*) (*c*) Photograph of an AFM (within its screening case), controlling computer and video display, and second monitor showing CCTV image of tip/specimen approach.
(d) AFM image of fixed and freeze-dried chick bone cell cultured on glass coverslip. Raw topographical image (*left*); image data processed to give shadowcast effect (*right*). (By courtesy of TopoMetrix Corp. Preparation and images in (*d*) by K. Piper, S. J. Jones, M. Arora & A. Boyde, University College, London.)

Scanning acoustic microscope

The velocity of ultrasonic waves in solids depends on the mechanical (elastic) properties of the solid. Waves from a point source are scattered by discontinuities such as grain boundaries or cavities, and images of internal inhomogeneities and surface modifications of bulk specimens can be obtained if waves from an ultrasonic transmitter are picked up by a detector (usually the same piezo-electric crystal) and the time lag between them measured. Scanning is usually achieved by physically moving the specimen under the transmitter, to which it is coupled by a liquid transducer. Different frequencies of oscillation in the GHz range are in common use, suited to the specimen type and the spatial resolution required.

The interpretation of acoustic images requires an appreciation of the processes and properties involved: they are not simply comparable with traditional light- or electron microscopy. The principles of the technique are explained by Briggs (1985).

7.11.2 Alternative methods for analysis
SIMS, SIPS and ion probe analysis.
Secondary-ion mass spectrometry, SIMS, is a destructive technique which slowly dismantles a surface and analyses the pieces. It has the rare capability of analysing the entire Periodic Table of elements, including hydrogen, and can detect elements at concentrations of parts per million and below.

When a surface is bombarded by a beam of positive gas ions in a high vacuum it is eroded ('sputtered') away; some of the atoms from it are themselves ionised. They can therefore be accelerated, focused, and sorted according to their energy and mass. Compared to electron probe x-ray microanalysis, this is an analytical technique which destroys the sample in the course of analysis, but which can produce information about compounds as well as elemental constituents. Thus, a SIMS spectrum from $CuSO_4$ will contain peaks for Cu^{2+} as well as SO_4^{2-}. The technique can be realised with different degrees of spatial resolution and complexity.

In the straightforward 'static SIMS' technique the primary-ion beam is several millimetres across and carries out an analysis of the first few atomic layers over a general area of surface. The ion gun and mass spectrometer may be fitted as accessories in the specimen chamber of a scanning electron microscope.

More localised analysis can be carried out using 'dynamic SIMS' on an *ion microprobe analyser* (Liebl, 1975) which uses a finely focused ($\leqslant 1$ μm) probe of positive ions to bombard and ionise a specimen. The secondary-ion current is analysed by a mass spectrometer, giving a composition analysis to a depth of about 1 μm. The ion probe may be scanned across the specimen, as in the electron probe x-ray microanalyser. Using high ion currents 'depth profiling' may be carried out, the specimen composition being measured as it is eroded away.

A very sophisticated ion–optical instrument was created by Castaing & Slodzian (1962) and subsequently further developed commercially by Cameca as the Imaging

Mass Spectrometer (Figure 7.9). This is a complex instrument which, in addition to carrying out static and dynamic SIMS with high mass resolution, can form microscope images showing the spatial distribution of a selected element over a field up to about 250 μm across, with a sub-micrometre spatial resolution.

With the use of modern digital imaging and storage techniques the lateral and depth information on elements of interest can be accumulated throughout a depth profiling experiment and reconstructed afterwards in a simulated 3-D display.

The sensitivity of element detection by the SIMS technique depends on the element considered. Mass concentrations between 1 part in 10^6 and 1 in 10^9 are detectable, compared with about 1 in 10^4 for electron probe microanalysis and 1 in 10^2 or 10^3 for EDX on the adapted SEM or TEM.

The ion probe analysers can use ion probes of O_2^+, N_2^+, Ar^+ or O^- which may be varied in diameter from 400 μm down to below 3 μm, as well as high-intensity sub-micrometre probes of oxygen, caesium or gallium ions.

The various forms of SIMS analyser all have a high sensitivity of detection, will separate isotopes, and will analyse hydrogen. Some of the particles emitted from a surface under ion bombardment are in an electronically excited state, and decay to their normal state in the vicinity of the surface with the emission of ultraviolet or visible light. The emitted wavelengths are characteristic of the atomic species con-

Figure 7.9. The Cameca Imaging Mass Spectrometer IMS 5F is a fully automated magnetic sector SIMS instrument capable of static and dynamic SIMS analysis, depth profiling, direct- and secondary-ion imaging of specimens including insulators. Very suitable for the analysis of semiconductors, the presence of as few as 3×10^{13} atoms cm^{-3} of P or As can be detected in Si, and 2×10^{16} atoms cm^{-3} H in Si, using Cs$^+$ primary ions. (Photograph and sensitivity data by courtesy of Cameca S.A.)

cerned, and can be used as a basis of a form of elemental analysis known as Sputter Induced Photon Spectroscopy (SIPS). This can be carried out simultaneously with mass analysis for SIMS or it can be used on its own. SIPS is particularly suited to the analysis and depth-profiling of non-conducting materials such as ceramics, glasses and minerals and thin films on silicon, since any specimen-charging which may occur has no effect on the analytical process.

SIMS and SIPS give an average composition over square millimetres of surface; the ion probe analyses smaller areas, below 1 mm², with a spatial resolution which can be sub-micrometre.

X-ray and UV photoelectron spectroscopy (XPS and UPS)

These are techniques for the non-destructive chemical analysis of the outermost surface layers of materials (Siegbahn *et al.*, 1967; Rivière, 1990; Briggs & Seah, 1990). All elements except hydrogen may be detected; the lowest detectable concentration of a particular element is about 1% but this may be prejudiced by interferences with other elements present in the specimen. The analysis is averaged over several square millimetres of surface and can be applied to solids, liquids and gases, although we shall only consider here the analysis of solid surfaces. The instrument commonly used for this form of analysis is known by the acronym ESCA (Electron Spectrometer for Chemical Analysis).

Photoelectron spectroscopy uses the property of x-radiation or ultraviolet light of causing the emission of photoelectrons when they fall on matter. The kinetic energies of these photoelectrons are characteristic of the elements present in the specimen. If the energies are analysed by an energy spectrometer, the spectrum can used as a basis of elemental analysis. An ESCA apparatus therefore consists of a source of monochromatic x-radiation or UV light (for XPS or UPS, respectively), an electron energy analysing spectrometer, and a specimen holder, all within an ultra-high vacuum chamber. Under the influence of the incident radiation photoelectrons are emitted from the specimen with a kinetic energy $E = h\nu - E_b - \phi$ where $h\nu$ is the energy of the exciting radiation, E_b the *binding energy* holding the electron within the atom, and ϕ is the work function of the energy spectrometer. Since $h\nu$ is a known constant for a particular type of source (e.g. 1486.6 eV for aluminium $K\alpha$ x-radiation, 1253.6 eV for Mg $K\alpha$, or 21.21 eV for helium I UV light), a plot of signal intensity from the energy analyser against analyser pass energy, as it is swept through the relevant energy range, provides a spectrum of binding-energy values for the sample (Figure 7.10(*a*)).

Each element has its characteristic binding-energy spectrum and an identification of elemental composition of a specimen can be made by inspection of the ESCA trace. The technique has wider capabilities, moreover, since the binding energies of electrons in a compound will be slightly different from those in free elements, and the characteristic binding-energy peaks are displaced by one or more eV by compound formation (Figure 7.10(*b*)). If the amplitudes and shifts of peaks in the spectrum are measured a unique form of surface analysis becomes possible using ESCA. The analysed region in the sample is restricted to a surface layer between 1 and 5 nm thick (i.e. only a few atomic layers), depending on the element and the kinetic energy involved. The incident x-radiation penetrates much more deeply into

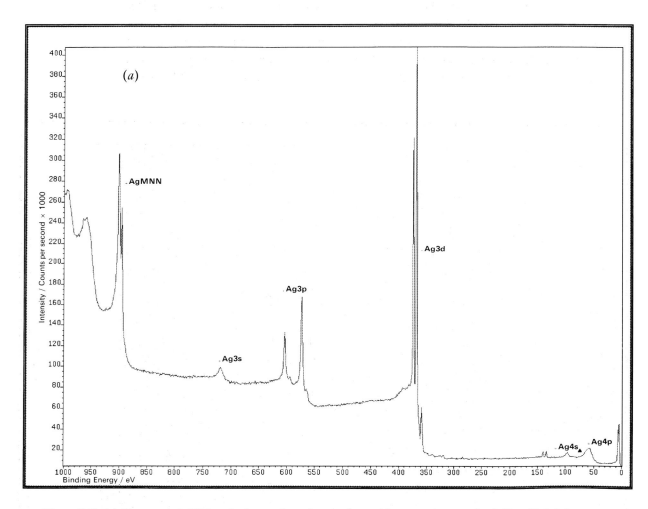

Figure 7.10. (*a*) Characteristic XPS peaks from a clean silver surface, with appropriate core levels identified. It has become conventional to plot these spectra with the origin in the lower right-hand corner, and the background rising towards the left. The reasoning is that the spectrometer is actually measuring kinetic energy, which has its zero on the left and increases from left to right. Thus, for Mg Kα exciting radiation binding energies 1000 to 0 eV correspond to increasing kinetic energies of 253.6 to 1253.6 eV.

[Note the presence of an Ag MNN Auger peak with kinetic energy 357.4 eV. This appears at a 'binding energy' of 895.6 eV when excited by Mg Kα x-radiation, but 1129.2 eV when excited by Al Kα radiation. So to check whether a peak in a spectrum is an XPS or an Auger peak, the operator would change the x-ray wavelength and see whether the peak appears at the same binding energy or not.]

the sample than this, generating photoelectrons, but those electrons originating more than one collision distance below the surface will have lost energy on the way out of the specimen and will no longer be characteristic of the sample. Instead, they contribute to the background on which the peaks are superimposed; a background which increases in intensity after each peak. This is why an ultra-high vacuum is needed in the analysis chamber; in a normal high vacuum the specimen surface would be covered by adsorbed gases and hydrocarbon contaminants and the intensity of photoelectrons from the specimen itself would be considerably reduced. A surface newly introduced into the ESCA apparatus will normally have pronounced peaks of C and O due to adsorbed CO_2 and H_2O present on the specimen. These may

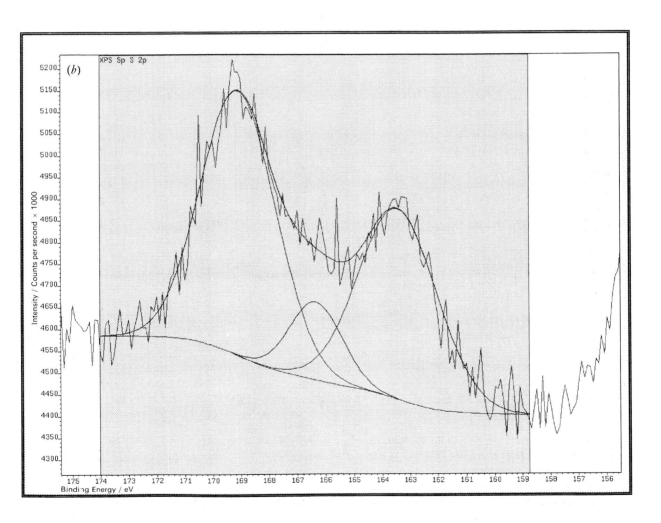

Figure 7.10. (*cont.*) (*b*) The noisy envelope of a sulphur 2p peak can be deconvoluted into sulphate (169 eV), sulphite (166 eV) and sulphide (163.5 eV). (Spectra by J. A. Busby, Johnson Matthey Technology Centre.)

be removed by giving the sample a gentle bombardment with Ar^+ ions. Commercial ESCA instruments are often provided with a preparation chamber in which the specimen can be processed by gas reactions, ion bombardment, heating or cooling, before being transferred into the analyser chamber.

Unlike the probe analysis methods, photoelectron spectroscopy gives an average analysis of several square millimetres of specimen surface, which is not actually imaged in the analysis. Smaller-scale analysis can be achieved if an electron lens is interposed between specimen and spectrometer, giving elemental analysis by XPS of regions below 1 mm (Drummond, 1992).

Auger electron spectroscopy and scanning Auger microscopy (AES and SAM)

This is a technique which, like XPS, gives an analysis of a very thin (1–5 nm) layer of the specimen (Rivière, 1990). Unlike XPS, however, it gives localised information because it uses a focused electron probe to stimulate the emission of Auger electrons whose kinetic energy is analysed and used for elemental analysis. Under electron bombardment in vacuo materials emit characteristic x-radiation and also comparatively low energy Auger electrons whose energies may be measured and used for elemental analysis (see also Chapters 2, 6 and Figure 2.14). A similar system results as for ESCA but there is an important difference in so far as the electron bombardment may be provided by a conventional scanning electron probe of very small dimensions, giving the possibility of very localised analysis ($\leqslant$50 nm) using Auger electrons. If secondary and/or back-scattered electrons are also detected the area being studied can be identified from a familiar SEM emissive micrograph.

As for XPS, the analyser chamber for AES must be evacuated to ultra-high vacuum. Although Auger electrons are emitted from deeper inside the specimen than the surface layer of atoms, only those from the surface will be emitted without change of energy due to collision on the way out. Hence the characteristic Auger peaks (Sekine *et al.*, 1982) are weak in comparision with the continuum of electron energies on which they are superimposed. In AES the background due to this and other causes is eliminated by modulating the voltage on the energy-analyser and using a 'lock-in' (tuned) amplifier to amplify only the modulated component of the signal from the analyser. In addition, it is usual to differentiate the output signal so that characteristic energy peaks in an AES spectrum are in fact double peaks, one positive and one negative. Although Auger electrons are emitted with a comparatively low efficiency (perhaps 1 for each 10^4 incident electrons) the efficiency increases at low energies (i.e. low atomic numbers) in contrast to x-ray emission which becomes less efficient at low energies.

By using more current in the scanning probe, e.g. 1 nA at 10 kV from a field emitter, sufficient Auger electrons can be obtained for elemental distribution mapping with spatial resolutions of 10 nm. Thus, combined SEM/Auger instruments are constructed, and may also incorporate facilities for ESCA. The analysis of insulating specimens may cause problems from charging; conducting coatings would of course be inadmissible.

Laser microprobe mass analysis (LAMMA)

This is a microanalytical technique in which the specimen is destroyed during the analysis (Kaufmann *et al.*, 1979). It has similar lateral and depth resolution to electron probe microanalysis but several orders of magnitude better sensitivity. It can detect the full Periodic Table of elements, but is not a quantitative technique.

A focused pulse of radiation from a laser melts, vaporises and ionises a very small volume of the specimen. Either the positive or the negative ions are analysed in a time-of-flight mass-spectrometer, giving a spectrum from which the original composition of the specimen can be inferred. Alternatively, a spectrograph can be used to record a line spectrum of the ionisation (Thibaut, 1992).

Ion-scattering spectroscopy (ISS)

In this technique a beam of low (3 keV) energy He^+ ions is reflected from a target surface. The final energy of the reflected ions gives a measure of the mass of the target atoms. The information depth of this technique is 0.3–1 nm, the lateral resolution 0.1 mm, the lowest detectable concentration 0.1%, and the analysis can be used for elements heavier than helium in the Periodic Table. Comparison of results by ISS, AES and XPS will be found in Hockje & Hoflund (1991).

7.11.3 Analysis of crystalline surfaces.

Low-energy electron diffraction (LEED)

This is a useful technique for studying the outermost atomic layers of crystalline material, and phenomena such as gaseous adsorption on these surfaces. The principle was originally established by Davisson and Germer (1927), but it is only with the recent advances in ultra-high vacuum technology that it has become a practical technique in surface studies.

LEED differs from the electron diffraction met earlier in the book in using electrons with energies of about 15–1000 eV, compared to the 20 000 eV upwards of the diffraction camera and electron microscope. Low-energy electrons have very limited penetrating power into the specimen, and wavelengths comparable in magnitude with the interatomic spacings of many materials (cf. Table 1.1). A broad beam of electrons is therefore diffracted by the surface layer of atoms of a crystal in the same way as a beam of light incident on a reflection diffraction grating. Normally incident electrons are preferentially reflected along paths making an angle α with the surface normal when the path difference $D \sin\alpha$ between electrons reflected from adjacent atoms is an integral number of electron wavelengths $n\lambda$ and where D is the interatomic spacing (see Figure 7.11(a)). Above about 100 eV the electrons penetrate further than the surface atomic layer and the intensity of diffracted beams behaves in a complex manner with electron energy.

Apparatus for LEED has the form shown schematically in Figure 7.11(*b*). Monoenergetic electrons are supplied from a gun in a direction normal to the specimen surface. The diffracted electrons move in a second, field-free space between the specimen and a set of three concentric hemispherical metallic grids held at specimen potential and at the electron gun filament potential (second and third grids). Electrons which have not lost energy in the specimen, but have only suffered diffraction, will pass through the second pair of grids to a region in which they are accelerated by a potential difference of about 5 kV on to a phosphor screen, where a diffraction pattern of spots is formed corresponding to the preferred directions α. The whole apparatus is in an ultra-high vacuum chamber to minimise specimen contamination. The pattern on the fluorescent screen is photographed through a window in the vacuum-chamber wall.

LEED phenomena are used mainly to study surface properties of solids and adsorption at these surfaces. The diffraction patterns are modified by adsorbed gas in such a way that the arrangement of the adatoms can be postulated. Surfaces may be treated *in situ* by heating, oxidation, reduction or ion bombardment, and the changes studied through the effects on the diffraction patterns.

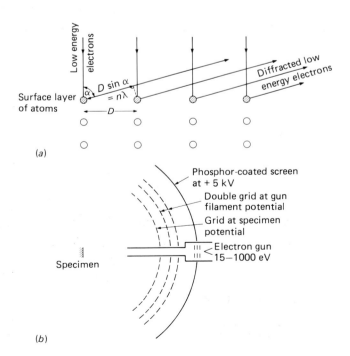

Figure 7.11. LEED geometry and apparatus. (*a*) Normally incident electrons are diffracted in a preferred direction from the surface layer of atoms in the sample. (*b*) Hemispherical retarding-field analyser filters out electrons which have lost energy to the specimen and accelerates the others to form a spot pattern on the fluorescent screen, enabling α to be calculated.

Reflection high-energy electron diffraction (RHEED)

This technique has been used for many years to study surface layers and thin films in electron diffraction cameras and electron microscopes, but has become a complementary technique to LEED when practised in ultra-high vacuum on the same specimen. A beam of electrons with energies of tens of keV is directed at grazing incidence on to a plane crystalline surface. The electrons are diffracted by atomic planes parallel to the surface and by any part of the specimen which projects proud of the surface. In effect one-half of a normal diffraction pattern is produced, which may be either rings or spots. The position of the central spot has to be inferred. Arcs and streaks in the patterns indicate preferred orientations of surface layers. The technique is illustrated schematically in Figure 7.12.

The combination of RHEED and REM in a transmission electron microscope is decribed in Yagi (1993) and Larson & Dobson (1988).

7.11.4 Methods for bulk analysis

X-ray diffraction (XRD)

Unlike a beam of electrons, a beam of x-rays cannot be manipulated by means of lenses. However, x-radiation is diffracted or preferentially reflected from crystalline materials in a manner similar to that already described in connection with electron diffraction (Birks, 1969). Since the x-radiation commonly used has a wavelength which is more than an order of magnitude larger than that of electrons used in transmission microscopes the angles of diffraction are correspondingly larger for x-rays, being tenths of or whole radians compared to the degrees or hundredths of radians

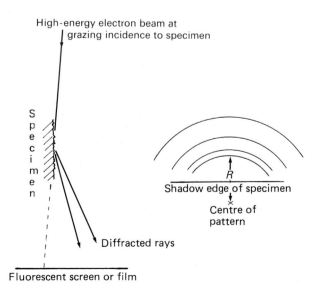

High-energy electron beam at grazing incidence to specimen

Specimen

Shadow edge of specimen

R

× Centre of pattern

Diffracted rays

Fluorescent screen or film

Figure 7.12. Schematic diagram of RHEED. High-energy electrons penetrate the outermost layers of crystalline surfaces and are diffracted according to the usual Bragg relationship $2d\sin\theta = n\lambda$. Much of the diffraction occurs in practice by transmission through edges and projections from the specimen surface, which is seldom absolutely smooth and flat on a microscopic scale.

in electron diffraction. Another important difference between electron and x-ray diffraction is that the x-radiation used in XRD is readily transmitted by air. An x-ray diffraction camera can comprise the x-ray source, the specimen – which may be a single crystal or a cylindrical tube of powder – and a detector – which may be an ionisation chamber or a photographic film moving or placed in a circle around the specimen – all in air.

In essence, the object of x-ray diffraction is to identify the crystalline nature of a specimen. A monochromatic source of x-radiation is used, the specimen giving diffracted beams in directions governed by the Bragg relationship $2d\sin\theta = n\lambda$. If λ is known and θ measured from a diffraction pattern the lattice plane spacing d can be calculated for each line in the pattern, and the crystal can be identified. It should be noted that, as in the case of electron diffraction, it is the actual chemical compound which is identified, not its elemental constituents. Since there are many thousands of compounds it may be difficult to make an identification starting from scratch, and it can be useful to start off by identifying the elements present by another method, e.g. electron probe microanalysis or x-ray fluorescence. As well as identifying compounds x-ray diffraction will also provide a figure for the average crystallite dimensions in the specimen, from measurements of the *line-broadening* effect in the diffraction pattern. Very small crystallites (e.g. down to 2 nm) give broad, diffuse, diffraction lines, whereas extended crystals (e.g. 100 nm or greater) form sharp, well-defined, lines. Even-larger crystals (~ 1 μm) give spotty lines.

X-ray diffraction can be used quantitatively to analyse the proportions of different substances in mixtures, by comparing the intensities of diffraction in certain lines with intensities from specimens of known composition or pure 'standards'. An accuracy of about 1% is attainable.

X-ray fluorescence (XRF)

XRD, just described, uses a beam of monochromatic x-radiation to study the nature of crystalline materials. XRF is a branch of x-ray spectrometry which uses the specimen itself as a source of x-radiation which is analysed to provide an identification of the elements it contains. The specimen may be amorphous or crystalline, solid or fluid.

In XRF primary x-radiation of high energy (short wavelength) is used to excite secondary x-rays of a lower energy in another material. The secondary or fluorescent radiation is characteristic of the element(s) present in the specimen and can be analysed by an appropriate spectrometer into its separate wavelengths. The pattern of these can then be used, as in microprobe analysis, to identify the elementary composition of the specimen. As in XRD the process does not require the use of a high vacuum except where very low energy fluorescent radiation is involved, as from elements of lower atomic number for which the absorption of radiation by air or the components of the spectrometer is excessive.

Primary radiation is provided by sealed, high-power (2–3 kW) x-ray tubes with tungsten or chromium anodes. The fluorescent radiation from the sample is analysed by wavelength- or energy-dispersive spectrometers of the types described in Appendix 3. One advantage of XRF over direct stimulation of characteristic x-radiation by electron bombardment is the elimination of the background continuum of *Bremsstrahlung*. This background can be an embarrassment when low concentrations of element are being looked for.

XRF is a technique which is sufficiently sensitive to detect a fraction of a percent of most elements in the Periodic Table, with micrograms or less of materials being measurable. It is therefore more sensitive for trace elements than electron-bombardment stimulated x-ray microanalysis, and does not require special coating procedures to render the specimens electrically conducting. On the other hand, it is not possible to see and select in microscopic detail which feature is being analysed, as in electron microscopes and dedicated EPMA. Trace analysis by XRF in the SEM was described in Chapter 6.

7.12 Suggested further reading
Since this chapter introduces so many different and very specialised topics the bibliography is equally specialised.

Examination techniques complementing electron microscopy are dealt with in numerous books, including Loretto (1993), Müller & Tsong (1969), Siegbahn *et al.* (1967), Woodruff & Delchar (1994), Williams *et al.* (1991), Smith (1994), Briggs (1985), Watts (1990), and Rochow & Tucker (1994). Surface analysis techniques are reviewed by Briggs & Seah (1990, 1992).

STM is reviewed by van de Leemput & van Kempen (1992) and Bonnell (1993). Some other books on scanning probe techniques are Chen (1993), Dinardo (1994) and Marti & Amrein (1993).

A bibliography of REM is reported in Hsu & Peng, 1993.

A series of conferences on x-ray microscopy have been held, and the proceedings published as *X-ray Microscopy I ... IV* (1994).

A short survey of biological applications of SIMS, including one of an imaging SIMS instrument, is given by Thellier *et al.* (1991). Murr (1991) describes ionic microscopy and microanalysis.

8 *Examples of the use of electron microscopy*

8.1 Introduction

In the fifty or so years since their commercial introduction electron microscopes have been used to study an ever-increasing variety of subjects. Nowadays, it is not unusual to see electron micrographs on television and in the popular press as well as in a wide range of periodicals and textbooks, even on postage stamps, illustrating subjects from abrasion to zoophytes.

The purpose of this part of the book is to mention some of the ways in which the instruments and technology which have already been described can be applied to practical problems in various fields and to illustrate a few of them. Some of the examples will be taken from the writer's own experience; the remainder have been provided by experts in other fields.

The basic techniques of specimen preparation and examination can be used, with suitable adaptation, for most examinations with the electron microscope. Working details can be found in a number of standard reference books, e.g. Kay (1965), Goodhew (1973 & 1985), Reid & Beesley, (1990), Hayat (1989a,b), Weakley (1981), Goldstein *et al.* (1992), and Robards & Wilson (1993). Variations on the standard themes can be found reported in the proceedings of national and international conferences on electron microscopy as well as scattered throughout the literature.

In a general survey it is convenient to divide the types of specimen into broad categories of *particles, surfaces of various kinds*, and *thin films, foils and sections*. There is biological and non-biological material in each category and it is hoped that every reader will find something in the chapter relevant to his or her own interests.

8.2 Particulate specimens

There is a very wide range of materials under this heading on which electron microscopy has been employed, including macromolecules, bacteria, viruses, cement, pigments, airborne dust and fragments of lunar surface. Because of the great variety of structures involved the whole range of electron microscope capabilities can be usefully employed in this field.

The size range of particles which can be studied by electron microscopy covers overall more than eight orders of magnitude, from below 1 nm to 10 cm! In addition to particle size distribution the surface topography, crystallinity and elemental composition can be studied, and in an ideal analytical laboratory this information could be complemented by other types of examination, such as surface-area measurements, x-ray diffraction and spectrographic analysis, to obtain a comprehensive picture of the nature of the specimen. In much work, however, one or two aspects will be of prime interest (e.g. size distribution and shape) and these will be studied routinely.

There is sometimes a tendency to forget that much useful work can be carried out on particles of micrometre and greater dimensions with the light microscope (McCrone & Delly, 1973). Refractive index and colour are two identifying characteristics which are lost to the electron microscope, and which can be useful even though not much of the surface of the particle may be in focus at any one time because of depth of field limitations.

Both types of electron microscope can be used to study particles, and specimen preparation may be one of the simplest operations for either instrument, involving spreading a single-particle layer on a support film on a grid for TEM or dusting a small quantity of powder on to a specimen stub, and metal- or carbon-coating it if necessary, for the SEM. On the other hand, the preparation may be more protracted, involving centrifugation, redispersion and deposition of the particles on a support film or substrate by spraying or some other technique. When particle sizes extend below 0.1 μm the TEM is preferred because it will produce clearer outlines to very small particles down to the sub-nanometre range. Also, errors in size measurement because of coating thickness for the SEM become appreciable around this size. The SEM is preferred for particles above about 10 μm across; these are inconveniently large for sizing in the TEM because parts of them are liable to disappear from sight behind the bars of the support grid and cannot be counted in the analysis. Particle dispersions are generally seen as outline shadowgraphs by transmission, but thinner particles ($\leqslant$0.1 μm for 100 kV or several times thicker for the Intermediate- and High-Voltage EM) may be penetrated by electrons and their internal structure and crystallinity studied using selected-area diffraction where appropriate (e.g. Gard, 1971). Interior structure and porosity can also be studied by encapsulating the particles in resin and cutting thin sections for transmission microscopy.

When surface shape as well as overall dimensions are required the 'pure' TEM user must generally resort to replication whereas the SEM produces very 'real'-looking micrographs showing the size and shape of particles. Stereoscopic microscopy can be carried out with very good effect on TEM replicas and in the scanning microscope.

The analytical TEM is particularly suitable for the examination of small particles. Clear dimensional information is obtained, and crystallographic data by electron diffraction. In addition there is the ability to examine surface shape and composition in the scanning mode and, for well-separated particles at least, to carry out elemental analysis of individual particles. In many cases insulating powders can be examined without being coated. One characteristic of the ATEM which can give unexpected results is that images in TEM and STEM modes show particles which are on both sides of the support film, whereas surface imaging techniques will only show those particles which are on the same side of the film as the secondary and backscattered electron detectors. Energy-dispersive x-ray spectra will always show peaks due to the material of the support grid. The various alternatives to copper grids, e.g. gold or nickel, may need to be employed if copper is being analysed in the particles.

One of the major requirements in particle examination is to achieve a good dispersion of the specimen during preparation; that is to say, to ensure that clusters of smaller particles are broken up or at least clearly identifiable as aggregates. In some cases, e.g. with magnetic particles, it may not be possible to overcome the cohesive forces holding a cluster together, but a replica or a scanning micrograph will enable the individual small particles to be identified as such and measured. The ability to see the shape as well as linear dimensions of a particle gives the electron microscope an interpretational advantage over volume-displacement methods of particle sizing. It can often resolve the apparent contradictions which may exist between indirect measurements based upon displacement and surface-area determinations. It is, of course, important that the size distribution of dispersed particles is not modified by the specimen preparation.

Finally, care must be taken to minimise the effect of the electron beam on specimens. Dehydration due to heating *in vacuo* can occur, and particles of a critical range of sizes between about 1 and 10 μm on thin support films can readily be heated to melting point in the concentrated beam of the TEM, since they absorb energy efficiently without having an effective means of dissipating it.

8.2.1 Industrial particles

The success or failure of many industrial manufacturing processes depends on the size, shape and composition of particles being correct. A number of electron microscopes are installed in manufacturing units to help in quality control of their products. The major manufacturers of photographic materials, for example, evaluate in this way the particle size distribution of each batch of silver halide emulsion which they produce (see Case Study 1); paint, pigment and polymer manufacturers do likewise (Cobbold & Gilmour, 1971). Because of the special requirements of electron microscopes in terms of specimen preparation and size, examination in high vacuum, etc., electron microscopy is not yet used in 'on-line' process control, but on a batch sampling basis. In this respect it can act as a backing-up facility to x-ray diffraction, β-ray scatter and transmission and various other continuous analytical quality control techniques. However, there is little doubt that further developments in this direction will occur in the future. Already the carrying out of particle size distribution analyses has been speeded up by sending a digitised image directly to an automatic image analyser, cutting out the need to develop and print micrographs before analysis (see Case Study 1).

In addition to production quality control there are, of course, requirements for particle studies in research and development, when the EM may provide a new viewpoint on a particular problem. The illustrations in this section show applications of electron microscopy in research and development and as an aid to production control.

8.2.2 Biological particles

The nature of a number of biological particles poses particular problems not encountered so frequently by the materials scientist. The organic, living, nature of bacteria, viruses, and blood cells, for example, requires that the preparatory techniques long used in biological light microscopy be carried into electron microscopy. Fixation, dehydration and staining are necessary preliminaries to examination. There may be particular difficulties to be overcome in achieving a uniform dispersion of suspension and negative stain on the carbon or plastic support film, and it may be necessary to experiment with procedures before selecting the most suitable one for a particular type of specimen (Harris & Horne, 1994). In the microscope contrast enhancement may be necessary by various means, e.g. strioscopy; post-examination image enhancement and analysis may be required to sort out regular but low-contrast structural details. Shadowcasting may enable more surface texture to be revealed, and electron diffraction will give information on the crystalline structure of macromolecules such as proteins. Quantitative mass determinations may be carried out on a range of natural particles. Over a period of many years the biologist has become accustomed to looking **through** his specimens, and all the comments above refer to examination by transmission electron microscopy, with its great gains in resolution and depth of field over the light microscope. More recently the potentialities of high-resolution scanning microscopy and x-ray microanalysis have begun to be appreciated, but the probable distortions of shape and composition by preparatory treatments make it necessary to be cautious before accepting the results of looking at and analysing biological particles. However, this is currently a very active area of research and the critical evaluation of preparatory procedures and cryo-techniques now taking place should bring increased confidence in the results obtained from these types of study (Nermut, 1994).

Much biological electron microscopy is of a routine nature. For example, the identification of viruses uses the electron microscope as a diagnostic tool and in times of emergency providing a rapid distinction between smallpox and chickenpox viruses can be a most important role for electron microscopes in public health laboratories. In addition a great deal of new work is being carried out, frequently stretching the present capabilities of specimen and operational techniques to their limit. It is particularly important in these cases to examine the results carefully for possible artefacts (Crang & Klomparens, 1988).

8.3 Surfaces

In non-biological fields probably more examinations are made of surfaces than of any other type of subject. Replica methods outlined earlier in the book were developed in order to make surface features suitable for study with the TEM. Many studies are still made by these procedures, which probably still give the most accurate and contrasty impression of surface shape, when they can be applied. But the real boon in surface studies was the arrival of the SEM, an instrument which makes it possible to examine all manner of surfaces over a wide range of magnifications without time-consuming specimen preparation and which, moreover, produces micrographs which are easy to interpret, in essence if not always in detail. The advantage is extended even further by the low-vacuum or environmental SEM in which moist and insulating specimens may be examined without any preparation at all.

A surface may be large or small in extent. It may be metallic, organic, inorganic, smooth or rough. It may be transportable to the electron microscope laboratory or it may be a permanent fixture or an indispensable component which cannot be removed from its present site. EM specimens result from the permutation of all these characteristics, and the competent microscopist will adapt his standard techniques of specimen preparation so that the end-product is suitable to go into his microscope or microscopes and gives the desired quality of image.

8.3.1 *Direct examination of surfaces*

When the body whose surface is to be examined is smaller than, say, a tennis ball it can be put directly into the SEM. Preferably it should be a lot smaller than this, because the smaller it is the more mechanically stable it is likely to be in the microscope. Also, it will be easier to examine it from a range of directions. Using secondary electron emission magnifications up to about $\times 250\,000$ can usefully be obtained, provided the surface is adequately conducting and has a good electron emission coefficient. Any doubtful surfaces, including metals which may have grown a very thin skin of oxide, will be improved with a thin conducting coating, e.g. 5–10 nm of sputtered platinum. It is most important that the specimen be firmly earthed to the stub, by painting a stripe of silver paste if necessary.

As well as topographical information the use of a solid-state backscattered electron detector will reveal atomic number differences within the surface of suitable (non-metal-coated) specimens. The image will contain detail from below the surface, however; the higher the primary electron energy, the deeper this will be. If surface shape is required the secondary electron image at low beam energy will give it most clearly of any scanning technique, particularly in one of the high-resolution field emission SEMs which are rivalling the ultimate in topographical imaging provided by the surface replica. Any ambiguities in interpretation of topographical details are readily elucidated from stereoscopic pairs of micrographs.

Surface imaging in the SEM makes it possible to examine many materials which, because of their optical properties, are very difficult to examine by light microscopy. These include glass surfaces and natural and synthetic fibres such as paper and woven fabrics, which are very difficult subjects because of their transparency to light and, often, because of their shape. In electron microscopy the electron emission occurs at or very close to the outer surface, and the interior is irrelevant. The enhanced depth of field and a eucentric tilt-and-rotate specimen stage, will overcome shape problems.

A great advantage of direct examination is the ability to analyse local regions of the specimen using wavelength- or energy-dispersive x-ray spectroscopy, bearing in mind that the spectrum accumulated by the spectrometer may have been modified by absorption and fluorescence effects between the point of generation and the detector.

Although requiring special accessories the technique of cathodoluminescence can be a very useful one in studying surfaces in many different fields of application.

Electronic components

Scanning electron microscopy has unique advantages when electronic and micro-electronic materials are to be examined. Voltage and magnetic contrast enable information to be obtained about microcircuits in static and dynamic situations, which would be unobtainable by any other means (Holt & Joy, 1989; Richards & Footner, 1992).

Surfaces in the ATEM

Whilst having very limited specimen accommodation and movements compared to the SEM, analytical transmission microscopes have particularly useful capabilities for surface studies, partly because of the wider range of gun voltages available, and partly because they have better illuminating and imaging capabilities at high resolution than many dedicated scanning microscopes. Whilst 20 or 40 kV enables surface imaging to be carried out at high resolution using secondary electrons, beam energies of up to 100 keV are very useful for backscattered imaging and x-ray analysis. In particular, the L lines and even K lines of heavy metals can be excited efficiently, which is not possible on the SEM with only 30 or 40 kV.

8.3.2 Replicas

When the surface to be examined cannot be spared, moved or chopped up there is no alternative to making a replica and studying that. Replication for the TEM is restricted to relatively flat, essentially two-dimensional, surfaces. The best techniques will reproduce detail of a few nanometres and run-of-the-mill methods will permit reproduction of 10 nm structures without much difficulty. However, the replica film has to be supported on a fine mesh, and the field of view is therefore periodically obscured. If a surface is to be studied continuously over a wider area, say 1 cm across, or if it is rugged or angular in topography, then it should be replicated for the SEM (Pfefferkorn & Boyde, 1974). The replica will be a robust reproduction of the original surface features with a conducting surface of good secondary emission capability. When magnifications of a few tens to a few thousand times are all that are required rough surfaces may be replicated in flexible materials which can be pulled off them. These materials may not reproduce the finest detail, and care and ingenuity are required to prepare the final, positive, replica from them.

One advantage of using a replica process for examining a surface is that a permanent record is available so that the original can be subjected to modifying treatments and any changes observed by comparison with the original specimen condition. One example of the use of this technique is shown in connection with the polishing of glass (Cornish & Watt, 1964 and Case Study 8). An example of the use of TEM replica techniques on a glass structure has already been shown in Figure 5.44, when it was possible to make depth measurements on features by the use of stereoscopic pairs of micrographs.

8.3.3 *Biological specimens*

In the natural world there are many surfaces which used to present a perpetual challenge to the transmission microscopist with his carbon replicas but which nowadays are taken in their stride by scanning microscopists. These include leaf and fruit surfaces, wood and other cellular structures, bone, teeth and blood cells. Before complete reliance can be placed on all the results from scanning studies on these subjects a number of preparatory techniques have to be critically assessed, as, for example, in the fixation and drying of the specimens. However, the main point to be made is that there is now an instrument which has a sufficiently long working distance and a great enough depth of field to be used to examine irregular structures. Many commercial questions, such as, for example, what happens to insecticides sprayed at growing vegetables, are being answered in detail for the first time.

Techniques which are proving of great help to the biologist are freeze-fracture and freeze-etching, and the examination of frozen specimens in the SEM. The former provides a unique form of structural information; the latter enables many subjects, from foams via chocolate to plant surfaces, to be examined which would otherwise have been difficult, if not impossible (see Case Study 5). More recently the arrival of the Environmental or Low-Vacuum SEM has forced a revision of what can and what cannot be examined by electron microscopy (see Case Study 12 and Griffith & Danilatos, 1993). There are disappointments, though, insofar as the electron microscope examines the outer surface of a hydrated sample, and cannot look inside it as the light microscope would.

As mentioned earlier the other operational modes of the SEM can be used to study biological specimens; the x-ray mode is obviously of great interest but distortions of composition by the processes of fixation and drying must be controlled reliably before biological structures can usefully be analysed. This aspect provides one of the strongest justifications for cryo-techniques.

8.4 Thin films, foils and sections

The final classification of specimen types involves mainly the type of study in which the transmission microscope is pre-eminent, just as the SEM has a great advantage in surface studies.

8.4.1 Thin films

Much of the present knowledge of the manner of growth and the resultant structure of thin films came from studies by transmission electron microscopy. High-resolution microscopes are necessary to observe the early stages of nucleation of films formed by evaporation, sputtering or chemical deposition. Much has been learned about the mechanism of formation of evaporated films from *in-situ* film depositions, for which a miniature evaporation source and a very thin substrate are incorporated in the specimen stage of the microscope. More recent work has been carried out with ultra-high vacuum in the film growing area (Valdrè *et al.*, 1970). One way of studying the growth phenomenon is to record the micrographs on ciné film or video tape so that it can be observed as a dynamic process.

Other work on thin films has involved growing or forming the films on massive substrates, e.g. glass slides or heated single crystals and then transferring the deposits to thin electron transparent supports for detailed study in the TEM. Whilst this is in general a useful technique it is desirable to check by another method (e.g. by replicating the surface of the film on its original substrate or using an Atomic Force Microscope on it) that the film structure has not been modified by the transfer operation. It may also be possible to form the deposit on a support film already on an electron microscope specimen grid so that it is only necessary to transfer it into the microscope for examination.

As well as looking at the shape of the deposit its crystalline nature can be studied by electron diffraction. A particularly sensitive technique used in studying the growth of thin films is scanning electron diffraction (Grigson, 1965), in which the diffraction pattern is deflected periodically over a slit behind which lies a detector. By this method the rings or spots in a diffraction pattern are distinguished from the background of scattered electrons with greater sensitivity. The early formation of crystals in thin films evaporated within the diffraction camera has been studied in detail using the Grigson technique.

By plotting a density profile across a micrograph of an amorphous film, (with a microdensitometer or by image analysis) the profile of the deposit can be obtained; something which could not be achieved by any other technique (see Case Study 4).

Thin films and foils can be examined by atomic resolution TEMs to give very detailed information about atomic arrangements within the materials and at surfaces and interfaces (see Case Study 13).

8.4.2 Thin foils

The preparation of thin metallic foils by electropolishing was a technique developed in the 1950s and early 1960s, and the study and understanding of the behaviour of crystalline materials took a great leap forward because of the technique. Work by Menter (1956), Bollmann (1956), Hirsch *et al.* (1965) and others showed that not only could the lattice planes of suitably thin crystalline specimens be imaged by microscopes with sufficiently high performance but the effects of lattice defects, e.g. dislocations, stacking faults, and precipitates, could be photographed for thicker specimens.

Observations on thin foils prepared from metallic specimens which have been subjected to various treatments have greatly improved our understanding of metallurgical processes. However, there has often been the suspicion that preparing a 0.1 µm thick specimen from an original many times that thickness has resulted in the loss or modification of some of the structure. High-voltage microscopy of thicker foils has confirmed that this can occur, and where possible metallurgical studies are now carried out on thicker foils in a higher-voltage microscope. The technique of STEM can also be helpful in giving maximum specimen penetration for a given accelerating voltage.

On specimens which are difficult to thin by electrochemical methods, e.g. refractory alloys, ceramics, or composite films, the method of ion-bombardment thinning (Barber, 1970, 1993) can be used to produce thin areas for conventional or HV transmission microscopy. Powders and dusts can also be thinned by this method, although the result seldom possesses an ideal parallel-sided profile and the transmission micrographs are often easier to interpret after the surface profile of the thinned material has been revealed by scanning microscopy.

8.4.3 Thin sections

The biologist has traditionally used sectioning techniques in his study of natural materials, whether of plant, animal or human origin. Embedding, sectioning and staining techniques were evolved long before the electron microscope was developed. The latter instrument has called for modification and refinement of these techniques to suit the different requirements of electron optics. The biological worker in electron microscopy now routinely uses epoxy or acrylic resins to embed his specimens, cuts ultra-thin sections and stains them with the salts of heavy metals. However, the wider availability of facilities for cryogenic preparation and examination has assisted many biological studies, particularly those involving elemental analysis, provided the necessary precautions are observed to avoid a new set of artefacts.

A new string to the bow of immunocytochemistry has been the use of colloidal gold and silver markers conjugated to specific stains for proteins and enzymes. Applied to ultramicrotomed sections they have proved valuable in both TEM and SEM, although the smallest, more sensitive, gold particles are beyond detection by all but the highest-resolution instruments.

Single and serial sectioning

Although for some investigations a single section or random single sections provide sufficient information, the realisation of the structure of a specimen in depth can be obtained most precisely by the painstaking collection and examination of a number of sections cut in a series from the same specimen. The same main structural features can be followed from picture to picture through the specimen and changes, e.g. branching, can be observed on the way. Some workers (Sjostrand, 1958; Pedler & Tilly, 1966) assembled the information so that it could be seen in three-dimensional form by printing the micrographs on transparent supports of thickness in scale with the lateral magnification of the micrographs (e.g. 1 mm separation corresponding to 0.05 μm sections at a magnification of ×20000). This could only be an approximation to the truth, however, since the thickness of a cut section is seldom known with anything like the precision of calibration of the magnification of the microscope. For that matter, the specimen may have undergone unknown dimensional changes during the preparation prior to sectioning.

The large number of serial sections required to examine even a small volume of specimen (e.g. 20 sections 0.05 μm thick to a micrometre) makes attractive the use of the high-voltage microscope to look directly at 'thick' sections several micrometres thick (Favard & Carasso, 1973). However, the difficulty is then to sort out the information in depth, and here the technique of stereoscopy can be of use.

As the older and more trusted technique, the ultra-thin section is used as reference when new or at least different methods of examination or preparation are being assessed for fidelity. Thus structural detail appearing in freeze-etched replicas is inevitably examined and compared with what a micrograph of an ultra-thin section would tell about the same sort of specimen. Likewise the success or otherwise of drying procedures is assessed by comparison of sections cut from normally prepared material with sections from material which has, for example, been freeze-dried, embedded and then sectioned.

Non-biological thin sections

Thin-sectioning techniques are often applied with advantage to industrial materials in order to find out more about their internal structure. Unlike the biologist, however, the industrial microscopist is frequently satisfied with a single scrappy-looking section, and would be embarrassed if expected to carry out methodical serial sectioning. One of the principal problems lies in the wide range of hardness of material which has to be cut, literally from putty to steel. The materials may also vary between solid metals and porous sponge-like structures. Each one therefore has to be treated on its own merits, embedded or left self-supporting, cut with a disposable glass knife or an expensive diamond knife, or frozen solid for sectioning, according to the circumstances.

In addition to the technique of looking through a parallel-sided thin section to study internal structure, there are many composite bodies for which all that is required is the ability to study the arrangement of different structural elements across the cross-section of the material; for example, to examine the penetration of a sprayed-on coating into a porous sheet of paper. In this case the required information is available from any fractured or sawn cross-section, but it is produced very conveniently for examination by either light- or scanning electron microscopy if the fractured face is trimmed off perfectly flat in a microtome or cryo-microtome. The specimen then becomes a surface to be examined and very well suited to electron probe x-ray microanalysis.

Microanalysis of thin foils and sections

The analytical transmission electron microscope is particularly useful for the examination of thin foils and sections (Weavers, 1971; Hren *et al.*, 1979). Whether the microscope is equipped for wavelength-dispersive or energy-dispersive x-ray spectrometry the thin parallel-sided specimen is very suitable for quantitative analysis with high spatial resolution, the latter being very approximately equivalent to the thickness of the foil or section. Small probes down to about 1 nm may be formed in the STEM mode, and illuminating spots down to 10 nm and below can be formed in the TEM mode on some microscopes, so that very localised identification can be carried out on a fully detailed static image.

Identification of selected crystalline features by electron diffraction is of course also possible; knowing the elements present from x-ray analysis is a useful start when trying to identify an unknown compound from its diffraction pattern.

An analysis with enhanced spatial resolution is also readily available on the SEM if the thin sample is appropriately mounted, but it may be difficult to determine which areas should be analysed without the advantage of a transmission image, e.g. if the SEM does not have a STEM detector.

8.5 Suggested further reading

Some textbooks describing applications of electron microscopy to a number of subjects have already been listed in earlier chapters. For the materials scientist Loretto & Smallman (1975), Wells *et al.* (1974) and Thomas & Goringe (1979) could be augmented by the more recent Forwood & Clarebrough (1992) and Holt & Joy (1989).

The biologist can find applications described in Wells *et al.* (1974), and Fujita *et al.* (1971). Both major fields of applications are to be found in Swann *et al.* (1974), in connection with the HVEM.

An intriguing picture-book edited by Williams *et al.* (1991) shows applications of a range of microscopic techniques, with explanations of how and why.

A unique collection of nearly 500 large-format TEM pictures illustrating biological applications will be found in Scanga (1964); SEM applications will be found in Kessel & Shih (1974). Geologists will find microfossils illustrated in Murray (1971). Soils and sediments are covered by Fitzpatrick (1993), Whalley (1978), Rutherford (1974) and Smart & Tovey (1981, 1982); mineralogy by Wenk (1976), Beutelspacher & Van der Marel (1968) and, more recently, McLaren (1991). There is, in addition, a wealth of material on all manner of applications reported in the proceedings of the international and regional conferences on electron microscopy, as well as other series of conferences organised by the Royal Microscopical Society (RMS), Microscopy Society of America (MSA), Electron Microscopy & Analysis Group of the Institute of Physics (EMAG), and other organisations throughout the world.

Biologists concerned about microanalysis would find of interest Roomans & Shelburne (1983), Gupta & Roomans (1993) and Sigee *et al.* (1993). Chemical applications of the TEM are described by Fryer (1979). The microscopy of cell structures is described by Gunning & Steer (1974), Tribe *et al.* (1975), and Hall (1978). Clinical studies are contained in Mandal & Wenzl (1979) and Johannessen (1978). Molecular biology is found in Sommerville & Scheer (1987). Microscopy of food is described by Vaughan (1979). Proteins are discussed in six volumes edited by Harris (1981–7). Immunocytochemistry is dealt with by Bullock & Petrusz (1982–9).

Readers interested in assessing the capabilities of Environmental SEMs would find articles on a range of different applications, together with a bibliography of earlier work, in Griffith & Danilatos (1993).

Case Studies

Case study 1

Electron Image Analysis in quality control of photographic emulsions

The properties of silver halide photographic emulsions are influenced by the size distribution of the sensitive grains. Batches of emulsions are evaluated using the procedures described below.

Sample Preparation (N.B. All preparation is carried out in safe light.)

The emulsion to be sized is diluted into water at 40 °C and centrifuged at 50 000 r.p.m. on to Formvar/carbon-coated TEM specimen grids. The grids are carefully removed from the centrifuge, dried and examined under a light microscope, using a green filter to minimise printout damage. Grids that are deemed suitable for image analysis are transferred to the TEM carousel in the CamScan SEM and examined in the STEM mode.

Examination

The grids are examined at a suitable magnification set for the analysis. The Image Analyser is calibrated using a cross-referenced traceable Sira 2160 lines/mm grid. The grid is imaged at 35 keV, the signal is fed to the Image Analyser and the 'true' magnification determined. The carousel is then rotated to the grid of interest and after suitable manipulation of brightness/contrast an image of the grains is sent to the image analyser. The image is thresholded, corrected for truncated grains and agglomerates, and the image measured. The results are classified into 4th root of two size classes and saved to disc. The results are plotted graphically and statistical information is printed. The results can be sent electronically to customers on the network. This mode of operating has resulted in faster turnaround of data to the customer. Compared with the previous mode of operating, direct imaging of grains, rather than carbon replicas, has resulted in a higher accuracy and the elimination of taking, processing and imaging TEM negatives. (By courtesy of Kodak Ltd, Research Division.)

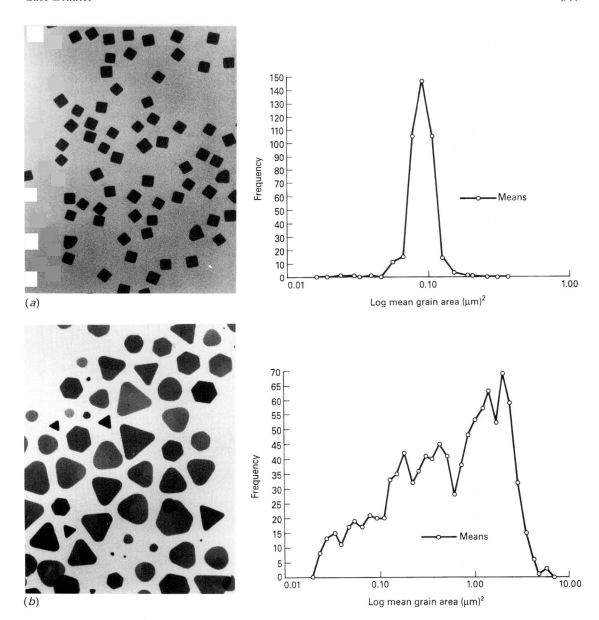

(*a*) Emulsion for Graphic Arts
(*b*) Tabular 'Gold' emulsion

Case study 2

Electron microscopy in the study of pigments and colours

From its earliest days the electron microscope has been used in production quality control and in the development of new pigments and colours. Although many pigments have particle sizes of micrometres and can be seen by light microscopy the sub-micrometre resolving power of scanning and transmission electron microscopes simplifies the recognition of agglomerates and enables the finest particles to be examined. With the addition of electron diffraction and x-ray microanalysis a much more detailed study of coloured particles is possible than hitherto. Care is needed, however, to ensure that the particles studied are representative of the original sample.

Dispersed pigments and fillers in surface coatings and glazes can be examined by using appropriate techniques of specimen preparation, including direct imaging with backscattered electrons, replication, cross-sectioning and thin sectioning, and ion-bombardment thinning. Selected-area studies using SEM or TEM can give a very detailed picture of ageing and weathering processes.

The illustrations show some examples of the use of electron microscopy in pigment research and development. (By courtesy of Johnson Matthey plc.)

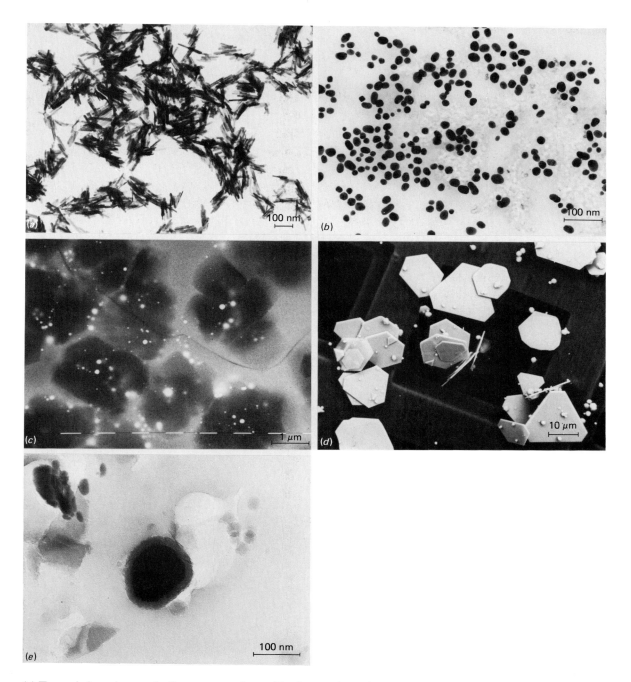

(*a*) Transmission micrograph of a transparent iron oxide pigment shows the particle morphology. Electron diffraction enables the compound to be identified.

(*b*) Micrograph of a gold colloid shows the particle size and uniformity in a pink ceramic colour.

(*c*) Backscattered electron scanning micrograph shows the distribution of gold particles in a pottery enamel. The 80 keV electron beam penetrates many micrometres into the enamel. Images of deeper-situated gold particles are diffused and unsharp.

(*d*) Scanning micrograph of gold flake used in pottery decoration.

(*e*) The stability of pigments may be greatly improved by encapsulation within materials such as alumina or silica. The completeness of the encapsulation can be proved with ESCA, and the thickness of the coating can be measured from a thin section, as in this micrograph.

Case study 3

Electron microscopy as an aid in virus identification.

Electron microscopy finds a means of differentiating between virus types by specific reactions and by morphology. Two examples are illustrated. (*a*) Virus particles can be identified by their reaction to specific antibodies. The micrographs show negatively stained papovavirus particles without anti-body (*right*) and with a specific antibody coating the particles (*left*). In (*b*), negatively stained dispersions of virus suspensions on a Formvar support film are examined at high magnification in the TEM. The morphology of viruses of the *varicella-zoster* group (chickenpox) (*left*) enables them to be distinguished from those of the *variola-vaccinia* group (smallpox) (*right*). A full identification within the groups would require the use of culture techniques. (Micrographs and information by courtesy of A. M. Field, Central Public Health Laboratory, Colindale, London.)

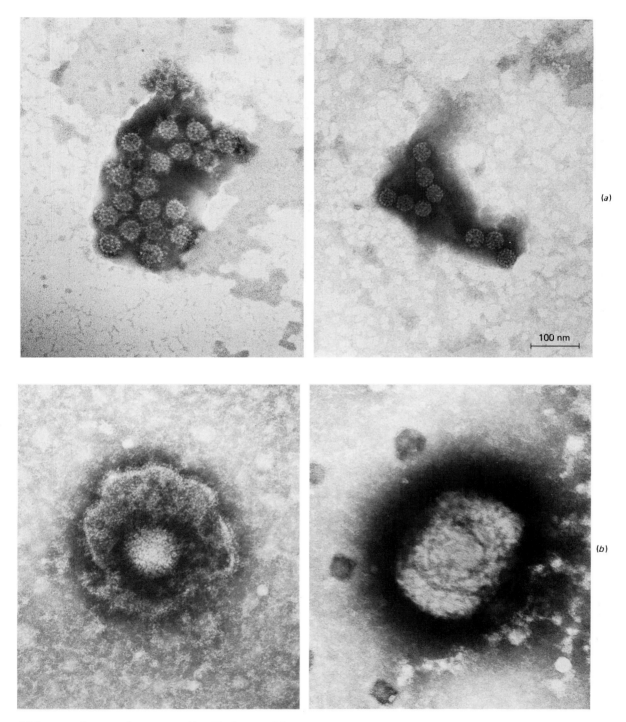

(*a*) Immune electron microscopy to identify virus particles
(*b*) Differentiation between smallpox and chickenpox viruses

Case study 4

Electron microscopy in the study of the growth of electroless nickel films

Films of metal may be deposited by vacuum processes such as evaporation and sputtering, or by wet processes such as electroplating or electroless plating. The last of these enables adherent metal films to be prepared on insulating substrates such as glass or plastics merely by immersing the surfaces in a succession of chemical solutions. The surface is first *sensitised* and *activated* in two baths and then *plated* with nickel, copper, gold or other metal in a third bath. The micrographs show some ways in which transmission microscopy was used to study the initial formation and characteristics of nickel films formed by the electroless process.

Thin (10 nm) films of evaporated silicon oxide on specimen support grids were used as substrates, being dipped into the necessary solutions for plating. Micrograph (a) shows a nickel deposit formed (after appropriate sensitisation and activation) by plating for 10 seconds at 65 °C. It is essentially a discontinuous single layer, although in some places a second layer of particles has started to form. The island structure is characteristic of this type of film, the gap between grains of nickel being approximately 1.5 nm. Quantitative microdensitometry on the photographic image shows that the individual granules are flat-topped. A cross-section, drawn to scale (b), shows the profile of three granules from a slightly thinner electroless nickel film. The electron scattering coefficient was established by measurements on micrographs of evaporated nickel films of known thickness.

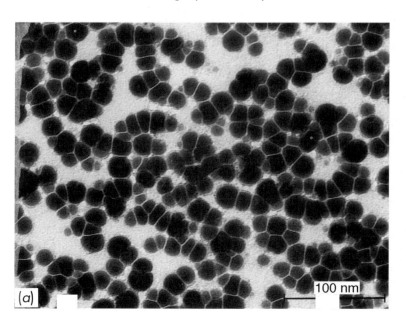

(a)

100 nm

(b)

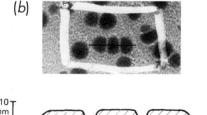

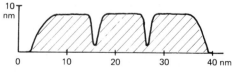

The thin electroless nickel films are shown to be amorphous by the diffuse haloes of their electron diffraction patterns (c)(i). Electroless nickel is plated from a solution containing a few per cent of phosphorus. Annealing the film shown in (a) at 450 °C in air resulted in identifiable rings of nickel phosphide (Ni_2P) being observed before the whole pattern became sharply defined as nickel oxide (NiO) after 30 minutes heating, as shown in diffraction pattern (ii). A transmission micrograph (d) of the same film in the final state showed no sign of the original structure, but strong diffraction contrast. (Work reported by courtesy of Sira Ltd.)

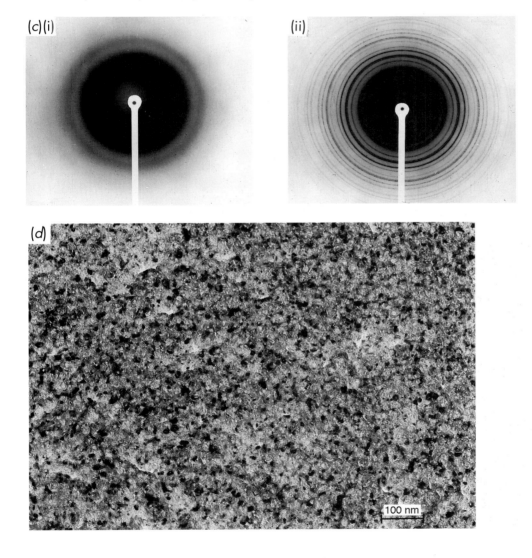

(c)(i)

(ii)

(d)

100 nm

Case study 5

Electron microscopy applied to food-processing research

A wide range of techniques of light and electron microscopy and analysis are used to study food processes. Three illustrations show the application of TEM and cryo-SEM.

(a) Micrograph made on a cryo-SEM of a frozen emulsion from the mixing tank of a pilot-plant producing a low-fat spread. Forty percent water is added to oil containing mono- and di-glyceride emulsifiers, in the mixing tank.

Cryo-SEM allows the structure of the emulsion to be examined whilst undergoing a dynamic process. Small amounts are removed, rapidly frozen to preserve structural features, and then fractured under vacuum. This allows successive stages in the development of structure to be identified.

(b) Scanning micrograph of frozen whipped cream. Adsorption of fat globules to the air/water interface of bubbles takes place during whipping, and globule bridges develop between bubbles to produce cream of good stiffness and stability. The effectiveness of this stabilisation can be assessed by directly observing the globules on the air/water interface and in the aqueous phase by SEM.

(c) Transmission micrograph of a section of resin-embedded cottage cheese. Most of the protein in milk exists in the form of particulate casein micelles. During manufacture, starter bacteria are added to the milk to produce lactic acid and to lower the pH. This causes the casein micelles to aggregate into a three-dimensional network. The quantitative relationship between processing parameters and texture can be established using microscopy of this sort in conjunction with image analysis.

(Text and micrographs by courtesy of BBSRC Institute of Food Research, Reading.)

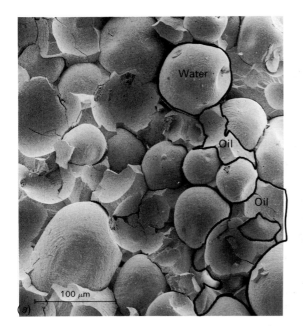

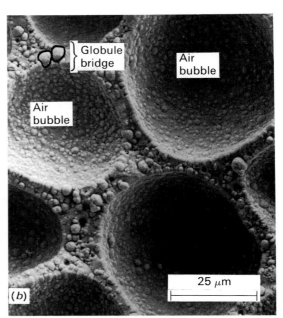

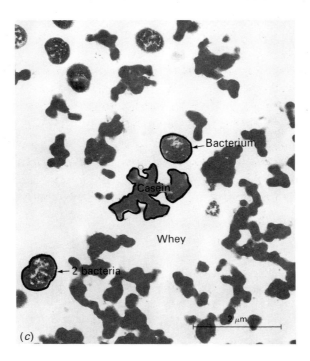

Case study 6

Examination of fibrous surfaces with the SEM

The scanning microscope has made possible and easy, even with the simplest of instrument, the detailed study of fibrous and interwoven materials, especially those of a translucent or transparent nature which can hardly be seen in the light microscope. These micrographs of a general purpose (No. 1) filter paper show the ease with which the general view can be magnified to reveal fine detail in individual fibres. At the highest magnifications local differences in the electrical state of the specimen are likely to be found, and it is important to achieve correct compensation of astgmatism for each field.

A small (about 1 cm) square was cut from the filter paper and stuck flat on a specimen stub using double-sided adhesive tape. Silver paste was applied around the edges of the square to earth the specimen to the stub. It was then coated with a gold/palladium alloy by vacuum evaporation in a rotary coater. 10 keV electrons were used and the specimen was tilted 30° from the 'square-on' position.

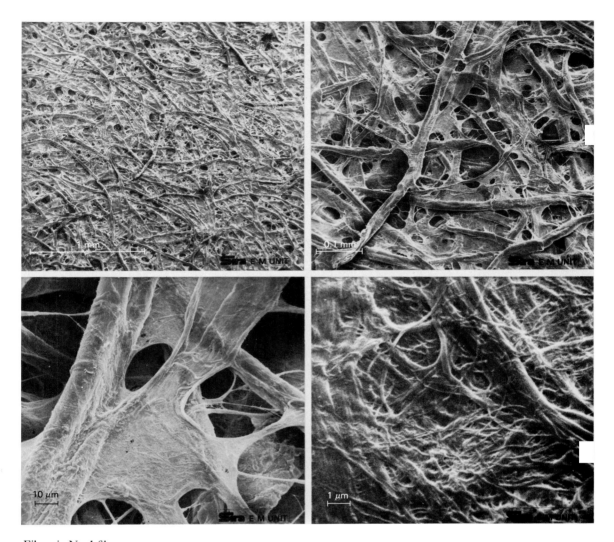

Fibres in No. 1 filter paper.

Case study 7

Study of the structure of opals using electron microscopy.

Precious and other opals are composed of spherical particles of silica formed over a very long period of time by the concentration into a gel of underground lakes of silica-rich water. Where the silica particles are uniformly sized and regularly arranged they form an optical diffraction grating giving coloured reflections which vary with the angles of incidence and viewing. The particle size of precious opal is too small to be seen except by electron microscopy. Micrograph (*a*) is a scanning micrograph of a chalky sample of opal, in which the layers of particles are not well cemented together. The regular hexagonal arrangement of the particles in their layers can be clearly seen. Micrograph (*b*) is a transmission micrograph of a replica of a similar chalky opal, in which small 'flats' or contact areas between adjacent layers of spheres can be seen. Micrograph (*c*) is a transmission micrograph of a replica of a fractured and etched surface of a sample of 'potch opal', a white glassy material in which the silica spheres have a range of diameters so that no regular layer formation is possible. The replica shows that each sphere is made up of shells of constant volume. The smaller spheres are so because the volume of each shell is less. Only spheres which fracture through the centre show the rings clearly. (Micrographs and information by courtesy of J. V. Sanders, CSIRO Division of Materials Science, Melbourne, Australia.)

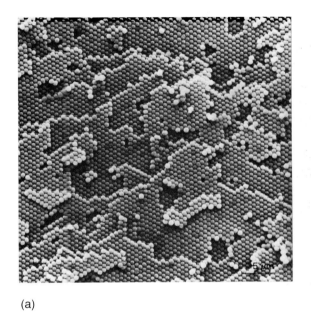

(a)

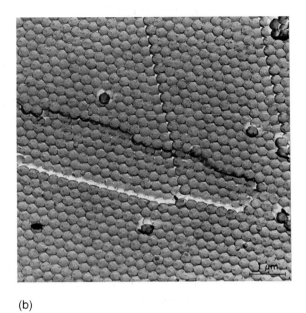

(b)

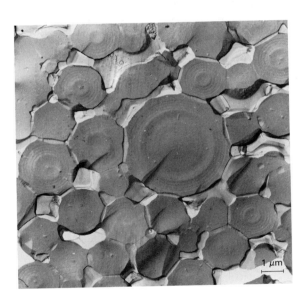

(c)

Case study 8

Electron microscopy in the study of glass polishing

Detailed studies have been made of glass surfaces to elucidate the processes occurring during the polishing of glass, e.g. for lenses for spectacles, cameras, etc. Replication is particularly useful in these studies as it enables the features of a chosen area on a glass specimen to be examined at intervals between the ground and polished states. Within certain limitations the exact area to be studied in detail can be selected in the final stages and then located and photographed in the earlier replicas to obtain an intimate picture of the surface changes occurring during polishing.

Four stages in the polishing of a 5 cm diameter flat blank of ophthalmic crown glass are shown here, starting in the 'greyed' state. Polishing was against a pitch lap with a slurry of cerium oxide and water. After 10 turns of the polishing lap, i.e. about 5 seconds polishing (*micrograph upper right*), the highest peaks of the surface have been removed and the sample has lost 0.4 mg in weight. A further 10 turns of the lap (*lower right micrograph*) develops the flat areas further and another 0.4 mg of glass has been removed. Arrows on the two right-hand micrographs point to one of the ways in which the surface changes during polishing, i.e. by flaking out of pieces of glass which were strained during grinding but not completely cracked out. The final stage shown (*lower left*) after a total of 320 turns of the lap, shows the development of an extensive area of flat surface by gradual removal of the high spots. It is significant that after the detailed study of a large number of micrographs the only polishing mechanism found was by a physico-chemical removal of glass. No sign of surface flow was observed under the conditions of polishing typical of normal commercial procedures. (By courtesy of Sira Ltd.)

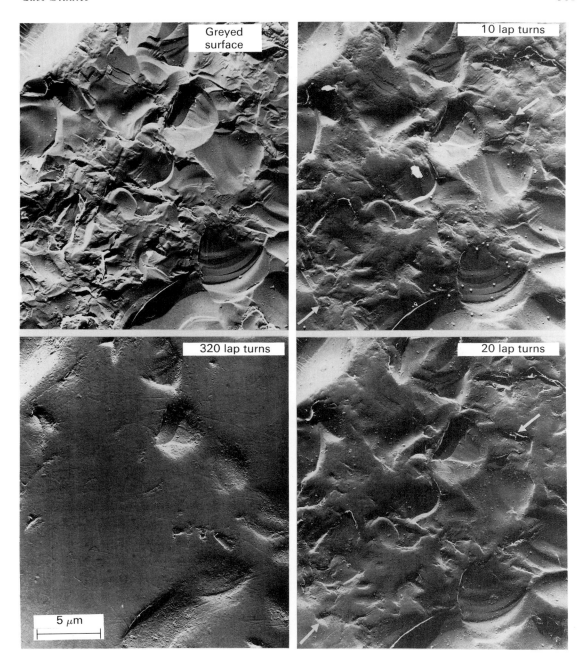

Case study 9

The xerographic copying process seen by electron microscopy

In this process a light image projected on to a selenium surface gives rise to an electrostatic charge pattern on the selenium. A fine insulating powder (the *toner*) is dusted across the selenium surface, where it adheres to the charged areas, forming a powder image corresponding to the original light image. This layer of powder is transferred to a sheet of paper on which it is fused by application of heat into a permanent image bonded to the paper fibres. There are a number of variants of this process, and the illustrations show how electron microscopy can be used in the study of the materials used in one of them.

1(a,b). The *toner* is a low-melting-point resin loaded with a black pigment, frequently carbon black. The particle size distribution in the toner powder can be seen from TEM shadowgraphs; the shape of the particles is perhaps more readily appreciated from SEM micrographs. Particle in (a) were dusted on to a Formvar support film and in (b) were held in a layer of silver paste on a stub and gold/palladium-coated *in vacuo*.

2. The *carrier* is the means of distributing toner over the charged selenium surface. In one process spherical steel shot coated with an insulating polymer layer is used as carrier. The carrier illustrated is some 25 times as large as the toner. Toner particles adhere to the insulating coating by electrostatic attraction; the toner-coated carrier forms the *developer* for the latent image on the selenium. The developer particles roll across the charged selenium surface and toner is pulled off the carrier to adhere to the local charges in the image. The SEM does not reveal the presence of the polymer layer on the carrier particles when operated at normal beam energies, since the layer is then transparent to electrons. At a much lower operating voltage (e.g. 3 kV) the extent of the polymer coating can be seen; thicker regions charge up brightly and thinner areas show up darker than the bare surface of the steel shot. The carrier particles were mounted on silver paste and examined uncoated.

3. The fused black resin image can be seen on the paper fibres in varying degrees of detail in the SEM. Bonding betwen adjacent toner particles and wetting of the paper fibres can be studied in this way; the effects of fusion temperature and surface additives can be studied. A portion of xerographic copy was mounted on a specimen stub with double-sided adhesive tape and coated with Au/Pd alloy in a rotary coater. 10 keV electrons were used; stub tilt was 30° in (a) and (b), and 70° in (c).

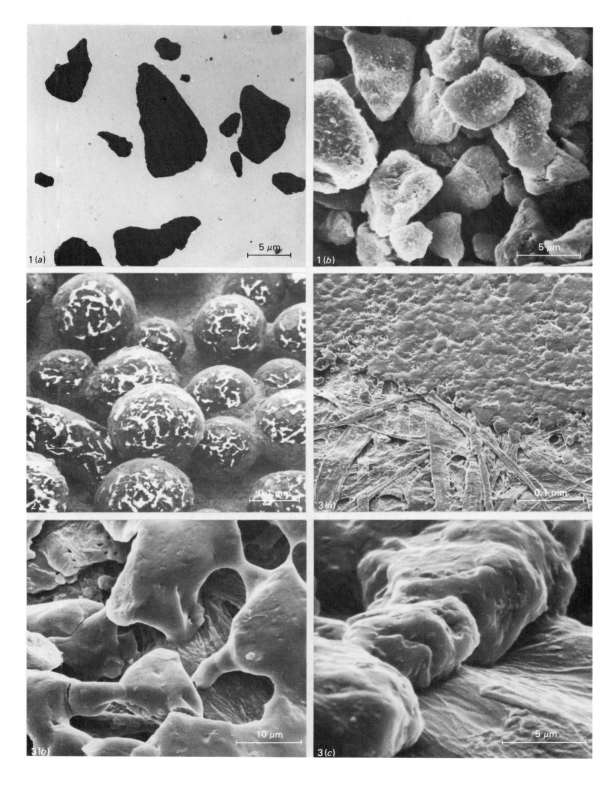

Case study 10

The electron microscope in pathology

The electron microscope, usually the TEM used to examine ultra-thin sections, provides the means of studying changes in tissue brought about by illness, accident or the use of drugs. Two such studies on skeletal muscle are illustrated here.

1. Changes in muscle resulting from a mild burn.

(a) Normal skeletal muscle fixed in glutaraldehyde followed by osmium and embedded in epoxy resin before being thin-sectioned. The muscle is totally relaxed and the Z-bands (Z) which make the striations seen under the light microscope are very regularly spaced. There is a large bundle of mitochondria (M) in the middle.

(b) Sample taken from the deep muscle after a burn, showing a small degree of injury. The Z-bands are less regular. The fibre on the left is more damaged.

(c) Severe disruption of the muscle fibre seen in a sample taken from muscle lying immediately beneath the skin. The striations can only be seen with difficulty in the fibre on the right.

2. Changes in muscle induced by treatment with a drug.

Muscle as in 1(a) was incubated in Ringer's bicarbonate solution for 40 minutes and in the presence of a drug similar to theophylline but much more pharmacologically active. The micrograph shows how the muscle is highly contracted and the mitochondria are swollen.

(Description and micrographs by G. R. Bullock, Ciba-Geigy Pharmaceuticals Division, Horsham.)

1(a)

M

Z

1 μm

1(b)

Z

1 μm

1(c)

1 μm

2

1 μm

Case study 11

Electron microscopy in the Plant Sciences

Ever since the first recorded light micrographs, studies of plant material have made use of all available microscopic techniques. Researches using SEM and TEM are reported below.

1 Tylose development in oak

Tyloses form in the vessels (tubular water-conducting structures with lignified cellulose walls) in the wood of many tree species. Living cells known as parenchyma lie in contact with these vessels and may absorb water and dissolved ions from them via unthickened regions of cell wall known as pit membranes. The tension in the water as it is drawn up the vessels counteracts the tendency of water to move into the living cell by osmosis. If the vessel is damaged, for example by cutting into the stem, the tension in the water is released, disturbing this balance, and water flows into the parenchyma cell from the vessel. As a result of the consequent increase in internal pressure, the pit membrane of the parenchyma cell begins to swell out into the vessel. The parenchyma cell prevents the membrane from bursting by actively adding more material to it, and the resulting tylose may swell like a balloon until it fills the vessel. The growth of many membranes in this way may result in the vessel being filled with a mass of membranous material. This prevents the entry of pathogens into the system, and limits the loss of water through the damaged region.

Micrograph (*a*) shows a transverse surface of a piece of oak wood (*Quercus robur*) in which the vessels are prominent. Micrograph (*b*) shows the inner wall of a vessel with the numerous pit membrane regions clearly visible. In micrograph (*c*) tyloses may be seen beginning to grow to fill the interior of a vessel close to a wound made four hours earlier, while in micrograph (*d*) a balloon-like tylose has almost blocked the vessel twenty-four hours after wounding.

2 Stylar transmitting tissue in tomato

Transmission electron micrographs illustrate cells from the style, or female part, of a tomato flower. The central core of the style, known as the transmitting tissue, consists of many elongated cells with thin cell walls (the lightly staining lines enclosing the cytoplasm of the cells in (*a*)). Between these thin walls is densely staining material which has been secreted by the cells, and this intercellular matrix provides the medium through which the pollen tube will grow during the process of fertilisation. The extent of this

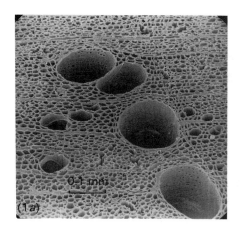

(1a)

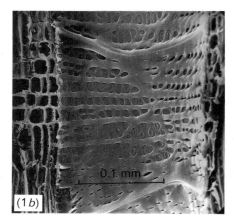

(1b)

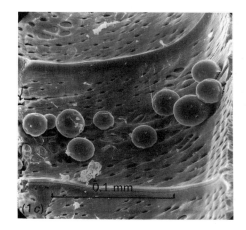

(1c)

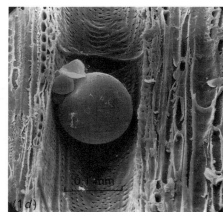

(1d)

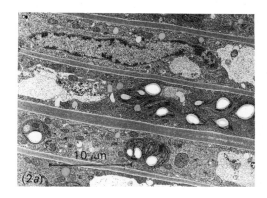

(2a)

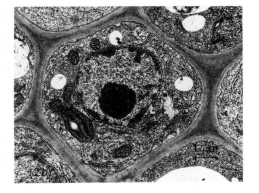

(2b)

intercellular material is apparent in transverse section (*b*), being thickest at cell junctions. Transmission electron microscopy established that it is through these intercellular regions that pollen-tube growth occurred, a fact which was not apparent from studies of light microscope sections.

(Text and micrographs by J. R. Barnett (1) and L. J. Bonner (2), School of Plant Sciences, The University of Reading.)

Case study 12

Environmental SEM for the study of hydrated specimens

Environmental scanning electron microscopy (ESEM) can be used to study biological samples in their hydrated state. By manipulating sample temperature and chamber water-vapour pressure, evaporation or condensation can be controlled by the microscope operator. It is often useful to compare images obtained using ESEM with those from low-temperature scanning electron microscopy (LTSEM).

Figure 1 (*overpage*) shows an ESEM image, taken at 20 kV, of fresh mushroom spores. The sample temperature was 8 °C with a chamber pressure of 8.5 mbar of water vapour.

Figure 2 shows an image from a similar area of the same mushroom obtained by LTSEM. Clear differences can be seen in the surface morphology. With ESEM, the surface is relatively smooth with the spores appearing turgid. In contrast, use of LTSEM shows partially collapsed spores with surface structures present. The LTSEM sample had surface moisture sublimed off and the surface structure may represent crystallisation of dissolved material. In some cases spores have become detached when preparing for LTSEM. Although this may illustrate the normal mechanism of dispersal, careful sample selection indicates that it is more likely to be mechanical stress during sample preparation.

Additional studies have been performed with artificial arterial grafts where endothelial cells have been seeded on to a gelatin-infused matrix of woven Dacron. In this study glutaraldehyde-fixed samples were used. Only ESEM adequately preserved the gelatin matrix, thus allowing an assessment of the endothelial cell coverage and morphology (Figures 3 and 4).

(Text and micrographs by C. J. Gilpin, School of Biological Sciences, University of Manchester.)

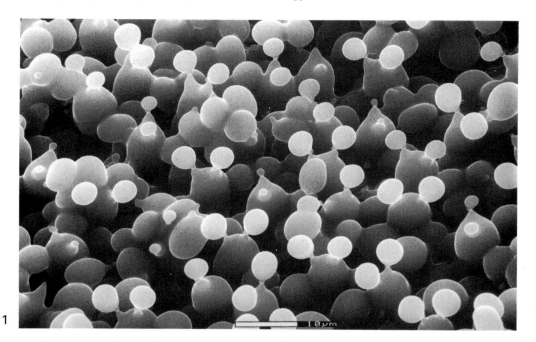

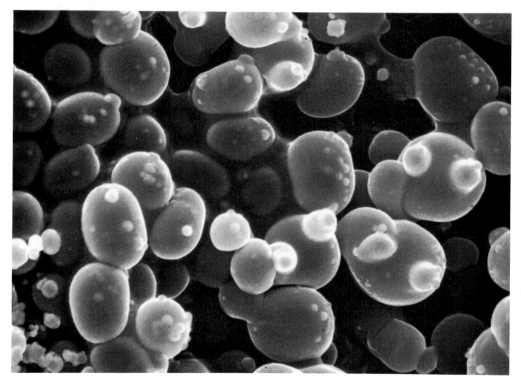

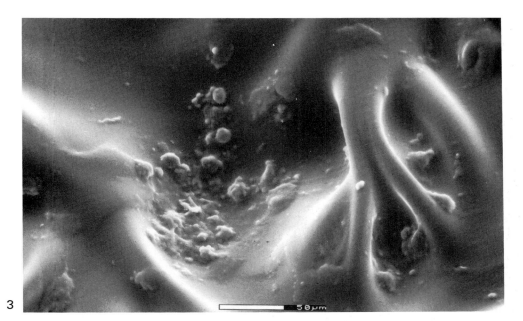

3

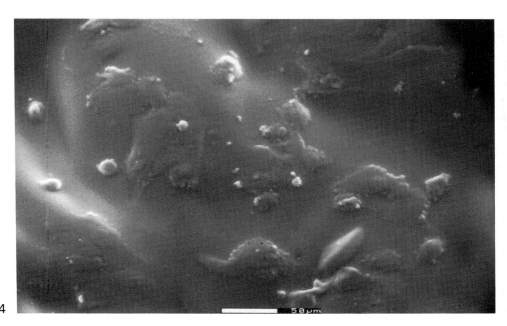

4

Case study 13

HRTEM study of defects in synthetic diamond films

The growth of thin films of diamond on various substrates is an important development, there being many potential applications for such materials. High-resolution transmission electron microscopy has made some valuable contributions to this field, particularly since instruments with resolving power better than 0.2 nm became available.

Sample preparation involves the chemical dissolution of the substrate (usually silicon), followed by careful argon ion milling, to leave free-standing diamond films, parts of which are thin enough (i.e. <10 nm) for successful imaging. Grains are tilted to bring a <110> axis parallel to the electron beam, since this orientation presents various sets of lattice planes edge-on.

Images reveal a wide variety of disorder within grains – Figure 1 illustrates simple *twinning*, i.e. reflection of the diamond crystal lattice across internal planes, with minimal disruption. Figure 2 shows more complex defect structures, some of which have been shown to arise from the interaction of different families of twin variants. Especially interesting are the features indicated by the arrow; they correspond to the intersections of five twin variants at a point, and are in fact planar, five-membered carbon ring structures. These can propagate to the surface of growing crystals, and account for the fact that many of the crystallites display apparent five-fold symmetry, surprising for the simple cubic crystal which is diamond. (Text and illustrations by J. L. Hutchison, Department of Materials, University of Oxford.)

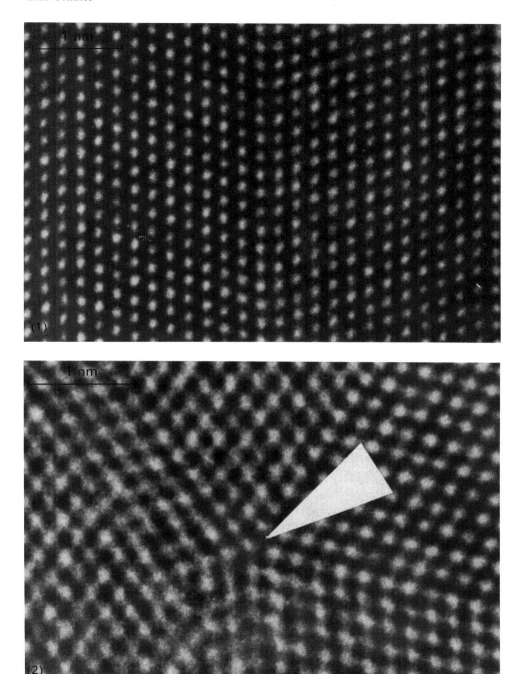

Case study 14

Electron microscopy as an aid in the development of car exhaust catalysts

In the USA, Japan and Europe emissions of noxious gases from motor vehicle exhausts are tightly governed by legislation. In order to meet this legislation, which governs the emission of unburnt fuel, carbon monoxide and the oxides of nitrogen, most vehicle manufacturers use emission-control catalysts to convert these toxic gases to harmless CO_2, N_2 and H_2O. Depending on the type of vehicle, up to 90% of the pollutants must be removed to satisfy the legislation.

In the catalysts for exhaust control various active ingredients, including metals from the platinum group, are applied to a 'washcoat' of microcrystalline alumina coated on a honeycomb block of ceramic which is incorporated in a metal canister in the exhaust system. The catalyst reaches temperatures of up to 1000 °C in operation and processes such as grain growth and phase transformation in the alumina, and sintering of the catalytic additive, can result in a progressive loss of efficiency of the catalyst. Commercial exhaust catalysts are therefore formulated and processed in such a way as to slow down the rate of change with use. The changes are microscopic in scale and are studied by a combination of high- resolution electron microscopy, x-ray diffraction, ESCA and EPMA, combined with physico-chemical techniques such as gas adsorption for surface-area measurements and differential thermal analysis. Examples of the microscopic studies are given here. (Reproduced by permission of Johnson Matthey plc.)

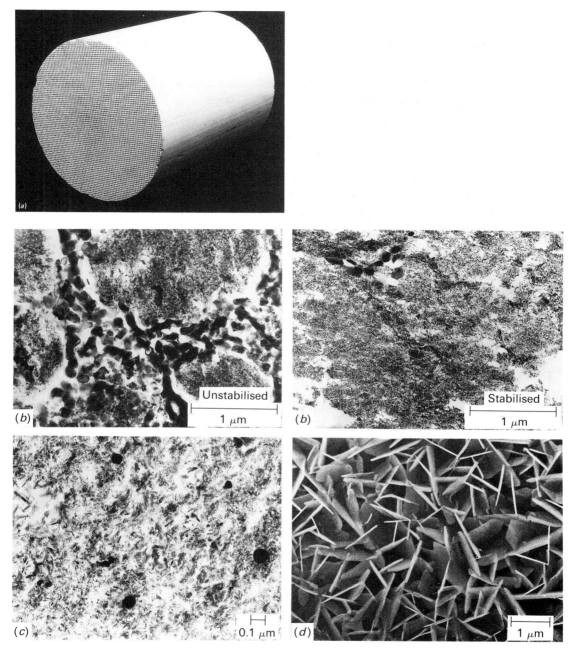

(*a*) Photograph of an uncoated block of honeycomb ceramic. This normally contains between 30 and 60 cells per cm² of cross-section.

(*b*) Transmission micrographs of thin sections of resin-embedded washcoat show growth of extended crystals of Al_2O_3 after heat treatment to 1100 °C compared with a similarly heat-treated alumina with stabilising additives.

(*c*) Platinum particles in a fresh catalyst are initially below 1 nm in size. After prolonged use they may sinter together into larger particles (thin section).

(*d*) Operation of a catalyst with leaded fuel can in certain conditions result in the formation of a surface growth of crystals of a lead compound which can be identified by electron diffraction. (SE image of a bulk specimen, carbon-coated).

Case study 15

Microfossils and the search for energy

Microfossils are the skeletal remains of microscopic plants and animals that abound in ancient and modern sediments. Terrestrial sediments, for example, are rich in the remains of spores and pollen and these may make up much of the bulk in some coal and peat deposits. Siliceous diatomites accumulate on the floors of tranquil mountain lakes while nearly one-half of the ocean floor is carpeted with the chalky remains of foraminifera and coccoliths.

These tiny organisms are so abundant and varied that a teaspoonful of sand from a tropical lagoon can contain tens of thousands of shells belonging to as many as 400 different species (see Figure). Many building materials such as chalk, Portland cement, refractory bricks and the white lines on roads owe their useful properties to the microfossils in them. But the main use of microfossils is for the relative dating of rock layers and the interpretation of their ancient environment of deposition. These are particularly important in the subsurface exploration of oil and gas fields since hydrocarbons tend to be formed in rocks of one age and type but accumulate in reservoirs of another rock type. Microfossils are excellent for this kind of analysis because they are small, abundant and varied, and occur in a wide range of rock types so that a story can be constructed from the tiny rock chips brought up to the surface by the drill of the oil rig.

The Figure shows a strew of microfossils upon an SEM stub from which a trained micro-palaeontologist could make the following rapid interpretation for a geologist searching for oil:

Age of sediment: Pleistocene (i.e. younger than 2 million years) as shown by the presence of *Globigerinoides ruber, Ammonia beccarii, Haynesina* sp.

Environment of deposition: shallow marine water, probably less than 100 m deep and within the photic zone, as shown by the dominance of bottom-dwelling (benthic) forms, especially miliolid foraminifera, reticulate ostracods and calcareous algae and by the presence of microborings in the shells. An open-ocean aspect is indicated by the presence of planktonic foraminifera. The water-current energy was moderate since the microfossils are not greatly size-sorted or fragmented.

Latitude: about ten species of miliolids, including striated forms, suggest warm temperate to tropical conditions; the planktonic forms are typical of tropical surface waters. Several species are restricted in their distribution to the Caribbean province.

Hydrocarbon potential: the microfossils in the sample indicate that the parent rock is too young and too near the surface for maturation of any

400 μm

hydrocarbons into oil and gas to have taken place; a warm, aerated tropical bay is unlikely to produce a good source rock, though shallow marine sediments such as this can provide good porous reservoirs for hydrocarbons. In this instance, the horizon is too young and too near the surface to trap any oil or gas satisfactorily. *Keep drilling down!*

(Micrograph and text by M. D. Brasier, Geology Department, Hull University.)

Appendixes

It is possible to prepare specimens for, and to use electron microscopes, by following established routines on commercial equipment without understanding what is going on inside the equipment. However, as with a great many practices, a knowledge of how and why things are happening can be of great assistance in making the best use of them and in appreciating the reasons for modifying them. For readers who would like some background information the following four appendixes provide introductions to the important topics of high vacuum, vacuum-deposition of thin films, the analysis of x-rays, and electron sources in electron microscopes. Each section can be no more than a précis of its subject; readers wishing to be more deeply involved are referred to more comprehensive publications.

Appendix 1: Production and measurement of high vacua

The operation of an electron microscope is dependent on electrons in the illuminating beam travelling along predictable paths from their source to their destination. They will only do this if they are not deflected by collision with any atoms or molecules other than those in the specimen. The inside of the microscope must therefore be kept sufficiently free from gas molecules, of which there are normally more than 10^{19} in each cm^3 of air at atmospheric pressure. It is only by reducing the gas pressure to less than one-millionth of atmospheric that a sufficiently long collision-free path is obtained for the microscope to operate. This is, however, only a statement of the upper limit of pressure which can be tolerated before electron microscopy becomes impossible. It is now appreciated that examination of specimens in electron microscopes is much more satisfactory if the column is clean, i.e. if the pressure of all gases and vapours in the instrument is well below the 'tolerable' level. This is in any case only one side of the story; at any higher pressures it would be impossible to maintain the electron gun in operation, and we shall see in a following section that improvement in vacuum by a further 10^6 times is necessary for some types of electron gun to operate reliably.

Although it would perhaps be possible to have an electron microscope column which was evacuated and sealed off in the factory, and operated thereafter as a vacuum-filled device like a television tube or radio valve, the complete outgassing of the column prior to seal-off would be a formidable operation. In any case the practical requirement for exchanging specimens, filaments and photographic recording materials means that a dynamic, continuously pumped vacuum system is used in all electron microscopes. A gas pressure at least as low as 10^{-3} mbar and preferably several orders of magnitude lower is necessary in the electron–optical column. In

Table A1.1. *Vital statistics of vacuum systems (rounded off to whole powers of ten)*

Pressure, mbar	1000 (atmospheric pressure)	1	10^{-3}	10^{-7}	$<10^{-7}$
		ROUGH VACUUM	FINE OR LOW VACUUM	HIGH VACUUM	ULTRA-HIGH VACUUM
Molecules cm^{-3}	10^{19}	10^{16}	10^{13}	10^{9}	$<10^{9}$
Wall collisions, cm^{-2} s^{-1}	10^{24}	10^{20}	10^{17}	10^{13}	$<10^{13}$
Molecular collisions, cm^{-3} s^{-1}	10^{29}	10^{23}	10^{17}	10^{9}	$<10^{9}$
Mean free path, cm	10^{-5}	10^{-2}	10	10^{5}	$>10^{5}$

addition the gases and vapours remaining in the column should be compatible with the efficient operation of the microscope.

In vacuum technology it is usual to divide the pressure range between atmospheric pressure and ultra-high vacuum into a number of bands. These are shown in Table A1.1 with rounded-off values for a number of relevant pressure-dependent characteristics. The pressures usually found in electron microscopes and specimen coating units are in the high-vacuum range, between 1×10^{-4} and 1×10^{-7} mbar being most common.

[N.B. Pressure on the Système International d'Unités (SI) is measured in *newtons per square metre*, N m^{-2}. This unit is named *pascal*, Pa. In this book references to gas pressures use the SI-approved alternative, *bar* (1.0 bar = 10^5 Pa), since the *millibar* is of the same order of magnitude as the previously used unit *torr*. This was itself the replacement for *mm of Hg*. Precisely, 1.0 mbar = 10^2 Pa = 0.750 062 torr = 1.450 38 $\times 10^{-2}$ lbf/in^2.]

Vacuum pumps

The essential characteristics of any type of vacuum pump are its pumping speed, i.e. the rate at which it will take in gas at the inlet side, and the limiting inlet and outlet pressures between which it will operate. There are three main divisions of vacuum pump commonly used in electron microscopy, each with its own particular characteristics which determine its field of use:

(1) Mechanical, including rotary (backing) pumps and turbo-molecular pumps
(2) Vapour diffusion pumps
(3) Sorption, ion and sublimation pumps

The majority of electron microscopes and vacuum-coating units use a combination of rotary and diffusion pumps, but in the ultra-high vacuum range turbo-molecular or ion pumps may be used as well. Systems for which the presence of even traces of hydrocarbon vapour are unacceptable (e.g. semiconductor processing plants) may use spiral pumps or diaphragm pumps instead of rotary pumps, but these will not be described here.

Rotary pumps

These are the simplest means of producing a 'fine', or high, vacuum and require only a source of driving power, usually a fractional horse-power electric motor. The rotary pump can be likened to a broom working in a cylindrical chamber, continuously sweeping gas molecules from the inlet port towards and out of the exhaust port. The schematic diagram (Figure A1.1(*a*)) shows how an eccentrically mounted rotor with sliding vanes traps gas between itself and the chamber walls at the inlet port and compresses it in further rotation until it is forced out through the exhaust port, which is sealed by a one-way valve. A second pumping chamber and rotor can be added within the same outside casing as shown, the second stage acting as a backing pump to the first one, giving an improvement in the ultimate vacuum obtainable. In practice, the two chambers would be coaxial, with the rotors mounted on the same shaft. The whole unit is mounted in a vessel of oil of relatively low vapour pressure.

The pumping performance of a rotary pump is such that it will reduce the pressure in a vacuum system from atmospheric pressure (1 bar) to below 10^{-1} mbar, usually 10^{-2}–10^{-3} mbar. Two-stage rotary pumps will reach vacua down to 10^{-4} mbar before the back-flow past the vanes reduces the pumping speed to zero. Easily condensible vapours, such as water and organic solvents, would condense to liquid form at the outlet port, and not be forced out of the pump chamber. This behaviour is prevented if a small volume of air ('gas ballast') is injected into the pump late in the compression cycle. The ultimate performance of the pump is impaired when it is operated with gas ballast, so this procedure is normally used intermittently to purge the pump of trapped vapours. Figure A1.1(*b*) gives typical performance curves for single- and two-stage rotary pumps suitable for use in electron microscopes or coating units. On its own a rotary pump would be unsuitable for producing the high

Figure A1.1. The rotary-vane vacuum pump. (*a*) In this two-stage pump, gas entering by the intake port (4) is compressed by the sliding vanes (3) in the first chamber and forced through the connecting duct (5) into the second chamber where it is further compressed until its pressure is high enough to open the exhaust valve (6) and allow it to escape. Key: (1) Cylindrical housing (stator); (2) Eccentric rotors; (3) Sliding vanes; (4) Intake port; (5) Duct connecting 1st and 2nd stages; (6) Exhaust valve. (*b*) Performance curves of commercial rotary vacuum pumps suitable for use in electron microscopy. The adverse effect of gas ballast operation on the ultimate vacuum can be seen. (By courtesy of Leybold AG (*a*) and Edwards High Vacuum International (*b*).)

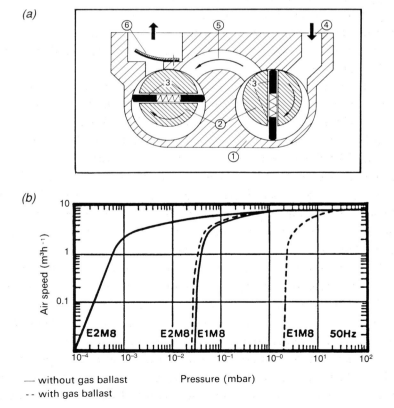

vacuum required in electron microscopes. It is used for the initial evacuation from atmospheric pressure to a level at which a faster pump can take over, and is then used to 'back' the outlet side of this faster pump.

Vapour diffusion pumps

In the same way as a leaf thrown on to the surface of a fast-moving river is rapidly carried off downstream, a gas molecule moving into the path of a jet of oil or mercury vapour molecules is swept along with them. This is the principle underlying the vapour diffusion pump. Gas molecules from the vessel being exhausted diffuse into dense fast-moving streams of vapour molecules from annular umbrella-shaped jets (see cross-sectional diagram, Figure A1.2(*a*)). The gas molecules are trapped in the vapour stream and carried to a region where the pump fluid (oil or mercury) is condensed out and the now liberated gas molecules are sucked away by a rotary pump, compressed and finally ejected into the atmosphere.

It is usual to arrange two, three or four vapour jets in series inside the same cylindrical pump body so that the gas pressure increases in several stages between the high-vacuum inlet and the low-vacuum outlet side of the pump. The oil or mercury is vapourised in a boiler at the base of the pump and requires an electrical or gas supply to heat it; the walls of the pump body are cooled to recondense the vapour to liquid, which runs back into the boiler for the cycle to start again. A supply of running cold water is usually required to provide the cooling although small diffusion pumps can be successfully cooled by natural or forced-air circulation past external cooling fins.

Vapour diffusion pumps can have very high pumping speeds, depending on the diameter of the body, number of stages and the fluid used. An oil diffusion pump with body diameter of 90 cm may pump 45 000 litres of gas per second, but the requirements of electron microscopy are usually satisfied by a 10 cm pump with speed of 300 litres per second. The trend is towards smaller pumped volumes in electron microscopes, for which smaller vacuum pumps are needed. The speed of an oil diffusion pump varies with pressure as shown in Figure A1.2(*b*). The pumping speed for gases is maintained to very low pressures, the speed becoming zero when the rate of back-diffusion of gas molecules against the jet equals the rate at which they are carried out of the high-vacuum chamber.

The total pressure in the vacuum chamber is the sum of the individual pressures of the various gases and vapours present. The pressure in the microscope or coating unit may be determined by the vapour pressure of the pumping fluid at the temperature of the vacuum system if this is higher than the residual gas pressure. A vapour diffusion pump using mercury, whose vapour pressure at room temperature is about 10^{-3} mbar, will therefore remove the residual gases to a low pressure and eventually fill the chamber with mercury vapour at 10^{-3} mbar unless a cooled surface (*cold trap*) is placed between the pump and the vacuum chamber. In this case the pressure of

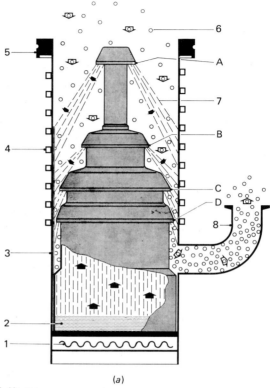

(a)

1 Heater
2 Boiler
3 Pump body
4 Cooling coil
5 High vacuum flange connection
6 Gas particles
7 Vapour jet
8 Backing vacuum connection port
A–D: nozzles

Figure A1.2. The vapour diffusion pump. (a) Diagrammatic explanation of the operation of a diffusion pump. (b) Performance data for an oil vapour diffusion pump suitable for use in electron microscopy. (By courtesy of Leybold AG (a) and Edwards High Vacuum International (b).)

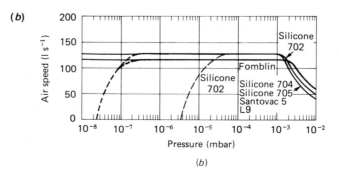

(b)

mercury vapour will fall to its value at the temperature of the trap, and the pressure in the system will be determined by the residual gases and the rate at which they are removed.

Although mercury was cited in the example above, and is still used in pumping systems on some earlier models of electron microscopes, diffusion pumps fitted to current microscopes all use oils whose vapour pressures at room temperature are 10^{-6} mbar or below. Some characteristics of a number of these oils are compared in Table A1.2. Any of these may therefore be used without a cold trap to generate a vacuum well below the 10^{-3}–10^{-4} mbar at which an electron microscope can be operated. With prolonged pumping oil vapour from the rotary and diffusion pumps does diffuse into the microscope or coating unit, where it forms a source of contamination and can give rise to a number of undesirable effects. (Visible effects of contamination were described in Chapter 5 (page 202). These can be reduced in electron microscopes by the use of cold traps, particularly if these are placed close to the specimen to reduce the vapour pressure of contaminants in its vicinity, or by using certain types of pump fluid, e.g. a fluorinated polyether, whose molecules are dissociated and dispersed rather than polymerised and 'fixed' when bombarded by the electron beam.

An operating diffusion pump takes in a large volume of gas at low pressure and compresses it into a smaller volume at a higher pressure on the outlet or backing side (the product *pressure* $\times$ *volume* is constant, by Boyle's Law). There is a critical backing pressure (typically about 0.6 mbar on modern oil diffusion pumps) above which the pump cannot operate, and the rotary pump must be capable of maintaining a pressure below this at all times. The rotary pump may operate continuously to back the diffusion pump or, in some systems, it may be switched on only when it is needed to remove the accumulated gas from a large ballast tank between it and the diffusion pump. The pressure in this is cycled repeatedly between the critical backing pressure and a value 50 or 100 times lower, which is still well within the capability of a two-stage rotary pump. Thus for long periods of time the possibility of vibrations from the mechanical pump disturbing the stability of the microscope image is eliminated, and at the same time the operator is left in peace, although pump noises these days are relatively inoffensive.

Boiling mercury and hydrocarbon oils are permanently damaged by exposure to oxygen and it is important not to admit air by accident or error into operating vapour pumps using these fluids. The first reaction of an automatic pumping system or a practised microscopist will be to close the inlet valve to the diffusion pump if for any reason there is an abnormal flow or air into the microscope column. Similarly, a protective device will turn off the diffusion pump heater should the flow of cooling water fall below that needed to remove the heat of condensation of the pump fluid.

Table A1.2. *Data on diffusion pump fluids*

(Extracted from Edwards High Vacuum International products catalogue, and reproduced with permission.)

Fluid propery	APIEZON			SILICONE		SANTOVAC	EDWARDS
	B	C	AP 301	704EU	705	5	L9
Ultimate vacuum achievable at 20 °C (typical), mbar	1.3×10^{-6}	1.3×10^{-7}	1.3×10^{-7}	6.5×10^{-8}	1.3×10^{-9}	1.3×10^{-9}	5×10^{-9}
True vapour pressure at 20 °C, mbar	5×10^{-6}	4×10^{-9}	4.8×10^{-8}	1.3×10^{-8}	2.6×10^{-10}	2.6×10^{-10}	7.8×10^{-10}
Boiling temperature at 1.3 mbar (approx.), °C	220	255	225	223	254	295	251
Molecular weight (average)	420	479	584	484	546	446	407
Viscosity, cSt, at 20 °C	142	283	23.7[a]	47	240	2400	71.3
100 °C	7.0	10.6	4.78	4.3	7.9	12	5.6
150 °C	3.0	4.0	—	2.2	3.3	4.5	1.2
Pour point, °C (approx.)	−9	−15	−40	<−20	−10	+5	−5
Specific gravity at 25 °C	0.869	0.872	0.965[b]	1.07	1.09	1.195	1.89
Energetic particle bombardment	Conducting polymers formed			Insulating polymers formed		Conducting polymers formed	
Oxidation resistance	Poor to fair		Good	Excellent		Very good	Good

Notes:
[a] 30 °C
[b] 20 °C

Turbo-molecular pumps

High and ultra-high vacua free from the vapours of pumping fluids can be achieved in closed systems if vapour diffusion pumps are replaced by turbo-molecular pumps or non-mechanical pumps of types which will be described later. The former are essentially high-vacuum gas turbines; one design is illustrated in Figure A1.3. The arrangement of blades on the rotor can be seen clearly from the photograph. The stator is a cylinder lined with a similar set of vanes sloping in the opposite direction. Spacings between the stator and rotor may be a millimetre or two. The rotor is driven at a speed of between 16 000 and 90 000 revolutions per minute, according to design, and gas molecules entering the pump at the high-vacuum side are knocked forwards by the vanes and gradually forced along the pump to the high-pressure end or ends of the casing, whence they are removed by a rotary backing pump. This type of pump is able to maintain compression ratios of up to 10^9 across itself, and hence is able to produce ultra-high vacua in sufficiently clean and leak-tight systems. The pumping action varies with the nature of the gas, however, being relatively ineffective for light gases such as hydrogen, as can be seen from the performance curves. With such high rotational speeds, the lubrication and cooling of the rotor bearings is of great importance. The bearings may be air- or water-cooled, and replacement at the annual service visit is a precautionary measure taken in some cases. The design shown has oil-lubricated steel bearings. Also used are grease-lubricated ceramic or magnetically levitated (Maglev) bearings.

Relatively few transmission microscopes as yet are evacuated by turbo-molecular pumps, but a number of scanning microscopes and coating units for specimen preparation provide them either as standard or as an option. In these instruments specimen changes involve a lot of pumping and the simplification of components resulting from having no changeover of pumps from atmospheric pressure to the operating vacuum is a practical incentive, in addition to the possible improvement in quality of vacuum.

Adsorption, sublimation and ion pumps

Several kinds of non-mechanical pump are available which will soak up gas molecules rather than move them from one place to another. Some require the assistance of a mechanical pump to reduce the gas pressure to a value at which they become operational, e.g. in the range 10^{-2} to 10^{-4} mbar. The different pumps and their characteristics are described briefly below.

(a) Adsorption pump: The operation of this type of pump depends on the ability of a molecular sieve (sodium- or calcium aluminium silicate) or any similar material with large surface-to-volume ratio to adsorb large quantities of gases or vapours when cooled to the temperature of liquid nitrogen ($-196\ °C$). The pressure–time graph

in Figure A1.4 shows that adsorption pumps can evacuate a vacuum system from atmospheric pressure to 10^{-2} mbar in a few minutes without assistance from any other pump. No power supply or cooling water is required – only a supply of liquid nitrogen and a vacuum vessel large enough to accommodate the body of the adsorption pump. The ultimate vacuum which can be produced by an adsorption pump is about 10^{-3} mbar.

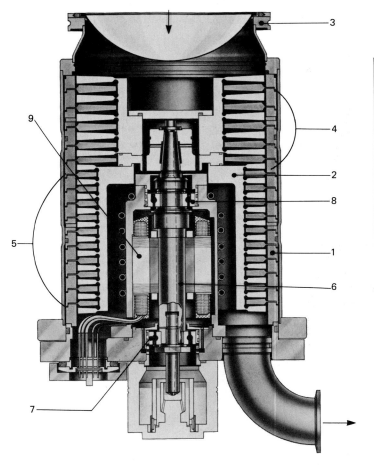

Construction of a turbo-molecular pump of single-ended axial flow design (vertical construction). 1 Stator blades; 2 Rotor body; 3 Intake flange; 4 Blades of the suction stages; 5 Blades of the compression stages; 6 Drive shaft; 7 and 8 Ball bearings; 9 High-frequency motor.

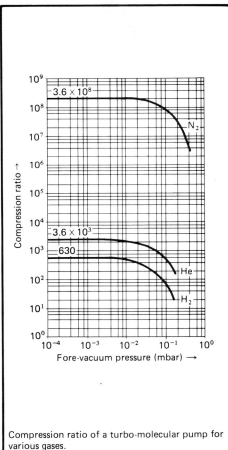

Compression ratio of a turbo-molecular pump for various gases.

Figure A1.3. Turbo-molecular pump for vertical operation. It can be seen that the rotor bearings are in the high-pressure (exit) end of the pump, reducing the possibility of vapour from the bearing lubricant reaching the high-vacuum chamber. The performance curves illustrate the degree of variation of compression ratio with the nature of the gas. (By courtesy of Leybold AG.)

Adsorption pumps are used to provide a completely oil-free starting vacuum for ion or sublimation pumps (see below).

(b) Sublimation pump: The metal titanium sublimes directly from solid to vapour without an intermediate molten stage. When the vapour condenses to form a film of metal, it is capable of pumping large volumes of gas by the combination of processes which are described as *gettering*. These comprise chemical combination (chemisorption), physical absorption within the bulk of the porous film, and adsorption or bonding to the outer surface of the condensed film. In the titanium sublimation pump, metallic titanium sublimes from heated filaments, such as those shown in Figure A1.5 and condenses on water-cooled surfaces adjacent to them. The pumping speed of the condensed film varies for different gases and is highest for hydrogen. Only chemically active gases such as nitrogen, oxygen, hydrogen, carbon monoxide and carbon dioxide are pumped; inert gases are not pumped.

The titanium sublimation pump has a very high pumping speed in the range of 10^{-2} to 10^{-4} mbar and is often used to back an ion pump. It is particularly useful when large gas bursts of short duration are expected. In view of the limited amount of titanium metal in the pump it is normally operated intermittently.

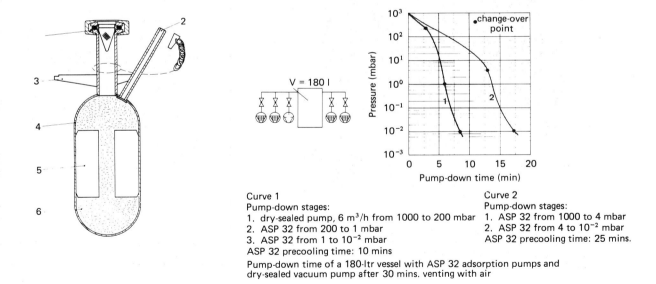

Curve 1
Pump-down stages:
1. dry-sealed pump, 6 m³/h from 1000 to 200 mbar
2. ASP 32 from 200 to 1 mbar
3. ASP 32 from 1 to 10^{-2} mbar
ASP 32 precooling time: 10 mins

Curve 2
Pump-down stages:
1. ASP 32 from 1000 to 4 mbar
2. ASP 32 from 4 to 10^{-2} mbar
ASP 32 precooling time: 25 mins.

Pump-down time of a 180-ltr vessel with ASP 32 adsorption pumps and dry-sealed vacuum pump after 30 mins. venting with air

Figure A1.4. Construction and performance of adsorption pumps. The section through an adsorption pump shows (1) Inlet port; (2) Degassing port (safety outlet); (3) Support; (4) Pump body; (5) Heat-conducting vanes; (6) Adsorption material, e.g. zeolite. The pressure–time curves are given for two unaided adsorption pumps (curve 2) and two adsorption pumps aided initially by an oil-free mechanical pump (curve 1). (By courtesy of Leybold AG.)

(c) Ion pump: A clean (this usually implies 'oil-free') ultra-high vacuum can be produced most economically by the use of one of several types of ion pump. These have no moving parts and rely on the phenomenon of ionisation (Figure A1.6) for their operation.

When a fast-moving electron collides with a gas atom the latter will become a positive ion if the collision knocks off one of its outermost electrons, or a negative ion if the electron becomes attached to the atom. The ions are denoted by the superscript $^+$ or $^-$ (or $^{++}$, $^-$ if multiple charges are involved) e.g. A^+, O^+, and being electrically charged their motion will now be influenced by magnetic and electric fields.

In a high vacuum the probability of an ionising collision occurring between an electron and a gas atom in a path of a few centimetres is slight, but increases if the distance travelled increases. The path length of an electron between two electrodes can be made many times the direct distance between the electrodes by modifying the electrodes or by superimposing an appropriate magnetic field. This is illustrated in Figure A1.7. Both these methods are employed in making ionisation gauges for measuring gas pressures in vacuum systems, as will be described later.

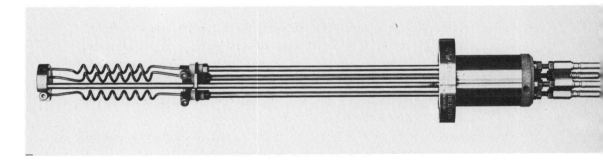

Figure A1.5. Titanium sublimation source as used in the production of a clean ultra-high vacuum. (By courtesy of Leybold AG.)

Figure A1.6. Ionisation and recombination processes. A moving electron with sufficient energy will ionise a neutral gas atom with which it collides, forming a positive ion and two electrons. The positive ion can combine with an electron to reform a neutral atom, releasing surplus energy as radiation, often visible light.

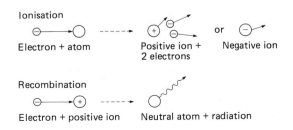

A positive ion which has been accelerated towards the cathode (negative electrode) expends its energy in ejecting material from the electrode surface by the process called sputtering. The sputtered material is mainly in the form of neutral atoms, which are free to travel through the vacuum uninfluenced by magnetic or electric fields which may be present.

The most commonly used ion pump is the *getter-ion sputtering pump* (Figure A1.8), generally shortened to *sputter-ion pump* or, less correctly, *ion getter pump*. This employs the Penning principle of increasing the ionisation probability between an anode and a titanium cathode by superimposing a magnetic field. Positive gas ions sputter metal from the cathode and produce a thin layer of titanium over all the walls and electrodes in the pump. Residual gas is gettered by this layer in the same way as in the sublimation pump described earlier. The ionisation principle ensures that additional titanium continues to be deposited as long as there is gas to be ionised. The pump can thus be used as its own vacuum gauge, after calibration of the discharge current against pressure measured with a normal gauge (see below). Although only active gases are removed by gettering, noble gases are pumped if they are first ionised and then accelerated towards the gettering layer, in which they bury themselves. Sputter-ion pumps are made in diode and triode configurations, each of which has its own strong and weak points in practical operation.

Getter-ion evaporation pumps rely similarly on the gettering effect of freshly formed titanium films, but the films are formed thermally as in the sublimation pump. Adsorption of gas occurs in and on the film, but noble gases (e.g. argon, neon, helium) would be unaffected without the further process of ionisation which occurs between a heated cathode and a helical anode. This results in noble gas ions burying themselves in the titanium films on the electrodes.

Getter-ion pumps can be constructed to pump gases at speeds of up to a hundred or so litres per second down to very low pressures, but must be backed by a rotary

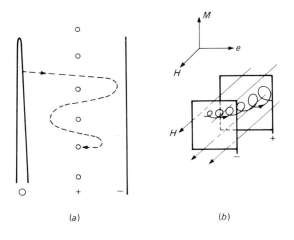

(a)   (b)

Figure A1.7. Methods of increasing electron path length between electrodes *in vacuo*. (*a*) Electrostatic. Electron from heated filament is attracted towards helical grid wire but passes between the turns and carries out several oscillations before finally being captured by the grid. (*b*) Magnetic (Penning effect). An electron moving across a magnetic field is deflected in a direction mutually perpendicular to its direction of motion and the magnetic field. An electron starting in the space between a plane cathode ($-$) and a wire-loop anode ($+$) with magnetic field direction between these two electrodes is attracted towards the anode but forced into a helical path by the magnetic field.

pump or an adsorption pump and a sublimation pump to reduce the pressure initially to 10^{-3} mbar, when the ion pump will take over.

Vacuum gauges

Pressure measurement in practical vacuum systems is based on one or other of two physical effects, heat transfer and ionisation. The former is used between atmos-

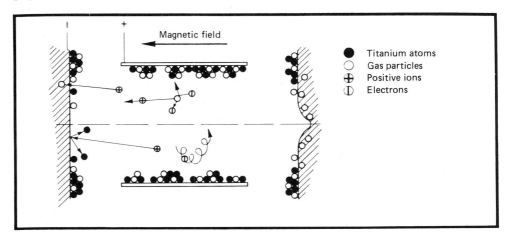

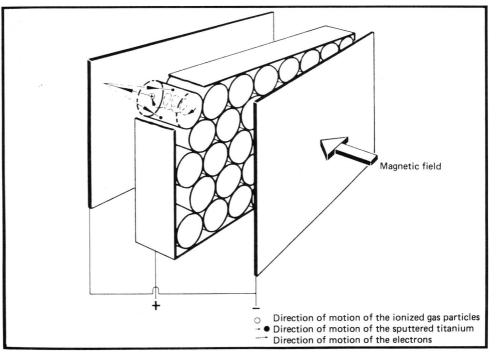

Figure A1.8. Schematic representation of the action of a diode sputter-ion pump. (By courtesy of Leybold AG.)

pheric pressure and about 10^{-3} mbar; the latter from 10^{-3} mbar downwards into the high- and ultra-high vacuum regions.

Vacuum gauges using heat transfer effects

An isolated hot body in a gas-filled enclosure loses heat by the combined effects of conduction, convection and radiation. Of these the last is independent of the gaseous environment but the first two depend on removal of thermal energy by impinging gas molecules and hence are influenced by the number of these, i.e. by the gas pressure.

Two different types of gauge are used to sense and measure the changes in heat loss from an electrically heated wire in the vacuum system. Both rely on the fact that for a given heat input to the wire the smaller the heat loss from it the hotter it will become, and vice versa. The gauges differ in their way of sensing the temperature reached by the wire.

The *thermocouple gauge* measures the temperature of the mid-point of the heated wire directly by means of a thermojunction welded to it. Two electrical circuits are required; one to provide a stabilised electrical input to the wire, the other to measure the output of the thermocouple. The gauge is calibrated against an absolute pressure measuring device using a known gas, since the heat loss at a given pressure will vary with the gas.

The *Pirani gauge* senses changes in wire temperature from changes in the electrical resistance of the wire – the higher the resistance the hotter it is. The wire is made one arm of a Wheatstone Bridge resistance-measuring circuit. The bridge is not balanced but the out-of-balance current is used as the gauge output and calibrated against the gas pressure around the heated wire.

Both of these types of gauge are rugged and trouble-free in normal operation. They will measure 'rough' and 'fine' vacua sufficiently accurately for most purposes. They are used in the pumping lines to rotary pumps to assess whether the pressure has been reduced far enough to permit the vapour diffusion pump to be used, but by 10^{-3} mbar they are relatively insensitive to further changes of gas pressure.

Ionisation vacuum gauges

Below 10^{-3} mbar it is usual to measure gas pressures in microscopes and coating units with gauges using the ionisation principle. The residual gas in the system is ionised under a fixed set of conditions; the magnitude of the positive ion current provides a measure of the gas pressure. A constant relationship between pressure and ionisation current is given over about seven orders of magnitude in the hot-cathode ion gauge and three or four orders in Penning gauges.

The *hot-cathode ionisation gauge* is the more accurate but more vulnerable type of gauge. It uses the ionising principle already described, and electrode arrangements shown in Figure A1.9(*a*). A constant and pre-selected electron current is drawn from a heated tungsten thermionic filament and collected by an open wire grid or cage held at a positive potential of a hundred volts or more relative to the filament. The current of electrons ionises gas molecules with which it collides in passage between filament and grid, and the positive ion current, which is proportional to gas pressure, is collected by a plate or wire electrode held a few tens of volts negative to the original filament. The electron path is lengthened over the direct physical gap between filament and grid because many electrons oscillate backwards and forwards through the spaces between the wires until they are finally collected. This gives an increased chance for the electrons to collide with and ionise gas molecules. At very low pressures the ion current will be very small and will require amplification before it will operate a panel meter. The ion current per unit electron current per unit gas pressure is determined by a preliminary calibration against an absolute pressure gauge of the McLeod or Knudsen types. (These are not suitable for continuous use on practical vacuum plants and are therefore not described here.)

The use of an incandescent tungsten filament as electron source in the hot-cathode

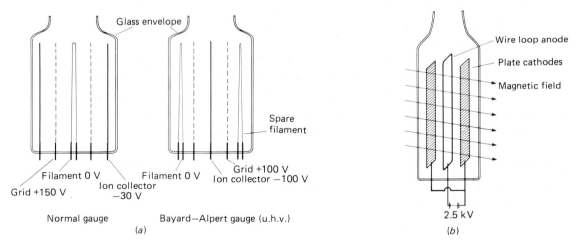

Figure A1.9. Ionisation gauges. (*a*) Hot-cathode ionisation gauge. Two forms of the gauge are illustrated. The normal version has a cylindrical ion collector of sheet metal or fine mesh. The lower limit of pressure measurement is set by the emission of photoelectrons from this cylinder, under the influence of x-radiation from the electron-bombarded grid. This photoelectron current is equivalent to the ion current at a gas pressure of about 10^{-7} mbar, and hence sets a lower limit to pressure measurement using this electrode geometry. The x-ray emission is very much reduced in the Bayard–Alpert design, in which the electrode arrangement is turned inside out, the filament being outside the grid and the ion collector being reduced in area to a fine wire on the axis of the grid cylinder. A second filament is often provided for use after the first one has eroded away. The low pressure limit of such a gauge is about 10^{-10} mbar. The electrode potentials in the diagram are approximate values only. (*b*) Penning cold-cathode ionisation gauge. This robust gauge consists of two flat plate cathodes with a wire-loop anode between them. A megnetic field perpendicular to the electrodes forces electrons to take a much longer path between the electrodes than the geometrical gap, and increases the ionisation probability.

gauge ensures a ready supply of ionising electrons at all pressures but makes the gauge vulnerable to filament oxidation and burn-out should there be any pressure rise above the 10^{-2}–10^{-3} mbar region, for example due to a leak in the vacuum system. Such burn-outs are less common nowadays with the incorporation of over-load relays into the gauge control boxes, but some gauges are provided with a second, standby, filament in case of failure of the first one.

The vulnerability of the hot-cathode ion gauge is avoided in the second type, the *Penning cold-cathode ionisation gauge*, which is tolerant of occasional abuses and per-forms adequately over the range of pressures most used at present in electron microscopy, i.e. 10^{-3}–10^{-7} mbar. This type of gauge, shown diagramatically in Figure A1.9(*b*), uses the principle already illustrated in Figure A1.7(*b*), in which a magnetic field causes electrons in the space between a wire-loop anode and a pair of flat plate cathodes (which can be the walls of the gauge) to move in a helical path towards the anode and hence to have an increased probability of colliding with and ionising a gas molecule. The potential difference between anode and cathode may be several thousand volts. The gauge relies on the release of electrons in ionising colli-sions to maintain and increase the supply of ionising particles. One spontaneous ion-isation is required in the anode/cathode space to initiate the ionisation process and 'strike' the gauge. This results from the 'natural' ionisation by cosmic radiation which is present continuously.

Commercial Penning gauges may differ from the plate and wire-loop geometry of the original gauges, in the interests of robustness, ease of dismantling for cleaning, and improvement of the operating pressure range. Their operation still requires a strong permanent magnet, however, which provides a source of attraction for tweez-ers and small tools which may inadvertently be brought near them!

One development has been to combine the pressure-sensing element of a vacuum gauge with a miniaturised version of the electronic control circuit in a single unit. In a further step a Pirani gauge and a Penning gauge are combined in the same enve-lope, and electrical supplies are integrated so that one meter reads pressure continu-ously from atmospheric pressure to high or ultra-high vacuum.

Practical vacuum systems

A conventional high-vacuum system has a similar layout whether the chamber is the column of a microscope or the work chamber of a coating unit for specimen prepara-tion. This is shown diagrammatically in Figure A1.10(*a*); a mechanical rotary pump backs a diffusion pump, with a Pirani gauge monitoring pressure in the backing line. The chamber is separated from the diffusion pump by a high-vacuum ('baffle') valve and perhaps a liquid-nitrogen-cooled trap for condensible vapours (e.g. water or hydrocarbons). A separate roughing line connects the chamber or column directly to the same or a second rotary pump. Taps or valves are placed in the pumping lines between all components. It is a natural consequence of Boyle's Law that the cross-sectional area of pumps, valves and connecting tubing is much greater in the high-vacuum region than in the backing line.

Nowadays, all components will be made of metal; frequently stainless steel for use in high and ultra-high vacuum, or aluminium, copper, brass or nickel-plated metals. Corrugated metal 'bellows' are used where flexibility is required in the pipe lines. Sections of piping are soldered to metal flanges which are bolted together with synthetic rubber 'O' ring gaskets to produce a gas-tight seal. Joints sealed with 'O' rings of 'Viton'® may be degassed by heating to 200 °C.

A typical pumping routine would involve first of all turning on the cooling water

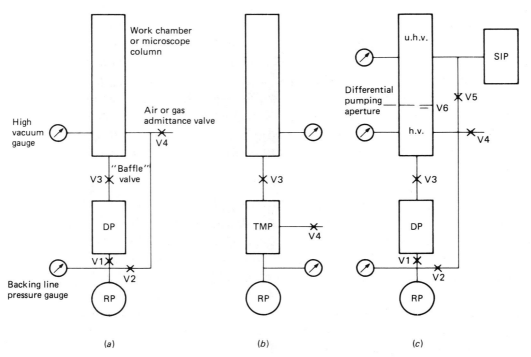

Figure A1.10. Vacuum systems in electron microscopy. (*a*) Traditional vacuum system in specimen preparation and microscopy. The high-vacuum chamber may be a work chamber for specimen preparation (carbon deposition, shadowcasting, etc.) or the column of the microscope itself. The chamber is evacuated from atmospheric pressure to operating high vacuum using rotary and vapour diffusion pumps and the valves V1,2,3,4 as described in the text. The whole procedure can be automated; valves operated by compressed air can be arranged to close automatically and protect the components in the event of an electrical power failure stopping the pumps. (*b*) Scanning microscopes may have a simplified pumping system consisting only of a rotary pump and a turbo-molecular pump. Both pumps are started simultaneously from atmospheric pressure and the microscope is brought to high or ultra-high vacuum without any valving operations being necessary. V3 is an optional high-vacuum valve which enables the microscope to be sealed off under vacuum with the pumps switched off to avoid unnecessary wear and tear. (*c*) A very clean environment (e.g. 10^{-7} mbar) for the electron gun and the specimen is pumped separately, frequently by a sputter-ion pump. The 'dirty' viewing and photographic chamber (pressure only 10^{-5} mbar) is pumped conventionally by a vapour diffusion pump. The two regions are connected only by a small diaphragm in the final projector lens, which can be closed by a sliding valve V6 when the camera chamber is let up to atmospheric pressure to change films.

to the vapour diffusion pump and heating the pump fluid, with a rough vacuum inside the pump. When the pump has warmed up (10–20 minutes) the backing line valve (V1 in Figure A1.10(*a*)) is closed and valve V2 opened so that the rotary pump evacuates the chamber/microscope column directly. When the pressure in this has fallen to 10^{-2} mbar valve V2 is closed, valve V1 is re-opened and baffle valve V3 opened. The chamber is evacuated until a sufficiently low pressure is obtained for the apparatus to be operated. After the operation has been carried out the baffle valve V3 is closed and air can be admitted directly to the chamber through the air inlet (V4).

In order to avoid having to evacuate the complete column of an electron microscope each time a specimen or photographic plate is changed, the better instruments are provided with vacuum airlocks by means of which a small volume of the instrument, e.g. the specimen stage, may be isolated by a valve from the main column. Letting this smaller volume up to atmospheric pressure in order to change a specimen and then rough-pumping it to 10^{-2} mbar so that the airlock may be re-opened is a very much quicker procedure (a few seconds in some cases) than evacuating the whole column and, in addition, keeps the column in a much cleaner state. It is also found beneficial if chambers being opened in this way are filled with clean, dry nitrogen to bring them to atmospheric pressure, rather than moisture-laden laboratory air.

The sequences of valve operations in a pumping routine may be carried out automatically, using servomotors controlled by a logic circuit supplied with pressure measurements from gauges in various parts of the system. This not only relieves the operator of the need to stand by the instrument whilst it is preparing for action but provides a safeguard against incorrect operation which could jeopardise the components. It has its drawbacks, of course, in so far as it works at its own pace and does not permit any short cuts or calculated risks to be taken, and the automatic control may prevent use of the microscope because of a faulty pressure-sensing component which in itself has no effect on the vacuum produced.

A further disadvantage of motor-operated valving systems becomes evident in the event of failure of the electricity supply. Electromagnetic backing-line valves can be arranged to close (or open) under gravity to safeguard the high-pressure side of the vacuum system, but large motor-operated valves in the high-vacuum side will remain open, to the possible detriment of the vacuum system. There is an increasing move in new instruments to use compressed air to operate vacuum valves. Pneumatically operated valves have the advantage of rapid, positive operation and, because they use stored energy, they are not rendered inoperative by electrical power failure.

Scanning microscopes may have a simpler pumping system, consisting only of a rotary pump and a turbo-molecular pump (Figure A1.10(*b*)). Both pumps are started simultaneously from atmospheric pressure and the microscope is brought to high or ultra-high vacuum without any valve operation being necessary. V3 is an optional high-vacuum valve which enables the column to be sealed off under vacuum with the pumps turned off, to avoid unnecessary wear and tear.

Differential pumping

This is a procedure whereby two parts of a vacuum system are maintained at different gas pressures, and flow of gas between them is restricted by a high-resistance diaphragm. In transmission electron microscopes the film camera/viewing chamber can tolerate a poorer vacuum than the specimen area and electron gun, especially if a brighter emitter such as LaB_6 or a field emitter is used. This pressure difference is produced by evacuating the two parts by separate pumping systems and linking the two chambers through a small diaphragm, e.g. a 1 mm or smaller pinhole which will admit a focused electron beam but restrict the gas flow (Figure A1.10(c)).

The principle of differential pumping is applied in all environmental and low-vacuum scanning microscopes, the specimen chamber being maintained at a higher gas pressure than the rest of the column. This is carried to the extreme in the ElectroScan Environmental SEM in which four separately pumped volumes are separated by diaphragms, and a pressure of up to 60 mbar in the specimen chamber is separated by diaphragms from 10^{-7} mbar in the electron gun. A schematic diagram of the vacuum system is shown in Figure A1.11.

Recent trends in electron microscope vacuum systems

In some analytical TEMs, differential pumping across a 1 mm diameter or smaller diaphragm enables the specimen and electron gun at high or ultra-high vacuum to be separated from the viewing and camera chamber which are separately pumped to a lower standard. The ultra-high vacuum in the gun and specimen chamber may be produced by a sputter-ion pump, with no pumping fluid to contaminate the clean areas, or by one or more vapour diffusion pumps, that nearest the column having a particularly low-vapour-pressure fluid. The preference between these alternatives differs between manufacturers.

'Clean' (meaning hydrocarbon-free) vacuum systems can be achieved more readily if the vacuum pumps are all oil-free. Oil diffusion pumps are therefore inadmissable. Ion and some turbo-molecular pumps are completely oil-free, but mechanical rotary pumps needed to cover the first few orders of pressure down from atmospheric are oil-sealed and hence potential sources of oil vapour contamination, even with oil-absorbing traps between them and the high-vacuum pumps. Certain types of mechanical pump, e.g. diaphragm pumps and spiral pumps, provide completely dry and oil-free pumping, but at higher cost than conventional rotary pumps.

The achievement of clean high vacua in electron microscopes has been greatly assisted by the concept of column liner tubes. These are thin-walled non-magnetic tubes which pass down the electron-optical column and separate the electron path, which has to be evacuated, from the pole-pieces, bores and windings of the lenses, which are better out of the vacuum. They have simplified both pumping and servicing at one and the same time.

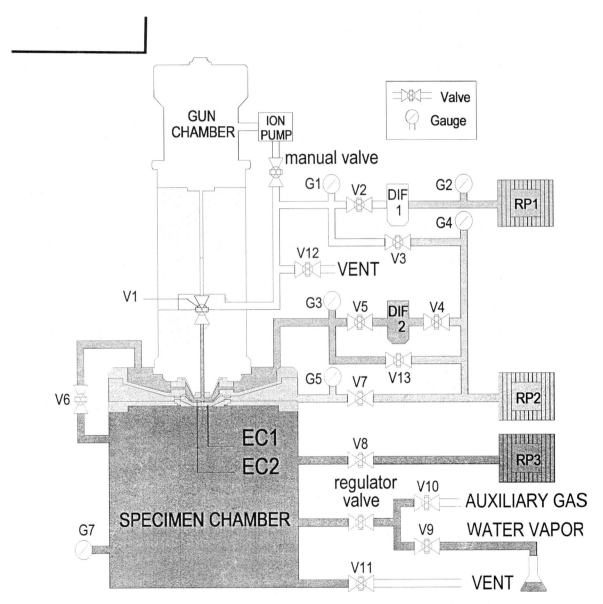

Figure A1.11. Schematic diagram of the vacuum system of the ElectroScan ESEM™ 2020 Environmental Scanning Microscope, showing the complex pumping and valving necessary to enable the specimen-chamber pressure to be controlled between 10^{-5} and 60 mbar. (By courtesy of ElectroScan Corporation.)

Suggested further reading

Several publications dealing with vacuum equipment are available. *Modern Vacuum Practice* by Harris (1989) provides a concise introduction to the production and measurement of vacuum, selection of pumps, and leak detection. Another publication, dealing not only with the principles of vacuum equipment but also with international standards, preferred units, and the properties of a wide range of materials used in vacuum systems, is Hucknall (1991). Bigelow (1994) deals in detail with vacuum technology specifically with electron microscopists in mind.

Appendix 2: Vacuum deposition of thin metallic and carbon films for electron microscopy

Vacuum deposition is an important process for a great number of users of electron microscopy. The uses described in the book have included the preparation of support films, replicas and shadowcasting layers for transmission microscopy, and conductive coatings for scanning microscopy. In general, when directional depositions are necessary thermal evaporation is used; sputtering is used for all-round depositions. Exceptions are carbon coatings and replicas, which are prepared by evaporation.

Evaporation and condensation processes

The processes involved in the vacuum deposition of thin films may perhaps be understood more readily by considering first of all a simple analogy. The reader will be familiar with the solid, liquid and gaseous states of H_2O in the forms of ice, water and steam, the transition from one to the other requiring the addition of a certain amount of heat.

Addition of heat to a block of ice at, say, $-10\,°C$, causes its temperature to rise to $0\,°C$. Further addition of heat results in some of the ice melting to form water, still at $0\,°C$. When all the ice has melted the temperature of the water rises steadily to 100 °C (its boiling point at normal atmospheric pressure) at which temperature it remains until all the water has vaporised into steam. Between the melting and boiling points a certain amount of water is continually vaporising (evaporating) and re-condensing – the higher the temperature the higher will be the rate of vaporisation and the higher the 'vapour pressure' above the water surface. Placing a cold surface above the warm liquid results in condensation of the vapour as liquid water. A person wearing spectacles entering a warm kitchen from outdoors is very quickly made aware of the presence of a large amount of water vapour in the air when his spectacles steam up with a multitude of tiny droplets of water, which run together into a continuous film if enough condenses; this remains there until his spectacles have warmed to the kitchen temperature and the process of condensation and re-evaporation have reached an equilibrium at the higher temperature.

Applying these observations to the behaviour of higher-melting-point materials, such as metals, we would expect that if a metal is heated above its melting point then some vaporisation will take place, the hotter the faster, and if a cool surface is placed adjacent to the liquid metal, small droplets of condensed metal would form and coalesce into a continuous film if enough vapour is collected. This is in fact the case, and anyone who has replaced a 'blown' electrical fuse of tinned copper wire will have seen the deposit of metallic copper which was formed on the fuse-holder by evaporation from the molten wire. It is much more satisfactory to carry out evaporations in a high vacuum however, and in a better-controlled manner, so that the nature and quality of the deposit will not be adversely affected by interaction with the atmosphere between the molten source and the substrate. Vapour pressure–temperature curves for a number of metals used in electron microscopy are shown in Figure A2.1. The evaporation temperature of a material is sometimes conventionally taken to be the temperature at which its vapour pressure is 1.33×10^{-2} mbar and evaporation becomes appreciable.

In the vacuum deposition of thin films the source and substrate are positioned opposite one another in a chamber which is evacuated to a gas pressure of below 10^{-4} mbar. The material to be evaporated is heated until it melts and then evaporates atom by atom, a proportion of the evaporant arriving at the cooler substrate, where it condenses. In general the atoms do not stick exactly where they land, but migrate along the substrate, collide with others and form clusters (nuclei) around which the final film is built up.

The droplet formation, although on a minute scale compared to the steamed-up glasses, is very similar to the condensation of water vapour which we considered earlier. Stages in the growth of a thin gold film are shown in Figure A2.2. The size of the droplet and thickness of deposit before all the droplets have merged together depend on the material being deposited and the deposition conditions, gas pressure, rate of deposition, cleanliness, temperature and nature of substrate; the experienced microscopist will choose materials and evaporation conditions most likely to give the

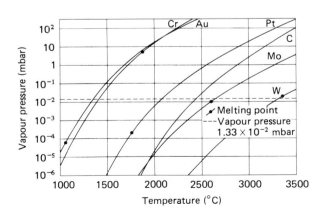

Figure A2.1. Vapour pressure–temperature relationships for some materials commonly used in electron microscope specimen preparation. Formation of vapour directly from the solid, i.e. below the melting point, is called *sublimation*. Chromium, carbon, and to a lesser degree tungsten form condensed deposits by this process. The temperature at which a substance has a vapour pressure of 1.33×10^{-2} mbar is arbitrarily called its *evaporation temperature*.

end product he or she requires. For example, in shadowcasting for transmission microscopy a very thin layer is required, with just sufficient mass thickness (= density $\times$ thickness) to provide contrast by electron scattering. Ideally the layer would be continuous and free from visible structure; in practice this seldom happens so it should be thick enough for most of the droplet structure to have been eliminated but not so thick that marked crystalline effects are visible. A shadowcasting film should be evaporated from as small a source as possible, so that the shadows have sharply defined boundaries. Shadowcasting films are generally made by evaporating gold, gold/palladium, platinum/palladium, chromium, platinum, molybdenum or tungsten, in order of decreasing droplet size or granularity; unfortunately, this is also the order of increasing difficulty of evaporation. Figure A2.3 shows shadow granularities for several of these materials. The shadow contrast from a given thin film depends on operational and recording conditions in the microscope, but a surface density of 5 μg cm^{-2} produces a useful contrast at low voltages, e.g. 60 kV gun voltage. This corresponds to a thickness of film 50/ρ nm, where ρ is the density

100 nm

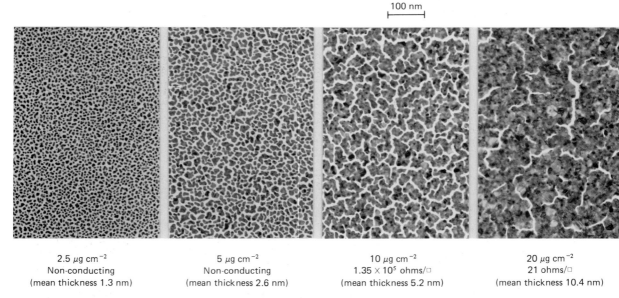

2.5 μg cm^{-2}	5 μg cm^{-2}	10 μg cm^{-2}	20 μg cm^{-2}
Non-conducting	Non-conducting	1.35 $\times$ 10^5 ohms/□	21 ohms/□
(mean thickness 1.3 nm)	(mean thickness 2.6 nm)	(mean thickness 5.2 nm)	(mean thickness 10.4 nm)

Figure A2.2. Four transmission electron micrographs show the structure of increasing amounts of gold condensed on carbon film substrates. Initially the gold atoms are mobile after landing on the substrate and join together to form very small droplets or aggregates. Further deposition results in these droplets extending sideways until they touch one another and join up to form a smaller number of larger islands. This is a random process and, as some channels between islands take longer to be eliminated than others, the films grow to a considerable thickness before the last channels are obliterated. A result of the early 'droplet' stage of growth is seen in the small crystallites still existing in the thicker and almost continuous films. When these are appropriately orientated in the microscope they appear as dark specks by diffraction contrast. Depositing the same amount of gold simultaneously on to glass substrates with metal electrodes enables the electrical conductivity corrsponding to the observed structures to be measured; these values are given with corresponding micrographs and the calculated average thickness of each film.

of the material in g cm^{-3}. Table A2.1 compares the important evaporation parameters for the materials listed above, as well as for carbon.

The granularity or self-structure in shadowcasting films can be the factor limiting the resolution and useful magnification of shadowcast replicas, since there is a danger of confusing fine structure in the specimen with the structure of the shadowcasting film when this is enlarged sufficiently to be visible. For example, the self-structure of gold/palladium shadowing films becomes obtrusive between 20 000 and 50 000 times magnification (i.e. resolutions of 4–10 nm), and if higher magnifications than these are to be used a finer-grained material should be employed.

Another reason for keeping the film thickness low in shadowcasting will be appreciated when it is realised that a shadowcast particle will have a build-up on one side of at least the thickness of the shadowing film (several times this at strongly oblique shadowing directions) and distortion of size and shape will therefore occur.

Coating films for normal scanning microscopy have to fulfil different requirements. It is essential that the film be thick enough to conduct electrical charges from the surface of an insulating specimen. It must therefore have passed from the droplet structure to, at worst, a joined-up lace-like structure, if not having become continu-

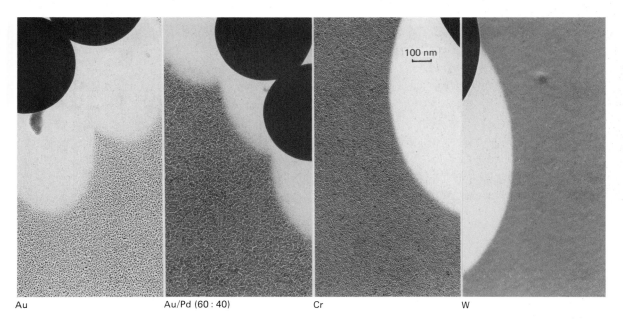

| Au | Au/Pd (60 : 40) | Cr | W |

Figure A2.3. Granularity of shadowcasting films. These films shown at the same magnification with shadows cast behind spherical polystyrene latex particles on carbon substrates, have a similar mass per unit area but very different degrees of self-structure. From left to right the materials are: gold, gold/palladium (60:40), chromium and tungsten. The degree of self-structure and the sharpness of the shadow edge are a result of the mobility of the metal atoms over the substrate. Gold and gold/palladium were evaporated from wire filaments; chromium and tungsten were sublimed from an electron-bombardment source.

Table A2.1. *Deposition parameters for shadowcasting films*

Material	Au	Au/Pd 80/20	Pt/Pd 80/20	Cr	Pt	Mo	W	C
Density, g cm^{-3}	19.3	17.7	19.4	6.9	21.4	10.1	19.3	2.2
Evaporation temp., °C (VP=1.33 × 10^{-2} mbar)	1350	—	—	1410	2090	2650	3310	2400
Average thickness at 5 μg cm^{-2}, nm	2.6	2.8	2.6	7.2	2.3	5.0	2.6	23

ous. This situation is readily attained on a flat surface, in which case a metal such as gold is very suitable at a mean thickness of 10–15 nm, but fibrous or particulate specimens required a conducting path from the top of the specimen over the sides and to the stub underneath. It is therefore essential that the film of gold be nowhere thinner than about 5 nm, which probably means that the coating on horizontal top surfaces will be several tens of nanometres thick. Alloys of gold with other metals become conducting at about one-half the thickness of gold on its own (depending on the alloy and the deposition conditions). Because of this they are to be preferred for irregular specimens, even though alloys such as gold/palladium are less convenient to evaporate on a routine basis.

High-resolution SEMs with resolving powers approaching that of the TEM place higher demands on applied coatings, since the irregular surface of a discontinuous coating layer may become visible in emissive mode images, even though the crystallinity of the film may not be noticeable in SEM, as it is in TEM. Coating materials which become continuous at small thicknesses are desirable, which means more refractory materials and also chromium and carbon. The behaviour will depend very much on the nature of the specimen: whether the arriving atoms stick where they land, or are mobile.

[N.B. Although the emphasis here is placed on the material of the deposit and its final film thickness, it must be remembered that the SEM specimen requires an all-over coating. The stubs should be rotated and tilted during the evaporation in the manner described in Chapter 4 (Figure 4.16).]

Evaporation sources for metallic films
These are basically of two kinds, differing in the manner of heating the charge being evaporated. Most straightforward is the resistance-heated metal wire or strip on to which the charge is loaded. For evaporation of refractory and other materials heating by electron bombardment is used.

Resistance-heated boats and filaments

These are made of suitably formed wire or sheet material which is clamped firmly between two electrodes in the vacuum chamber and heated by passage of a current of tens of ampères, a.c. or d.c., through it. By appropriate design the heating effect can be concentrated at the region of the boat or filament where the charge is to be placed.

A metallic 'boat' is formed from a thin (e.g. 0.05 or 0.1 mm) sheet of refractory material such as molybdenum, tantalum or tungsten. The area of the boat may be a few square millimetres or as large as several square centimetres. Material to be evaporated may be in wire or granular form and is first of all melted into the bottom of the boat before being evaporated in a controlled manner. Obviously, since the charge is held in the boat by gravity the boat will be situated near the base of the coating unit and evaporation will be in an upward direction. This may rule out the use of this type of source for SEM specimens, unless the stubs are held facing downwards in the rotary coating jig.

The alternative is a filamentary source. This may be a length of 0.5–1.0 mm diameter molybdenum or tungsten wire with a small V section at the middle in which can be hung a measured length of wire of the material to be evaporated. This type of source is very suitable for shadowcasting, since evaporation only takes place from a restricted portion of the wire. Evaporation may be in all directions, upwards, downwards or horizontally. Chromium can be sublimed in all directions from chips of metal heated in a helical wire basket. For coating SEM specimens a broad source is preferred and a multi-stranded filament (e.g.three strands of 0.5 mm wire twisted together) is used which can be loaded over a centimetre or so with a greater weight of gold or other wire than can a single filament. In all cases it is good vacuum practice to clean the boat or filament by heating to white heat and to fuse the charge on to the source in separate pump-downs before the specimens are loaded for coating. Typical filaments and boats are illustrated in Figure A2.4.

Electron-bombardment sources

Refractory metals such as molybdenum, tantalum and tungsten form shadowcasting films with very fine texture. Deposition of the materials from resistance-heated filaments is very slow since the filaments must obviously be run below their melting points, and the vapour pressure–temperature curves show that the evaporation rate will be low in this region. However a focused beam of electrons accelerated through some 2.5 kV may be used to melt pieces of the refractory metals resting on a 'hearth' of carbon or water-cooled metal and hence to provide a faster source of evaporation. Simple small electron-bombardment evaporators are now available for use in electron microscope preparation units (Moor, 1970). At the temperatures involved there is appreciable electron emission from the source in addition to the desired atomic beam, and if this is not suppressed damage can be caused to the specimens being shadowcast. Electron-bombardment evaporators are illustrated in principle in Figure A2.5.

Deposition of carbon films

Carbon is an important material in electron microscope specimen preparation, being used for replicas and support films in transmission work and for electrically conducting coatings for scanning microscopy and analysis. For support films it is first deposited on a flat substrate such as cleaved mica, and then transferred to the specimen grid. As a coating material it has two very desirable properties. Firstly, at all thicknesses used in microscopy (about 2 nm upwards) the films are continuous and essentially free from self-structure apart from a uniform speckled appearance visible only at the highest magnifications; secondly, carbon depositions produce a deposit around the sides and back of specimens without any specimen rotation being given. The reason for this effect is not universally agreed but it is probably associated with deflection of the light carbon atoms out of their straight-line paths by collisions with gas molecules and reflection from the walls of the vacuum chamber. The effect can be demonstrated impressively by making a carbon replica of chains of magnesium oxide smoke particles (Figure A2.6). After dissolving away the magnesium oxide the cubes are seen to have all sides complete, even though the carbon was evaporated from a point source over the top of the original specimen.

Examination of the vapour pressure curves in Figure A2.1 shows that carbon vapour sublimes from the solid at temperatures of 2400 °C or higher, well below the melting point of 3500 °C. Because of the difficulties of working at such temperatures carbon was little used in electron microscopy (if at all) before 1954 when Bradley introduced a technique for carbon sublimation from resistance-heated rods (Bradley, 1954a). A practical carbon evaporation source is shown in Figure A2.7. Two rods of spectrographic graphite or carbon are used, one fixed, the other sliding in an insulated bush and pressed against the fixed one by a light spring. The contact region is

Figure A2.4. Sketches of typical evaporation sources for shadowcasting or coating films. (*a*), (*b*) are made from molybdenum or tungsten wire 0.5–1.0 mm diameter. (*a*) is suitable for evaporating material in wire form, e.g. gold, gold/palladium, and platinum/palladium, which is hooked over the 'V' and pre-fused *in vacuo*. (*b*) is suitable for evaporating chromium chips. (*c*) is a boat bent up from molybdenum or tungsten sheet and suitable for evaporation, upwards only, of materials in wire, chip or pellet form. (*d*) is a boat made of molybdenum or tungsten sheet with a dimple pressed into it; it is used similarly to (*c*).

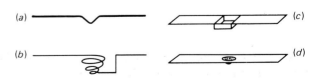

tapered to a reduced cross-section so that passage of an electric current of some tens of ampères results in resistive heating of the contact area sufficient to cause carbon to sublime. The heating current may be switched on and off by an electronic timer so that depositions of a second or less may be used for support films, replicas, etc.,

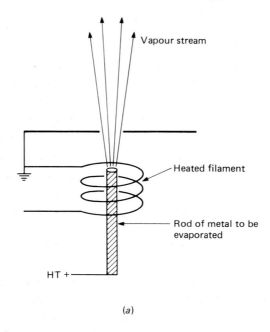

(a)

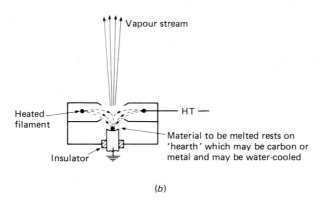

(b)

Figure A2.5. Operating principles of two types of electron-bombardment evaporator. (N.B. Not to scale.) (*a*) Heated tungsten filament emits electrons which bombard the central rod, melting the tip and providing a stream of metallic vapour. The temperature and, hence, rate of evaporation depends on the voltage applied to the central rod and the electron current available from the filament, i.e. on its temperature. (*b*) The electron current from a heated tungsten filament is used to bombard and melt a charge of the material to be evaporated. In this type of source the charge cannot be contaminated with material evaporated from the filament and it is, therefore, used where a deposit of high purity is required. The rate of evaporation is controlled by varying the HT voltage and/or the filament temperature.

or tens of seconds when a number of specimens are being coated for scanning microscopy. The deposition takes place in a vacuum chamber at a gas pressure of 1 $\times 10^{-4}$ mbar or below.

The deposition process from heated rods is characterised by the emission of a shower of small glowing particles of carbon or graphite, whose tracks may be seen on photographs of carbon depositions (Figure A2.8). In spite of their prominence

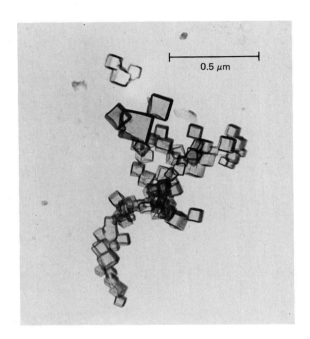

Figure A2.6. Transmission micrograph of carbon replicas of magnesium oxide smoke particles.

Figure A2.7. Practical source for evaporation of carbon by the Bradley process. R1 and R2 are graphite rods 6.35 mm in diameter tapered at one end, e.g. in a pencil sharpener. S is a compressed spring to hold the rods in contact as the tips sublime away. I is an insulating bush to permit the rod-holder to slide relative to the main body of the jig. A heavy current is passed through the rods, heating them to the sublimation temperature of carbon. Other diameters of graphite rod are sometimes used, e.g. 3 mm.

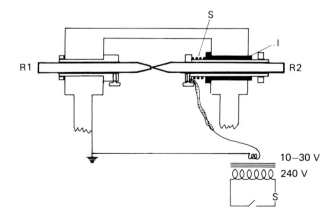

(START)

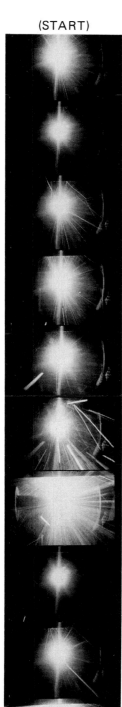

Figure A2.8. Ten consecutive frames from a cine film of a Bradley-type deposition of carbon from resistively heated graphite rods in a glass-domed vacuum chamber. Tracks of fast-moving incandescent graphite particles can be seen originating from the heated region. Reflection of particles from the rod-holders can also be seen, providing one explanation for the effective all-over coating property of carbon when the geometry of the source and the chamber are suitable. The film was exposed at 64 f.p.s. whilst the rods were heated with 50 Hz a.c. Brightness differences between successive frames emphasise that the rods are fluctuating in temperature at 100 Hz, and the deposition of carbon is consequently intermittent. Heating the rods with d.c. provides a more efficient deposition, as well as certain other advantages. A single frame shows the rod profile after the deposition.

these particles are seldom noticed on or in specimens, the majority of the carbon presumably subliming in atomic form. The carbon rods at deposition temperature are also hot enough to emit thermionic electrons, and these may bombard the specimens being coated unless the electrical connections to the evaporator are suitably chosen, preferably using d.c. for the rod heating (Watt, 1974).

Carbon may also be sublimed from rods heated by electron bombardment. This method gives a lower rate of deposition than the Bradley technique but is capable of a finer control since the power input to the rods can be precisely adjusted.

Carbon films from string

An alternative procedure for attaining the sublimation temperature of carbon in a small chamber is to use a length of carbon 'string', a carbonised viscose filament developed for the carbon reinforcement of resins. Either a single strand of fibre or a length of woven braid several centimetres long is clamped between electrodes and heated by passing an a.c. or d.c. current through it. The centre of the strand or braid heats up to incandescence and sublimes until a thin hot-spot develops and the fibre melts and breaks. Enough carbon can be sublimed to prepare support films and conductive coatings at short range of a few centimetres. As with evaporation from rods the effects of emission of heat and electrons should be borne in mind, especially if alternating current is used for heating.

Monitoring carbon depositions

It is not easy to determine just how thick a carbon deposit is; sufficient to have a routine procedure which results in carbon films which are adequate for the purpose in hand. The optical density of a carbon deposit on a glass slide can be used for monitoring purposes. A small droplet of oil on a glass slide provides a visual indication during deposition, since carbon will condense on the glass but not on the oil drop. Electrical resistance of a film between a pair of electrodes is another property which can be monitored, but the resistance should be reduced by the use of an interdigitated design of electrodes (Watt, 1991). An oscillating quartz-crystal film-thickness monitor can be used, provided its operation is unaffected by radiant heating or electron bombardment.

Platinum/carbon shadowcasting

Very fine textured, high-resolution shadowcasting films are formed by the co-deposition of platinum and carbon. Bradley (1959) demonstrated that resolution of surface features not much greater than 1 nm in extent is possible with this combination of materials.

It appears that when the proportions are right, individual platinum atoms arriving at the specimen are prevented by carbon atoms arriving at the same time from clustering to form platinum crystallites. Hence, the good electron-scattering properties of the heavy platinum atoms are used to form thick shadowing layers with minimum granularity. The co-deposition is achieved either by forming a carbon rod with a core of platinum powder or by winding a measured amount of fine platinum wire around the narrow neck at the contact point between the two carbon rods. (see, for example, Steere *et al.*, 1977). (N.B. If the conditions are not right, the Pt/C film may be found to be almost entirely C with minimal observable self-structure but low contrast or secondary emission, and excessive layer thickness.)

Deposition of thin films by sputtering

Cathodic sputtering is a process whereby atoms are ejected from a surface by bombardment with positively charged gaseous ions in a vacuum chamber. The sputtering process in independent of gas pressures, but the generation of the positive ions and the subsequent behaviour of the sputtered material are strongly influenced by the pressure. Its most familiar form, diode sputtering, is illustrated schematically in Figure A2.9. Two plane electrodes, which may typically be circular discs of metal 5–10 cm in diameter, are separated by a parallel-sided gap of several centimetres in a vacuum chamber filled with an inert gas, usually argon, at a pressure of between 10^{-1} and 10^{-2} mbar. When a potential difference of 1000 volts or more is applied across the electrodes, a glow discharge is set up between them, and a current can be measured on a sensitive meter. Some of the argon is ionised into electrons and positive ions; the former are attracted towards the positive electrode (anode), while the latter move towards the negative electrode (cathode) where their arrival causes neutral atoms of cathode material to be ejected by the sputtering process. These atoms cross the discharge, suffering collisions with other atoms on the way, since the gaseous mean free path at 10^{-1}–10^{-2} mbar pressure is only 1–10 mm. The sputtered

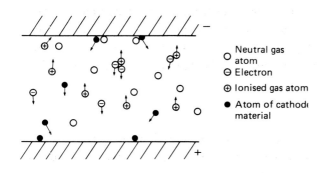

Figure A2.9. Schematic diagram (not to scale) of processes occurring in the cathodic sputtering process. Neutral gas atoms in the space between the anode (+) and cathode (−) are ionised by electron impact to positive ions and electrons. The ions are attracted to the cathode where they (i) give up their energy and cause emission or 'sputtering' of atoms of cathode material and (ii) regain an electron from the cathode and become neutral atoms once more. The electrons are attracted towards the anode with increasing velocity. On the way, if they encounter a neutral gas atom they may knock off an electron and create a new positive-ion/electron pair. The sputtered atoms cross the gap and condense on the cathode; their paths are deflected by numerous collisions with gas atoms and they arrive at the cathode from a wide range of directions.

○ Neutral gas atom
⊖ Electron
⊕ Ionised gas atom
● Atom of cathode material

atoms arrive at the anode from a range of directions and condense to form a thin film of cathode material. Some electrons reach the anode; others give up their energy in ionising collisions with gas molecules in the gap, thus sustaining the glow discharge.

(N.B. If the anode plate contains a central hole a beam of electrons would emerge through it from the discharge. The first electron microscopes used this type of source in their electron guns, see Appendix 4, 'Electron sources for electron microscopes'.)

Diode sputtering can be used for depositing thin films of any material which can be prepared as a sheet or a surface coating on the cathode. Surfaces which it is desired to coat are placed on the anode. The heavy mass of Ar^+ ions gives a higher sputtering yield than would a lighter gas, e.g. N^+ or He^+. If oxygen or air is used for sputtering the process of reactive sputtering results in the deposition of oxides (except in the case of noble metals).

Materials may be deposited by sputtering which are difficult or impossible to evaporate. However, the rate of deposition is in general slower than can be achieved by thermal evaporation. Increasing the current through a glow discharge by increasing the voltage across the gap or altering the gas pressure results in an increased sputtering rate but at the same time an increased heating of the cathode surface by the bombarding ions. To reduce this the cathode may sometimes be water-cooled. Any alteration in gas pressure affects the collision mean free path and the structure of the deposited film.

The random arrival direction of sputtered material and the fact that platinum and gold can be deposited very economically makes this form of deposition attractive for coating SEM specimens, since all-over coatings are obtained on stationary specimens in a fraction of the time, over-all, that it takes to prepare the boat, and evaporate metal coatings. Moreover, simple sputter-coaters can be pumped satisfactorily by two-stage rotary pumps alone. The only elaboration provided commercially is a permanent-magnet system to deflect electrons away from bombarding the specimen on the anode.

The deposition of gold at high gas pressures results in specimen coatings which can show visible lumpiness at high magnifications in high-resolution SEMs. This can be reduced by working at lower gas pressures (with appropriately increased sputtering times) or by using platinum in place of gold (Watt, 1978). Further reduction in the visibility of sputtered coatings depends very much on the nature of the surface being coated, and a certain amount of trial and error is advisable to find the simplest satisfactory procedure. Reduction of coating thickness and/or cooling the specimen should be tried first. Evaporation or sputtering of refractory metals such as W, Mo, Ta, and ion–beam sputtering from a target in a high vacuum (Franks, 1980) are recommended. Depositions in a high vacuum will not have an all-over coating effect, and specimen rotation will be necessary during the deposition. Sputtered chromium has been found satisfactory by some workers, but an oxygen-free atmosphere is essential.

Suggested further reading
For a number of years the standard text on this topic has been Holland (1956). There is also a certain amount of relevant information in Kay (1965).

Appendix 3: X-ray spectrometry

The principles behind the generation of characteristic x-radiation in electron-beam instruments were described in Chapter 2. This appendix gives the background to the two types of spectrometer used for x-ray analysis. Their construction and the strong and weak points of each type will be described, so that the reader can see how they can be used in electron microscopes and microprobe analysers.

Wavelength-dispersive x-ray spectroscopy (WDS or WDX)
Beams of x-rays and electrons are diffracted, i.e. preferentially reflected in particular directions, by the regular arrays of atoms in crystalline materials. If θ is the angle between a set of atomic planes in a crystal and an incident beam of x-radiation of wavelength λ there will be a strongly reflected beam making an equal angle to the crystal planes when θ, λ, and the separation d between the atomic planes satisfy the relationship $2d\sin\theta = n\lambda$ (Bragg's Law, Figure A3.1(*a*)). Here n is an integer, 1, 2, 3, 4, etc., giving the order of the reflection; unless otherwise stated it is assumed that $n = 1$. When θ does not satisfy the Bragg condition the reflected intensity is very low and the fall-off on either side of the exact angle is rapid. Figure A3.1(*b*) illustrates schematically the different orders of diffraction of bromine Kα radiation ($\lambda = 0.104$ nm) by a crystal of lithium fluoride ($2d = 0.4027$ nm). It is possible, therefore, to use a crystal of known lattice plane spacing with a suitable detector to separate and measure a range of different wavelengths emitted by an x-ray source. Figure A3.2 shows diagrammatically the arrangements of components in a wavelength-dispersive spectrometer – also familiarly known as a *crystal spectrometer*. In order to analyse the x-radiation from the source S the reflecting crystal X of known lattice spacing d is rotated through a range of angles θ whilst the detector C is rotated simultaneously through an angle 2θ. Measurement of the values of θ at which the output from the detector passes through a maximum enables the corresponding values of λ to be calculated from the Bragg equation.

The selectivity or resolution of the spectrometer is a measure of the precision or certainty with which one wavelength can be recognised from another. In the arrangement illustrated the detector will actually receive radiation of a range of wavelengths which are diffracted from different parts of the crystal at slightly different angles. The selectivity of the spectrometer would be improved if the radiation from source to crystal were collimated into a very narrow parallel beam, all of which would have the same angle of incidence at the crystal. However, this would reduce the proportion of the emitted x-ray photons reaching the detector and would result in a lower sensitivity of detection.

An instrument which provides both selectivity and sensitivity is the focusing crystal spectrometer, shown in principle in Figure A3.3. If the analysing crystal X is curved into a cylindrical form with radius $2R$ then x-radiation from a source S will be diffracted at the crystal and brought to a focus at C, where S, X, C all lie on the circumference of a circle of radius R, called the Rowland circle. This property was used by Johann in a focusing x-ray spectrometer. Because of the curvature the whole surface of the crystal could be used for diffraction at the same angle θ, and a greater proportion of the radiation emitted by the source could be utilised for spectrometry than in the case of a plane crystal. However, the focusing effect is not quite perfect: although the centre of the curved crystal may be placed precisely upon the Rowland circle the extremes will leave the circle because of the different radii of curvature. Johansson (1933) remedied this defect by bending the plane crystal to a curvature of radius $2R$ and then grinding the surface to a smaller radius, R.

In the focusing spectrometer the angle θ is varied by moving both the crystal and detector along the circumference of the Rowland circle. In practice, the greater the

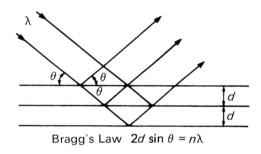

Bragg's Law $2d \sin \theta = n\lambda$

(a)

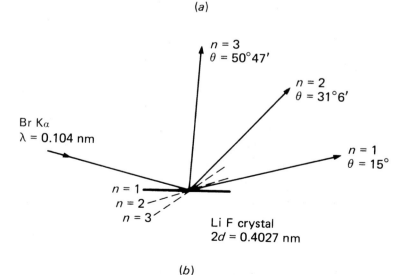

Figure A3.1. Schematic diagrams illustrating (*a*) diffraction of x-radiation by crystal planes, and (*b*) diffraction of monochromatic x-radiation by a crystal rotated to positions where the integer *n* in Bragg's equation equals 1, 2, and 3.

Br Kα
$\lambda = 0.104$ nm

$n = 3$
$\theta = 50°47'$

$n = 2$
$\theta = 31°6'$

$n = 1$
$\theta = 15°$

$n = 1$
$n = 2$
$n = 3$

Li F crystal
$2d = 0.4027$ nm

(b)

diameter of the circle the better the wavelength resolution of the spectrometer. It can be seen from Figure A3.3 that the geometrical relationship $D = 2R\sin\theta$, (where SX = CX = D) has to be satisfied as well as the Bragg relationship for reflection at the crystal surface. Thus, as θ varies to accommodate changes in λ, the distances R or D must also vary to maintain the focusing action. To achieve this, either the crystal-to-source distance can be variable with a fixed radius of crystal or the curvature of the crystal can vary and with it the Rowland circle diameter. Spectrometers which fulfil either of these requirements over the whole of the analysed spectrum are termed fully-focusing spectrometers. If the requirements are only partially satisfied the design is termed semi-focusing.

The fully-focusing crystal spectrometer may be constructed to follow the circular

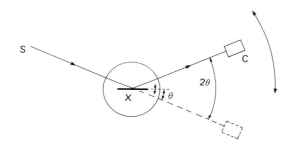

Figure A3.2. Components of the x-ray wavelength-dispersive spectrometer. Key: S = x-ray source; X = diffracting crystal; C = x-ray detector (counter).

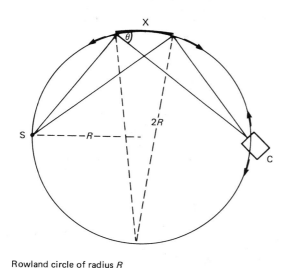

Rowland circle of radius R
Crystal X curved with radius $2R$

Source and counter positions at large θ values

$$\sin\theta = \frac{D}{2R}$$

Figure A3.3. Geometry of the focusing crystal spectrometer.

geometry literally. However, when the x-ray source is the electron-bombarded specimen in an electron microscope there is a disadvantage in using a fixed Rowland circle. This is that the x-ray take-off angle, or direction of the analysed x-ray beam relative to the specimen plane, varies as the crystal moves around the circle. Thus, the proportion of the generated x-radiation which is intercepted and absorbed by the specimen itself varies throughout the spectrum being analysed, which is an undesirable feature in quantitative analysis. To overcome this defect the linear fully-focusing spectrometer was developed; this is the design which is used on electron-beam instruments where accurate quantitative analysis is desired. In this spectrometer the analysing crystal moves along a straight line path at constant take-off angle, rotating as it goes to give variation in θ. The detector is moved simultaneously by a mechanical linkage so that its position at all times satisfies the requirements of Bragg's Law and the R, D, θ relationship. Figure A3.4 outlines the principle of this spectrometer and shows how the movements may be achieved in practice.

Since D is varied, the curvature of the crystal can be constant throughout the spectrum and still maintain the spectrometer in focus, provided the x-ray source position remains constant. The accuracy of positioning must be within a few micrometres or the sensitivity and selectivity of the spectrometer are adversely affected. The direction in which the spectrometer is built on to the electron microscope or microanalyser can influence the ease of maintaining the exact focus on a practical specimen. Since the curvature of the crystal is cylindrical, the source of x-radiation may be moved parallel to the axis of curvature without affecting the focus. In the SEM, therefore, where it may be desirable to be able to vary the working distance along the axis of the probe-forming lens and to analyse rough-surfaced specimens, a horizontally placed Rowland circle (Figure A3.5) gives the desired degree of freedom to specimen level. However, the specimen plane must be tilted towards the spectrometer, since a horizontal specimen would present a zero x-ray take-off angle. Also at low magnifications, the extent of the scanned field will be so great that focus will only be maintained at the centre.

If the spectrometer is mounted with its Rowland circle vertical the specimen plane must be accurately located within micrometres of the focus of the spectrometer (Figure A3.6). This is only possible on the SEM and EPMA with a horizontal, flat polished specimen whose plane can be precisely adjusted within the shallow depth of field of a light microscope.

Latitude of specimen height with an untilted specimen is obtained with an inclined Rowland circle, i.e. at an intermediate angle between a horizontal and a vertical spectrometer (Figure A3.7). Vertical and inclined spectrometers can be combined usefully on a single SEM or EPMA column if the angle of tilt of the inclined spectrometer is the same as the x-ray take-off angle of the vertical spectrometer (e.g. 40°).

If one particular wavelength is to be examined repeatedly, it is possible to use a

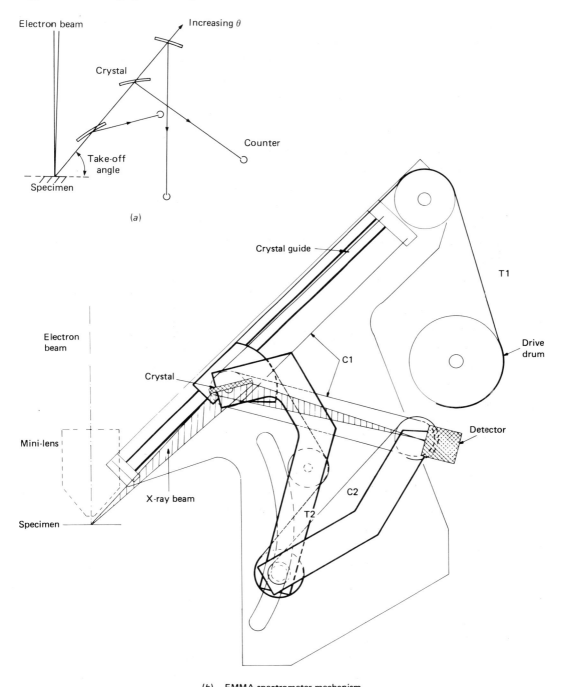

Figure A3.4. (*a*) Principle of the linear fully focusing spectrometer and (*b*) an outline of the mechanism for achieving the linked movement of crystal and detector. The drive is transmitted by flexible metal tapes. (Spectrometer detail from EMMA-4; by courtesy of Kratos Ltd.)

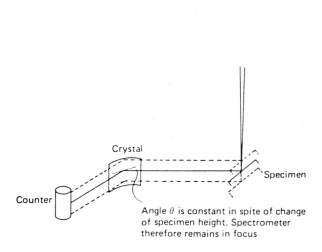

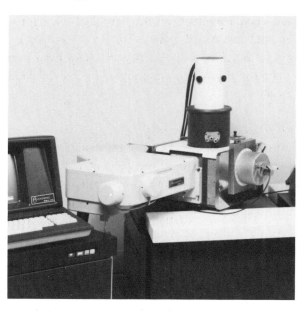

Crystal

Specimen

Counter

Angle θ is constant in spite of change
of specimen height. Spectrometer
therefore remains in focus

Figure A3.5. With its Rowland circle horizontal the focusing of a crystal spectrometer is insensitive to specimen height, as shown in the diagram. The photograph shows a focusing spectrometer fitted in the horizontal configuration to a Stereoscan 100 scanning microscope. (Photograph by courtesy of Microspec Corporation.)

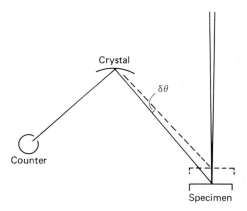

Crystal

$\delta\theta$

Counter

Specimen

Change in specimen height results in change
in angle $\delta\theta$ and reduction in diffracted signal.
The magnitude of the effect is shown in the
accompanying graph, compared with the effect
on an inclined or horizontal spectrometer.

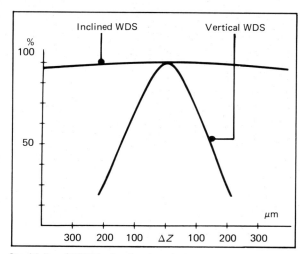

Sensitivity of WDS inclined and vertical
against ΔZ

Figure A3.6. When its Rowland circle is mounted in the vertical plane the performance of the spectrometer is critically dependent on the constant and correct setting of the specimen height. (Graph by courtesy of Cameca S.A.)

fixed-crystal monochromator, in which an appropriate crystal and a detector are mounted in a fixed relationship. The cost of the complex and precise mechanical linkages required for a tunable crystal spectrometer is thus avoided.

The two essential components of a monochromator or spectrometer will now be described.

(a) X-ray diffracting crystals.

These are essentially sheets of crystalline material with sets of atomic planes parallel to the surface. Reinforcement of reflected x-radiation from successive planes within the crystal produces the strong reflection when x-rays are incident at the Bragg angle. The range of wavelengths which can be analysed with one crystal depends on the geometry of the spectrometer; in commercial instruments the range of $\sin\theta$ covered lies between 3.25:1. and 4.2:1 Since the range of x-ray wavelengths excited by a 30 keV electron beam extends well over two orders of magnitude (30 kev = 0.04 nm, beryllium K wavelength = 11.40 nm) it is obvious that several analysing crystals providing a range of lattice spacings (*d*-values) will be required on a practical spectrometer. Commonly used diffracting materials are listed in Table A3.1 along with the *d*-values of the lattice planes used in spectrometry. Approximate values are given of the wavelength and atomic number range which would be analysed on an

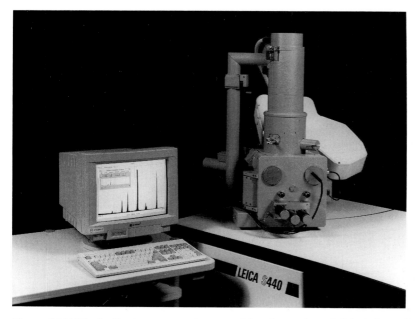

Figure A3.7. The inclined spectrometer can combine the specimen height independence of the horizontal spectrometer with a positive x-ray take-off angle from an untilted specimen (e.g. 40°). An inclined spectrometer is shown fitted to the column of an SEM. (By courtesy of LEO Electron Microscopy Ltd.)

Table A3.1. *Crystals and multilayers used in diffracting spectrometers on electron microscopes*

| Formula or abbreviation | Chemical name | d(nm) | λ(nm) | Approximate range covered[a] | | | |
				K	L	M
LiF(200)	Lithium fluoride (200)	0.2013	0.097–0.365	20–36	50–89	–
LiF(220)	Lithium fluoride (220)	0.1423	0.069–0.258	23–43	58–92	91–92
PET	Pentaerythritol	0.4371	0.212–0.792	14–24	36–63	71–92
ADP	Ammonium dihydrogen phosphate	0.5321	0.257–0.964	13–22	34–57	66–92
TAP	Thallium acid phthalate	1.287	0.623–2.33	9–14	24–39	53–76
RAP	Rubidium acid phthalate	1.306	0.632–2.37	8–14	24–39	53–76
LSM-60	Layered synthetic microstructure	3.0	1.451–5.44	6–11	20–28	–
	W/Si					
LTD	Lead tetradeconate	3.9	1.890–7.07	5–8	17–25	–
LSM-80	Layered synthetic microstructure	3.9	1.890–7.07	5–8	17–25	–
	Ni/C					
STE	Stearate	5.0	2.420–9.06	5–7	17–23	–
LOD	Lead octodeconate	5.016	2.440–9.13	5–7	17–22	–
LTE	Lead lignocerate	6.5	3.150–11.8	4–7	17–20	–
LSM-200	Layered synthetic microstructure	9.85	4.765–17.85	4–5	17–20	–
	Mo/B_4C					

Notes:

[a] Calculated for a spectrometer with 'average' range of 2θ coverage, $28°–130°$

'average' spectrometer. Crystal spectrometers normally have a rotatable turret on which up to six crystals are mounted so that they can be brought into position as and when required to cover a particular wavelength range. Crystals on commercial spectrometers vary in size between 10×30 mm and 20×50 mm. If all other factors are equal, larger crystals would intercept more x-radiation and permit a greater sensitivity of analysis.

Synthetic crystals: Several of the 'crystals' in Table A3.1 are man-made diffracting multilayers (e.g. layered synthetic microstructure, LSM, or multilayer device, MLD) formed by the alternate deposition by sputtering or evaporation of layers of heavy metal and a light spacing material transparent to the x-rays being analysed. These devices may be tailor-made for particular wavelengths in the light-element range, and have a marked improvement in peak-to-background ratio over conventional crystals (Nicolosi *et al.*, 1986).

(b) X-ray detectors (counters).
X-ray photons are normally detected and counted by a proportional counter, consisting of a gas-filled cylindrical chamber with an insulated central wire held at a high electrical potential. An x-ray photon passing through the chamber ionises the gas and a pulse of current passes between the wire and the body of the chamber. Current pulses are processed to give measurements of count-rate, which is proportional to the intensity of x-radiation, and total counts over a given duration of analysis. Pulses can also be energy-analysed to discriminate between first and second orders of diffracted radiation ($n = 1$ at wavelength λ (energy E), and $n = 2$ at $\lambda/2$ (energy $2E$) which both satisfy the Bragg equation). Count rates of up to 10^5 per second are obtainable with gas proportional counters. Two types of proportional counters are used in wavelength-dispersive ('crystal') spectrometers. A gas-flow proportional counter (FPC) is used to detect soft x-radiation. This has a continuous flow of an argon/methane gas mixture (P10 gas – Ar 90%, CH_4 10%) passing through it. X-radiation enters the counter through a very thin plastic membrane. A xenon sealed proportional counter (SPC) detects harder x-radiation, e.g. up to 0.4 nm wavelength or down to 3 keV energy. This has an admittance window made of beryllium up to 50 μm thick. When the spectrometer has a turret of crystals to enable a wide range of wavelengths to be covered, two counters will be fitted in tandem (FPC first). A fixed-crystal monochromator specifically intended for analysis of light elements, e.g. carbon, nitrogen or oxygen, will have an FPC only.

The high selectivity of a fully-focusing crystal spectrometer is preserved by admitting the x-radiation to the counters through a narrow slit placed on the Rowland circle and aligned with the sharply focused x-ray line. Because of the high sensitivity of the detection system great care is taken to eliminate extraneous

Table A3.2. *Performance data for Microspec fully focusing crystal spectrometer*[a]

Spectral line	Wavelength (nm)	Elemental standard	Operating conditions[b]	Resolution FWHM (eV)	Peak intensity (c.p.s.μA^{-1})	P/B ratio	Sensitivity (ppm)[c]
BeKα	11.40	Beryllium	LSM-200, FPC, 10	8	3.0×10^4	40	300
BKα	6.76	Boron	LOD,FPC,10	5	2.5×10^4	140	175
BKα	6.76	Boron	LSM-200,FPC,10	15	1.0×10^6	30	60
CKα	4.47	vitreous C	LOD,FPC,10	10	1.2×10^5	130	80
CKα	4.47	vitreous C	LSM-80N,FPC,10	14	5.0×10^5	50	65
NKα	3.160	Boron nitride	LSM-80N,FPC,10	16	1.0×10^4	3	1900
NKα	3.160	Boron nitride	LSM-80F,FPC,10	13	6.0×10^3	10	1350
OKα	2.359	SiO$_2$	TAP,FPC,10	3	6.0×10^3	350	230
OKα	2.359	SiO$_2$	LSM-80N,FPC,10	17	1.2×10^5	50	135
AlKα	0.834	Aluminium	TAP,FPC,20	9	3.0×10^6	800	7
SiKα	0.7126	Silicon	PET,FPC,20	2	6.0×10^5	2600	8
TiKα	0.275	Titanium	PET,FPC,30	20	3.0×10^6	500	8
FeKα	0.1937	Iron	LiF(200),FPC/SPC,30	25	1.3×10^6	550	12
CuKα	0.1542	Copper	LiF(200),FPC/SPC,30	40	1.5×10^6	330	14
GeKα	0.1255	Germanium	LiF(220),FPC/SPC,30	48	7.0×10^5	250	25
MoKα	0.0711	Molybdenum	LiF(420),FPC/SPC,50	150	5.4×10^4	60	185

Notes:

[a] Data measured with normal electron incidence and 35° x-ray take-off angle. Other x-ray take-off geometries will result in different intensity data.

[b] Key to operating conditions: crystal type; detector type (FPC = flow proportional counter; SPC = sealed proportional counter); accelerating potential (kV).

[c] Sensitivities for standards were determined using the relationship:

$$C_{DL} \geq 3.29/(t \times P \times P/B)^{1/2} \text{ (Ziebold 1967)},$$

where t = total count time (1000 s), P = peak count-rate (0.1 μA specimen current), P/B = peak to background ratio.

Source: (Reproduced with permission from Microspec Corporation.)

x-radiation such as that caused by electrons backscattered from the specimen striking parts of the spectrometer.

Analysis by wavelength-dispersive spectrometry

Faced with a specimen of unknown composition the user of a crystal spectrometer would first of all scan through the whole of the accessible wavelength range to look for peaks in the x-ray spectrum. Having located the peaks and calculated their wavelengths, he or she would know which elements are present in the specimen; if required, the user could then examine and measure selected peaks in greater detail in order to determine how much of each element was there, by comparing the count-rate on the unknown with a count on a pure element or known compound under the same conditions of beam energy and current. Given an appropriate set of crystals all elements down to beryllium ($Z = 4$) can be analysed. The whole operation can be very much simplified by allowing a computer to carry out the settings and measurements.

As a result of the deficiences in the diffraction and the effects of finite widths of slit in the spectrometer, monochromatic x-ray lines are broadened into Gaussian peaks with a finite breadth, although slender compared to those we will meet in connection with energy-dispersive spectrometry. Peak-height to background ratios of hundreds or several thousand can be obtained. The lower limit of detection can be below 0.01% for wavelength-dispersive spectroscopy; hence it is frequently more conveniently quoted in parts per million, ppm. Against this must be put the fact that the efficiency of collection of x-radiation by the spectrometer is very low, and high beam currents of 10^{-9}–10^{-6} A are required to obtain a reasonable count-rate of x-rays and so reduce the time taken for analysis. Table A3.2 reproduces performance data for a modern crystal spectrometer with a four-crystal turret, as fitted to a number of scanning microscopes.

Energy-dispersive x-ray spectroscopy (EDS or EDX)

This very successful method of x-ray analysis is often popularly known by the acronym EDAX (Energy-Dispersive Analysis of X-rays). However, this is now the registered trade name of one of the major suppliers of EDS systems, and the EDS equipment on a microscope may be supplied by Ortec, Kevex, Oxford Instruments (Link), PGT or Noran (Tracor), to mention the most prominent alternatives to EDAX.

The technique uses a semiconductor detector to classify x-radiation according to energy rather than wavelength. [N.B. Energy (keV) $= 1.24/\lambda$ (nm) $= 12.4/\lambda$ (Å)]. The detector, a single-crystal disc of lithium-drifted silicon (Si(Li)) or high-purity (intrinsic) germanium (HPGe or IG) some 3–5 mm thick and with an active area between 10 and 30 mm^2, (Figure A3.8), converts the energy of incident x-ray photons into pulses of current proportional to the photon energy. These pulses are amplified, digitised and fed into a multi-channel analyser or the memory of a computer, which stores them in a location appropriate to the pulse height. The memory may have 1000 channels, each 10, 20, 40 or 80 eV wide, so that the total energy range covered by the analysis would be 10, 20, 40 or 80 keV, respectively. The progress of an analysis and its final spectrum may be displayed on a TV monitor, which may be in monochrome or colour (nowadays usually the latter). Thus the ED spectrometer is able to analyse a whole spectrum simultaneously instead of peak-by-peak as in WDS.

An example of such a display is shown in Figure A3.9. Computer software has been developed to enable the displayed spectrum to be manipulated, e.g. expanded horizontally and vertically, and analysed using a bank of data stored in the computer memory. With a little experience it is possible to recognise at a glance the 'finger-prints' of the more common elements. It is found in practice that the energy range 1–10 keV is particularly useful, containing K lines of elements with Z from 11 to 32, L lines of Z from 30 to 80 and M lines from $Z = 62$ upwards.

The detecting crystal with its associated pre-amplifier is mounted at the end of a cylindrical tube which is inserted through a side port into the specimen chamber or column of an electron microscope so that the crystal can 'see' the specimen. There is no focusing requirement, but in order to make effective use of the x-radiation generated in the specimen the detector should subtend a large solid angle at the specimen. The largest angle currently in use is 0.3 steradian, which means that more

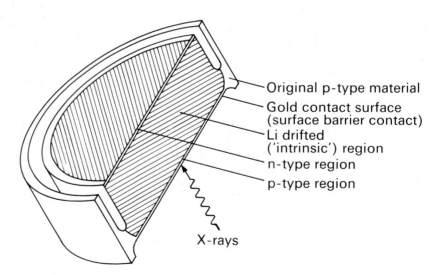

Figure A3.8. Schematic cross-sectional diagram of a lithium–drifted silicon crystal x-ray detector. This type of detector is made in various sizes, a typical crystal for an SEM having a detecting area of about 10 mm². The analytical TEM will use a 30 mm² detector placed very close to the specimen to make the most of the weak signals from very small areas of specimen. (By courtesy of Kevex Corporation.)

than 4% of the x-radiation is detected. Most EDX systems collect much less than this, but the efficiency is still an order of magnitude better than the wavelength-dispersive spectrometer. The detecting crystal and the first stage of the pre-amplifier are cooled with liquid nitrogen in order to improve the electronic signal-to-noise ratio of the spectrometer. The detector holder is therefore built on to the bottom of a dewar vessel which will contain up to a week's supply of liquid nitrogen. This dewar, which may be spherical, cylindrical or even flat-sided in shape, provides the most obvious outward sign that a microscope is equipped for x-ray analysis (e.g. see the detector on the SEM in Figure A3.7 and on the analytical TEM, Figure 3.47). A schematic drawing of a complete EDX detector is shown in Figure A3.10.

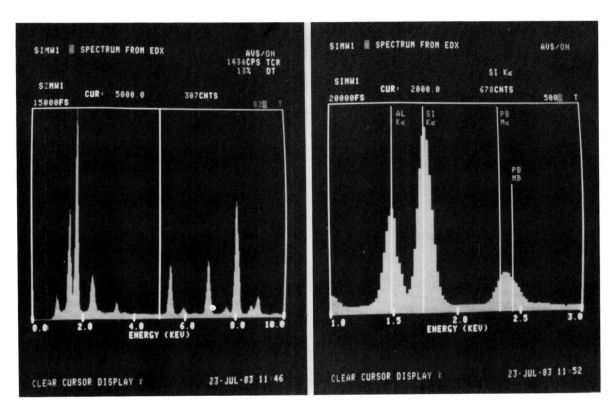

Figure A3.9. Energy-dispersive analysis in operation. Photographs from a display on the monitor during and after acquisition of a spectrum illustrate some of the features of this form of analysis. These include: all peaks within the chosen energy range are measured simultaneously; total count-rate (c.p.s.) and dead-time (see below) are displayed during the counting; the energy value and number of counts at the position of a movable cursor are displayed; identification of peaks can be confirmed using stored data in the computer memory. The automatic-vertical-scale (AVS) device (top right) ensures that the highest peak is always on the screen, by changing the vertical scale appropriately. The content of the spectrum remains essentially the same during an extended count, but becomes less ragged and statistically more reliable as the number of counts increases. To look for the presence of very small peaks the AVS would be switched off and the spectrum expanded vertically.

A cooled detector crystal in a conventional electron microscope high vacuum would quickly become coated with ice and any other condensible vapours present in the column. It is usual, therefore, to screen the crystal from the vacuum with a thick (7.5 μm) beryllium 'window', which seals the visible end of the detector. Although this window is quite transparent to most of the x-radiation, its transmission falls rapidly around 1 keV. Transmission of the 1.04 keV Na Kα line is 60%, but at 0.52 keV (oxygen K) it is less than 2%. Hence the normal EDX system is only usable for atomic numbers from 11 (sodium) upwards. It is, in fact, the detector's blindness to

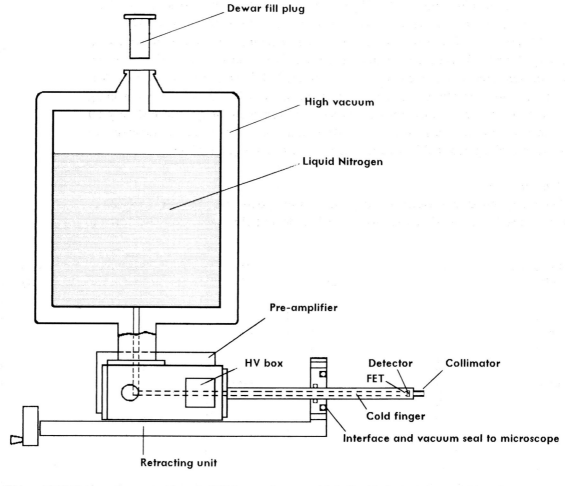

Figure A3.10. Schematic cross-section of a Si(Li) x-ray detector with its liquid–nitrogen dewar and interface to a microscope. The detecting crystal and FET (field effect transistor) of the pre-amplifier are mounted close together at the end of a cooled rod. The collimator is, in essence, a metal screening tube lined with carbon to prevent the detector from receiving x-radiation arising from regions other than the electron-bombarded area of the specimen, e.g. from pole-pieces and the specimen stage. The effectiveness of the collimator can be maximised if the specimen is at a constant position when it is to be analysed. (Based on a drawing supplied by EDAX Ltd.)

carbon x-radiation which makes it possible to employ that material for charge-dispersing coatings on insulating SEM specimens which are to be microanalysed.

Windowless and thin window EDX

If the microscope vacuum is sufficiently 'clean', i.e. free from condensible vapours such as water and pumping fluids, semiconductor x-ray detectors can be used without the protection of a window. A safer, intermediate, stage is to use a less absorbent window material than beryllium foil. Certain polymers and sub-micrometre thicknesses of diamond-like material or boron compounds are described by such terms as ultra-thin windows (UTW) or super atmospheric thin window (SATW). They must be vacuum-tight and strong enough to withstand 1 bar pressure over 10–30 mm² sensitive area when the specimen chamber is opened for specimen change. The UTW is usually coated with a thin, visually opaque, metal film to screen the crystal from light. A choice of Be window, UTW or completely windowless operation may be offered on a rotating turret mounted in front of the detecting crystal. Whichever window is used, it is necessary to mount a magnet near the crystal to deflect stray electrons away from it. An example of the extension of detection to lower energies permitted by a thin window is shown in Figure A3.11.

Spectral resolution

The ED spectrometer has a resolving power rather poorer than the WD spectrometer. The lack of resolution is shown in the broadening of response to sharp spectral

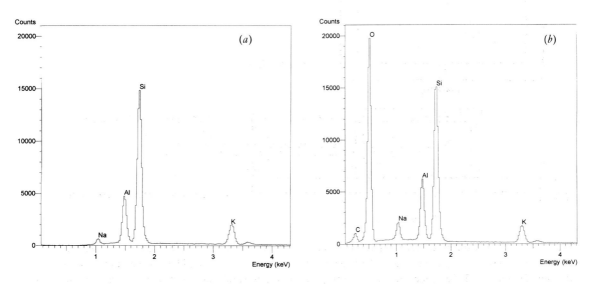

Figure A3.11. Energy-dispersive spectra of an orthoclase mineral from a Si(Li) detector with (*a*) a Be window and (*b*) a thin window (SATW). The improved detection of C, O and Na will be noted. (By courtesy of Oxford Instruments Microanalytical Group.)

lines to give a bell-shaped envelope extending over a number of channels on either side of the peak value. The usual way of comparing resolutions is to state the full-width at half-maximum (FWHM) value at 1000 counts per second for the Mn Kα peak at 5.9 keV (see Figure A3.12). Resolutions down to about 135 eV are currently found for Si(Li) detectors under these conditions, and 115 eV for HPGe; for theoretical reasons the lower limit is unlikely to be improved on. By comparison a fully focusing wavelength spectrometer can show a resolution of less than 10 eV by the same test. Figure A3.13 shows part of the spectrum of a superalloy containing Ta, W, Re as it is analysed by the two types of spectrometer. In this part of the spectrum there are overlaps between the M lines of the three elements, which are clearly resolved in the WD spectrum whilst being within a single broad 'envelope' in the ED spectrum.

It can be seen from Table A3.3 that it would be necessary to go to higher-energy lines (by increasing the electron gun energy if necessary) in order to be able to analyse these elements separately by EDS.

This type of overlap may often occur in ED spectrometry. Reference to another

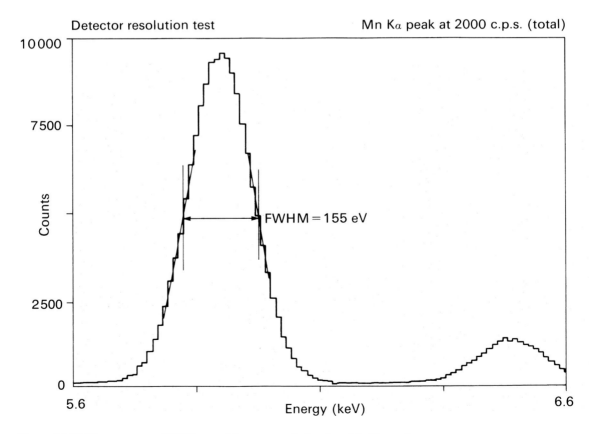

Figure A3.12. Measurement of Si(Li) crystal detector resolution from Mn Kα peak.

part of the energy spectrum usually provides a solution to such a problem. Alternatively, if the statistics of the spectrum are good enough, the presence and position of extra peaks can be revealed by a spectral deconvolution programme. In WD spectrometry the analogous situation results from every spectral line having the possibility of several different orders of reflection.

There are several peculiarities of ED spectrometry which should be noted. One of these is the phenomenon of *escape peaks* in the spectra due to the loss or escape of Si Kα or (Ge Kα) radiation generated in the detector itself during the collection process. The input energy E is then recorded as $E-1.74$, the energy of Si Kα being 1.740 keV. The proportion of lost silicon radiation is small, about 0.5%, but extended counts can give rise to observable small peaks 1.74 keV lower in energy than the main lines being counted. In the case of HPGe detectors the escape peak is 9.885 keV below the parent peak, with a higher probability of occurrence, but escape peaks will only be present if the beam energy used is greater than 11.1 keV.

There can also be *summation peaks* at high count-rates, in which two photons of

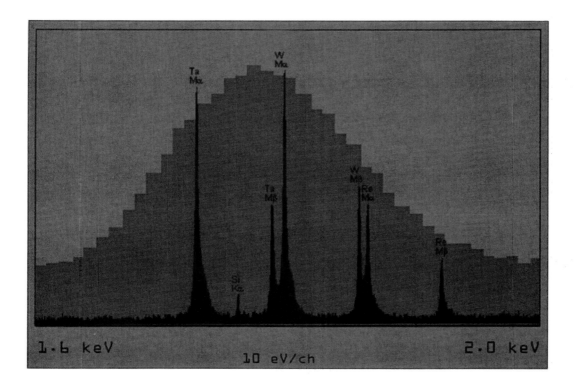

Figure A3.13. Comparison of ED (*dark grey*) and WD (*black*) spectra from the same sample. (Spectra by courtesy of Microspec Corporation; sample from Cannon–Muskegon Corporation.)

Table A3.3. *Characteristic spectra for tantalum, tungsten and rhenium*

Element	Energy (keV) for given spectral line					
	Mα	Mβ	Lα	Lβ	Kα_1	Kα_2
Tantalum Ta	1.709	1.765	8.145	9.342	57.52	56.27
Tungsten W	1.774	1.835	8.396	9.671	59.31	57.97
Rhenium Re	1.842	1.906	8.651	10.008	61.13	59.71

energy E are counted as a single photon of energy $2E$. However, the distortion and peak displacements which used to occur when the detector produced pulses faster than the amplifier could accept have been eliminated in present day EDX systems. If x-ray photons arrive faster than the system can deal with them the detector operates with a proportion of 'dead time' during which it will not accept a new pulse until the previous one has been cleared. Some differences may be expected between different systems, but in the best ones the pulse-processing has been improved to such a degree that count-rates of tens of thousands per second are achieved before dead-time limitations are met. This, and the reduction of FWHM for energy resolution, particularly with Ge detectors, has gone a long way towards overcoming the greatest disadvantage of EDS. Since the limiting count-rate is that of the whole spectrum, counting from minor constituents cannot be speeded up above the rate set by the presence of the major peaks and the Bremsstrahlung. However, even the best EDX systems processes x-radiation one or two orders of magnitude more slowly than WDX.

Because of the count-rate limitation and the higher efficiency of x-ray collection it is possible to analyse with very much lower beam currents in EDS than in WDS; a range of 10^{-11}–10^{-9} A for EDS against 10^{-9}–10^{-6} A for WDS. The lower current is associated with a smaller-diameter electron spot and slightly better spatial resolution of analysis. With very small spots, e.g. tens of nanometres in the analytical TEM or 1 nm or less in FESTEM, count-rates are usually on the low side, and the high collection efficiency is essential.

Comparison of important characteristics of wavelength-dispersive and energy-dispersive spectrometers

The important characteristics of the two types of x-ray spectrometer are shown in Table A3.4, with particular reference to usage on electron microscopes.

Light-element analysis

The elements carbon, nitrogen and oxygen are widespread in nature, particularly in geology, and it is often desirable to know whether a compound is an oxide or a

Table A3.4. *Summary of characteristics of WD and ED spectrometers in x-ray analysis*

Property	WD	ED
Elements covered, Z range	4 to 92 (Be to U)	11 to 92 (Na to U): 4 to 92 (Be to U) windowless
Resolution, peak width at half-maximum, eV	~10	~130–150 (115 for HPGe)
Min. detectable concentration, % (ppm)	≤0.01 (100) (varies with element & crystal)	≤0.1 (1000)
Accuracy of quantitative analysis, %	0.5–1	0.5–1[a]
Method of analysis	One wavelength at a time[b]	Whole spectrum simultaneously
Ease of fitting	Bulky, strict focusing requirements,– SEM, EPMA only[c]	Easily fitted to TEM, SEM, EPMA. No focusing requirements
Probe current, A	10^{-9}–10^{-6}	10^{-11}–10^{-9}
Running costs	Argon/methane gas for FPC	Liquid N_2 for cooling[d]
Field of usefulness	Very good for precise quantitative analysis and line-scans	Very good for quick, readily interpreted qualitative analysis; quantitative analysis possible

Notes:

[a] Accuracy lower with overlapping peaks

[b] Multiple elements if more than one spectrometer fitted

[c] Restrictions less severe with non-vertical spectrometer but space limitation worse.

[d] Not necessary if the detector has an alternative means of cooling

carbonate, for example. Detection and measurement of the light elements is therefore a desirable facility for electron microscopy. The x-ray energy range is 1.00 keV and below, or $\lambda = 1.2$ nm and above.

In WDS, natural crystals and synthetic multilayer crystals with wide d-spacings are available (see Table A3.1). These latter may be tailor-made for particular wavelengths and provide a marked improvement in count-rate and/or peak-to-background ratio over conventional crystals (Nicolosi *et al.*, 1986). The windows of the flow proportional counter must be transparent to these very soft x-rays, and a thin film of a polymer such as polypropylene is usually chosen. Simplified WD spectrometers with up to four pre-set single-element monochromators can be usefully employed to augment EDX if much work is to be done in this part of the spectrum.

In EDX, elements down to carbon can be detected if the 7.5 μm Be window in front of the detecting crystal is replaced by a thinner material. The thinnest of windows, or none at all, allows beryllium to be detected. The improved resolution possible with Ge detectors is helpful in this part of the spectrum (Figure 6.8).

It is in the analysis of light elements that EELS (see Chapters 2, 6) proves to be a more sensitive analytical technique than EPMA, subject to the restrictions that the specimen must be a transmission specimen thin enough for single scattering only to have occurred.

Suggested further reading

The following books will fill in details of x-ray spectrometers and their use: Woldseth (1973), Birks (1971), Reed (1993), and Scott *et al.* (1994).

Appendix 4: Electron sources for electron microscopes

The source of the illumination is one of the most important components of an electron microscope, although it is very often taken for granted until it begins to give trouble. Yet the correct operation of this part of the column is crucial to the performance of the whole system. If the electron wavelength fluctuates, or the intensity is inadequate or unsteady, the remainder of the instrument is disadvantaged. It is worth spending a little time, therefore, in considering the principles involved in providing a source of electrons adequate for the needs of present-day scanning and transmission electron microscopes.

The early experimental transmission electron microscopes derived their illumination from a gas-discharge tube situated at the top of the microscope column. Electrons formed by ionisation of the gas (fundamentally the same process as that described in earlier sections of Appendixes 1 and 2) were extracted through a hole about 0.1 mm in diameter in the anode of the tube. This small hole restricted the flow of gas and allowed a higher gas pressure to exist in the discharge tube than in the main body of the microscope. (It is interesting that the situation is now reversed; in many microscopes a diaphragm is placed in the column so that the electron source can operate at a *lower* gas pressure than the remainder of the microscope.)

In the late 1930s a gas–discharge tube was replaced by a heated (thermionic) cathode, a form of electron source which has remained the basic standard on electron microscopes ever since, although it is now better used and understood. More recently the demands of high-resolution microscopy and microanalysis have resulted in a change in the material used for the thermionic cathode and in the exploitation of the quite different principle of field emission. Both of these developments have become practicable because of recent advances in vacuum technology and the construction of electron microscope columns. The electron emitter together with one or more other electrodes for accelerating and shaping the illuminating beam constitute an electron gun.

For electron microscopes the important characteristics of an electron source are its size and its brightness. (For some purposes there are additional requirements, to be mentioned later.) We shall see that for most microscopy, i.e. until nearing the limits of performance, the simple thermionic electron gun can be used, which projects its electron stream into a bright 'crossover', which may be 10–100 μm across and can be regarded as the actual electron source. Although high electron currents can be obtained from such a source, the large size means that the actual brightness (current density per unit solid angle) is relatively low, and inadequate for forming very small electron spots for microscopy or microanalysis, especially at low accelerating voltages. For these purposes the more elaborate field emission electron guns are used, in which the electron source is effectively a region 1000 times smaller. The different forms of field emitter and their associated advantages and limitations will be outlined later in this Appendix. First we will describe the most common form of electron gun, which depends on the thermionic emission of electrons.

Thermionic emission

Metallic bodies contain large numbers of conduction electrons which are in constant thermal motion within the body and free to move under the influence of any potential difference which may be established between one part of the body and another. If the metal is heated up, the kinetic energy of the electrons becomes greater and a certain proportion are sufficiently energetic to penetrate the boundary of the metal and escape. (N.B. Here, and in the remainder of this Appendix, we assume that the electrons are emitted into a vacuum.)

As the temperature is raised, the number of electrons with sufficient energy to escape increases rapidly. The process is called *thermionic emission* and the electron emission current I amperes escaping from each square centimetre of a metal surface at a temperature T kelvin ($= {}^\circ C + 273$) is given by the modified Richardson equation:

$$I = AT^2 \exp(-\phi/kT)$$

where A is a constant (theoretically 120 A cm^{-2} K^{-2} for pure metals)

          ϕ is the work-function of the metal in electron volts (eV)

          k is Boltzmann's constant 8.62×10^{-5} eV K^{-1}

The *work-function* is the additional energy needed for an electron to escape from within the metal. Materials with low work-function (e.g. caesium and barium, for which $\phi = 1.9$, 2.5 eV respectively) emit thermionic electrons at relatively low temperatures, whereas metals such as tungsten ($\phi = 4.54$ eV) require to be heated to white heat before they emit electrons strongly.

Experimental values of the 'constant' A are invariably lower than 120; as low as 10^{-2} for so-called 'oxide cathodes' of mixed Ba, Sr, Ca oxides. However, the influence of the exponential term $\exp(-\phi/kT)$ is much greater than AT^2 in determining the electron emission from a heated body. (N.B. the energy needed for emission can be supplied by other means, as in the photoelectric effect, when it is provided by absorbed electromagnetic radiation, e.g. light, UV or x-rays.)

It is useful to visualise the effects of temperature and work-function from a consideration of the electron theory of metals. Fig. A4.1(*a*) shows the number of free electrons $N(E)$ in a metal plotted against electron energy E. At absolute zero ($T = 0$) the curve falls abruptly to zero at the energy E_F, the Fermi energy. At a higher temperature T_1 some electrons have energies greater than E_F. The higher the temperature the further upwards the energy 'tail' extends. Fig. A4.1(*b*) shows the electron energy change across the interface between the metal and the vacuum outside it. There is an energy barrier ϕ_0 above the Fermi level which has to be surmounted before a free electron can escape into the vacuum. Heating the metal results in a few electrons in the metal being given sufficient energy to be emitted over the barrier as thermionic emission. Reducing the height of the barrier ϕ (i.e. the work-function) allows thermionic emission to take place at a lower temperature.

In practice, the surface barrier is not square-cut, but rounded. Also, the presence of a positive electrode to attract electrons results in the electron potential-energy falling with increasing x. These two effects combine to result in a lowered energy barrier, shown dotted. The reduced work-function ϕ_F is the effective work function in the presence of the applied electric field.

In a thermionic diode, when a second electrode (anode) at a positive potential is placed opposite a thermionic emitter, the current being collected by the anode may be reduced below the full thermionic emission because some of the electrons are repelled and turned back into the emitter by the negatively charged cloud of electrons (the *space charge*) already in the space between the electrodes. The collected current *I* and applied potential *V* are related by the *space-charge equation*, or *three-halves power law* $I \propto V^{3/2}$, as shown in Figure A4.2. When *V* is increased to a sufficiently high value such that the applied electric field overpowers the effect of space-charge the current saturates at the value determined by the emitter temperature and work-function. Above the 'knee' of the $I - V$ curve the current increases in proportion to $V^{1/2}$, according to the Schottky Law. Thus below the knee, *I* depends on voltage; above it, the current is temperature- and voltage-

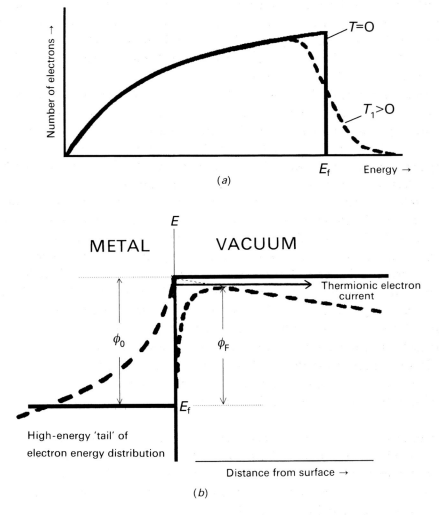

(a)

(b)

Figure A4.1.(a) Free-electron energy distribution in a metal. (b) Energy diagram for thermionic emission from a metal. With zero applied field thermionic emission will occur when the tail of the electron energy distribution reaches the height ϕ above the Fermi level. In practice, the combination of several effects including an applied electric field results in the rounding-off and lowering of the surface barrier (shown dotted) to an effective ϕ_F above the Fermi level.

dependent.

Finally, it should be noted that the constant of proportionality in the three-halves power relationship contains a factor $1/d^2$ where d is the anode–cathode separation in the plane-parallel two-electrode geometry we have considered. Thus if the electron current alters as a result of changing V, the situation may be restored by appropriately altering the inter-electrode spacing.

The relationships described above apply in this form only to the two-electrode situation; however, the important principle for the reader to grasp is that distance and voltage are related parameters when determining currents between electrodes.

Choice of emitter materials for electron microscopes

The total emission current drawn from the electron source in an electron microscope is quite small, several hundred microamperes at the most, and the emitter operates in a vacuum which may be as poor as 10^{-4}–10^{-3} mbar and can become worse than this if the microscope is shut down overnight. Under these conditions, pure tungsten is the most practical thermionic emitter for use in the electron gun. The most common arrangement is the 'hairpin' filament made by bending a length of fine (e.g. 0.1 mm diameter) tungsten wire back on itself into a 'V' or parallel configuration and welding or soldering the two ends on to posts of thicker metal embedded in an insulating

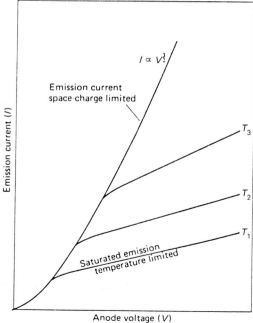

Figure A4.2. Relationship between current and voltage in a diode. The current reaching the anode follows the ½ power relationship until the emission from the cathode is *saturated*. The current which can be drawn then depends on the temperature of the emitter. In the figure T_1, T_2 and T_3 are emitter temperatures: $T_3 > T_2 > T_1$.

block (Figure A4.3). This forms an easily replaceable unit which can be plugged into a holder in the electron gun. The filament is heated by passing an electric current through it until the tip of the 'V' reaches the desired electron emitting temperature, which lies in the range 2600–2900 K. The other electrodes of the electron gun are described later.

Pure tungsten continues to be used as filament material in most electron microscopes because of its ability to provide high beam currents in comparatively poor vacua with very little trouble. At the operating temperature there is a continuous evaporation of metal from the hottest regions of the wire, thinning it down. The end of life comes when a local 'hot-spot' develops and the wire melts, breaking the circuit. Filament lifetimes vary from instrument to instrument, and even between different operators of the same microscope. The range from 10–150 hours would probably include most microscopes. Tungsten filaments are eroded more rapidly by operation in oxygen or water-vapour; filament lifetimes can be increased markedly if a microscope is not put back into use immediately after the gun has been opened to the atmosphere for a filament change, but allowed to pump out for a few hours. Continuous pumping without overnight shut-down and the provision of refrigerated surfaces within the vacuum ('cold fingers') to condense out water and other vapours are also beneficial to filament lifetime.

In radio and television valves and tubes, pure tungsten filaments were super-

Figure A4.3. Hairpin filament assembly from the electron gun of an electron microscope. The filament is mounted on a ceramic base (*left*) with current leads which plug into a larger block (*right*) which is, in turn, mounted within the Wehnelt cylinder (see later, page 439) of the triode gun.

seded many years ago by thoriated tungsten ($A = 3.0$ A cm^{-2} K^{-2}, $\phi = 2.63$ eV), thoria (similar constants to thoriated tungsten) and by oxide-coated cathodes ($A = 0.05$, $\phi = 1.0$). These operate at a higher efficiency than pure tungsten, in terms of emission current per watt of power needed to maintain them at operating temperature, a consideration that is more important in the case of electron tubes than in electron microscopes. Unfortunately, these emitters require an *in-situ* activation to achieve their best emission and are more sensitive to the quality of vacuum and the degree of positive-ion bombardment than is tungsten. However, apart from the work reported by Uyeda and others (cited in Tochigi *et al.*, 1962) on oxide-cored cathodes and a simple TEM made earlier by Akashi, there has been little evidence that they have been tried seriously in electron microscopes. In the improved vacuum of modern microscopes the objections may not still be valid, and comparable emission to tungsten over very much longer lifetimes, or higher emissions than tungsten over comparable lifetimes, may be possible. (But see later, Schottky emitters, where ZrO/W appears to be a robust substitute with usefully reduced work-function.)

More recently, the hexaborides of a group of alkaline-earth and rare earth metals have been found to give high and stable thermionic emission without any special activation procedure. Of these lanthanum hexaboride, LaB$_6$ ($A = 29$, $\phi = 2.66$) is most commonly used, although CeB$_6$ is also available. Cathodes made of these materials are now available for use on a number of commercial microscopes. Lanthanum hexaboride is used in the form of a pointed rod, indirectly heated to a temperature of about 1600 °C. Provided the vacuum around the emitter is maintained better than 10^{-6} mbar at all times, the LaB$_6$ cathode supplies a stable electron current coming from a smaller area than the tungsten hairpin. The brightness can be up to 10 times that of tungsten and will be maintained for a longer period (a minimum of 500 hours is guaranteed by one supplier). This is particularly useful in any application where a small bright source is needed, e.g. microanalysis and scanning microscopy near the resolution limit (Verhoeven & Gibson, 1976). There has also been reported (Harada *et al.*, 1991) the use of a sharpened LaB$_6$ emitter at 800 °C, combining the lower work-function with high-field effects (see later), giving an increase of 300 times in brightness at 20 kV over a normal LaB$_6$ emitter at 1500 °C.

Figure A4.4 gives the emission current density–temperature relationship for the four thermionic emitters discussed above.

The triode electron gun

It is possible to produce a beam of accelerated electrons with a two-electrode or diode gun consisting of the electron emitter or cathode and a positively charged plate with a hole in it, forming the anode. Electrons will be accelerated towards and out through the anode hole. No further electrodes are needed to keep the beam moving, and the arrangement is literally an electron gun in so far as it projects electrons in a chosen direction. In electron microscopes the cathode is maintained at a negative potential whilst the anode and all the other metalwork of the column are at zero or earth potential. This arrangement means that the only high-voltage insulation in the microscope column is the cathode support, which is usually a robust ceramic body.

The electron current from a diode gun with fixed dimensions and anode potential can only be varied by altering the filament temperature; moreover the constancy of the beam current depends on the stability of the filament heating current.

In electron microscopes the diode electron gun has been superseded by a triode or three-electrode design which allows the beam current to be varied independently of filament temperature and anode voltage. The third electrode or *Wehnelt grid* is introduced between cathode and anode. It usually takes the form of a metal cylinder (the *Wehnelt cylinder*) surrounding the filament assembly. The cylinder is closed at one end by a metal plate with a small hole at its centre, the tip of the filament being placed just behind or level with this hole and centred within it. The filament tip and the centres of the holes of the grid and anode lie in a straight line, which is the axis of the electron gun. The Wehnelt grid is held at a negative potential relative to the filament; this negative bias may be a hundred or more volts in a gun whose anode voltage is tens of kilovolts.

The effect of the grid is to deflect the electrons emitted by a filament into a small crossover between grid and anode (Figure A4.5(*a*)). This is the effective source of the illumination which is directed onto the specimen by the condenser lens system of the microscope. The bias voltage on the grid may be a fixed potential supplied by a battery but is more usually produced by passing the emission current through a bias resistor ('*self-bias*'), shown schematically in Figure A4.5(*b*). This has the added effect of stabilising the emission current against fluctuations in filament temperature, since any tendency for the emission to increase results in a more negative bias voltage which acts against the increase. Several values of bias resistor are normally provided and can be switched in and out according to the current which is required

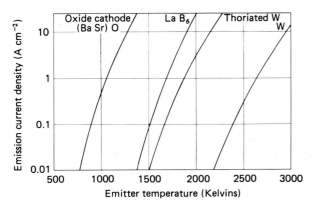

Figure A4.4. Current density–temperature relationship for saturated emission from four types of thermionic cathode. It will be seen that a range of current densities is available from each material depending on the temperature at which it is run. The latter is normally chosen to give a reasonable lifetime before failure due to evaporation, i.e. 10 to 100+ hours in an electron microscope, more than 1000 hours in an electron tube.

from the gun.

The characteristic behaviour of a triode electron gun is illustrated in Figure A4.5(c), in which the current drawn from it is plotted against the power input to the filament. As the latter is increased the beam current increases to a saturation value (3) at which the crossover has the smallest diameter. Heating the filament beyond this point shortens its lifetime (since it increases its rate of evaporation) without resulting in any increase in beam current. If more beam current is necessary the value of the bias resistor is reduced.

The electron guns from two generations of high-resolution TEM are shown in cross section in Figure A4.6. It can be seen that the construction of the more recent gun has been simplified in order to make the alignment procedures more convenient, and to enable a better and cleaner vacuum to be provided in the gun chamber.

Brightness of an electron gun

The brightness of an electron gun is a very important factor which sets the limits of performance of both the SEM and TEM. It is defined as the current density per unit solid angle (A cm^{-2} sr^{-1}) and has the theoretical value, originally calculated by Langmuir in 1937, of

$$\beta = \frac{AT^2 \exp(-\phi kT)}{\pi} /(1+V/kT)$$

where V is the accelerating voltage and A, T, ϕ, k are as defined earlier. For a tungsten hairpin, filament brightnesses up to 10^6 A cm^{-2} sr^{-1} at 120 kV are possible.

Important points to note are that the expression is for maximum brightness which can be obtained from an electron gun and that for a given gun this is dependent on the operating voltage. If a gun is designed to give the theoretical brightness at, say, 120 kV, it will only achieve a lower brightness when operated at a lower voltage unless the anode–Wehnelt grid separation can be reduced to compensate. Although earlier electron microscopes provided this adjustment it is only rarely possible on present-day instruments. However, maximum brightness is only needed under certain conditions of operation; namely very high magnification in the TEM and scanning microscopy and analysis with very small focused probes.

At magnifications of hundreds of thousands the tungsten filament on a 100 kV TEM becomes unable to provide a bright enough image for focusing and photography. (Very high magnifications impose greater demands on specimen stability, and short recording times, requiring brighter images, are more necessary than at lower magnifications.) Similarly, when operating the SEM or analytical TEM with spot sizes of 5 nm or below the image becomes grainy (noisy) and the count rate in x-ray analysis very low, because of the smaller number of electrons in the

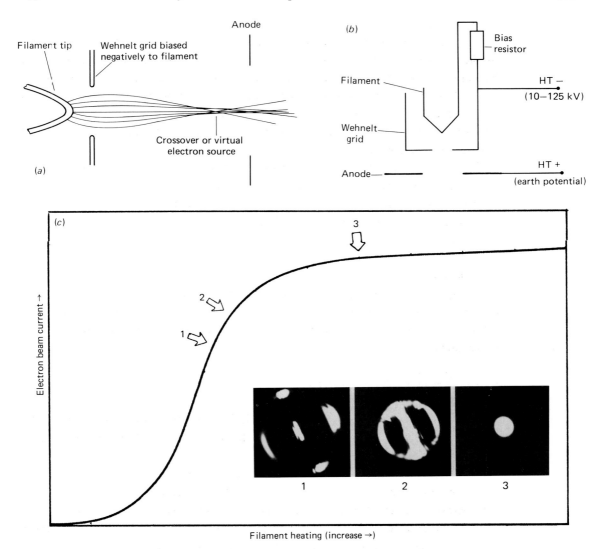

Figure A4.5. Self-biased triode electron gun. (*a*) The effect of grid bias is to concentrate the electrons emitted by the heated filament into a small crossover which becomes the effective illumination source for the microscope. (*b*) Passing the emission current through a bias resistor gives the filament a positive potential relative to the grid. For example, a current of 100 μA will develop a bias of 100 V across a resistor of 1 megohm. (*c*) Characteristic emission–temperature relationship for the triode gun. As the filament heating is increased the negative bias also increases and the shape of the virtual source tends towards the eventual small spot. Micrographs (*inset*) show the appearance of the source at points 1, 2, 3 on the curve. Position 3 is the normal operating point; making the filament hotter than this does not increase the emission current but increases the rate of evaporation of the filament and hence shortens its lifetime before failure.

spot.

Fortunately, the lanthanum hexaboride emitter has now been developed to such a state that it can provide a practical replacement for tungsten. The physical size of the emitting tip is smaller than the emitting region of a wire hairpin; a smaller crossover results and a brightness up to ten times as high. This allows smaller electron probes to be formed, and hence a higher resolution in scanning microscopes. For example, commercial microscope specifications may contain a statement such as 'resolution 6 nm (tungsten cathode), 4 nm (lanthanum hexaboride cathode)'. As mentioned earlier, it is essential for the vacuum around the LaB_6 cathode to be better than 1×10^{-6} mbar whenever it is hot, or full brightness will not be obtained. It may be necessary to provide the electron gun with its own localised pumping system, e.g. a sputter-ion pump, in order to meet the vacuum requirement, and it is important also that the cylinder of the Wehnelt grid has adequate pumping holes in it.

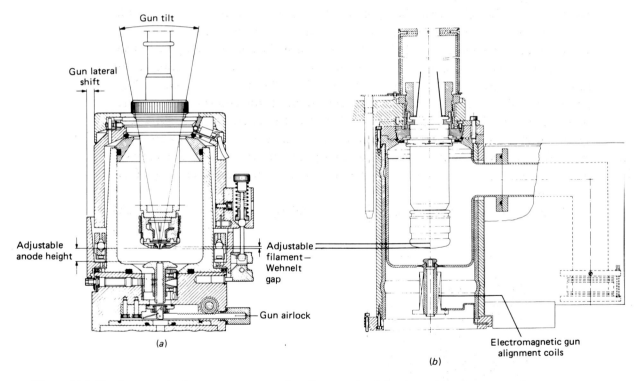

Figure A4.6. Developments in electron gun construction. Cross-sectional drawings of the electron gun from two generations of high-resolution TEM (Philips EM301 (*a*) and EM400 (*b*)). Mechanical adjustments of gun tilt and shift in (*a*) have been replaced by electromagnetic controls at desk level. The only mechanical adjustment still retained is the setting of the filament/Wehnelt cylinder separation, which does not involve sliding vacuum seals. The result is a very clean vacuum which assists in assuring bright, stable emission with maximum filament lifetime. (By courtesy of Philips Electron Optics.)

Other methods have been used to obtain brighter and smaller electron sources using tungsten filaments. These include various means of increasing the temperature of the tungsten, e.g. by grinding flat faces on either side of the tip (Bradley, 1961) or by heating a tungsten bead close to its melting point with a laser beam (van der Mast *et al.*, 1974). It is possible to purchase so-called 'pointed filaments' which consist of a very short needle-point of tungsten welded to the tip of a tungsten hairpin. All of these methods will result in smaller and brighter emitting sources in the hands of a skilful operator, but are unlikely to combine the merits of high brightness, small source size, long lifetime and direct interchangeability with conventional filaments (albeit at a higher price) which are found with lanthanum hexaboride. The pointed filament is a compromise between a thermionic filament and a field emission source; this, the ultimate in bright emitters, which is not interchangeable with a tungsten hairpin, is described in a later section.

High-voltage electron guns

100 kV used to be considered the upper limit of accelerating voltage which could be applied to a triode electron gun in an electron microscope, without the risk of an insulation breakdown. The 100 kV has been 'stretched' successfully to 120 kV on a number of instruments and even to 150 kV on one. The more recent Intermediate Voltage microscopes, operating at 200–400 kV, have applied the high tension in multiple stages, so that the full gun voltage is not applied between adjacent electrodes. The accelerator tube of a 200 kV microscope is illustrated schematically in Figure A4.7, which shows the electrical connections to the seven electrodes involved. Freon gas at twice atmospheric pressure is used as external insulation to reduce the chance of electrical breakdown within the housing of the accelerator.

The principle of multi-stage acceleration is similarly applied at the higher volt-

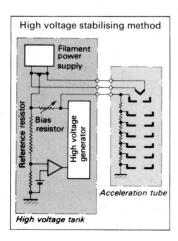

Figure A4.7. Schematic diagram of multi-stage electron accelerator from 200 kV transmission electron microscope, JEOL JEM 2010. (By courtesy of JEOL Ltd.)

ages of 600 and 1000 kV. In these microscopes the electron gun has considerable length, and together with the high-voltage generator, occupies a large volume, as was seen earlier (Chapter 3). An experimental compact 1 MV instrument with a 'folded' accelerator tube was reported by le Poole *et al.* (1970).

Field emission

If a very strong electric field is applied at the surface of a metal it is possible to extract electrons from the metal by the process of field emission – sometimes referred to, more accurately, as high-field emission. Field strengths of 10^9 V m^{-1} are necessary for this with an unheated emitter; the easiest way of achieving very high electric fields without using abnormally high voltages is to give the emitting surface a very strong curvature. Etching a fine single-crystal tungsten wire to a sharp needle-tip (figure A4.8(*a*)) of radius below 1 μm and applying a voltage between the tip and an apertured metal anode results in field emission apparently coming from a small area within the tip. This 'virtual source' may be only tens of nanometres across, compared with the tens of micrometres of a thermionic cross-over, and hence there is the possibility of achieving higher brightnesses. In fact, brightnesses of 10^9 A cm^{-2} sr^{-1} are achieved from a gun consisting of a needle-tip and two electrodes, the first having the extraction voltage V_1 and the second the final accelerating voltage V_2 required for the electron gun (Figure A4.8(*b*)). The effect varies with crystallographic orientation of the tungsten, and wires with <100> and <310> orientation are used.

(a)

Figure A4.8(*a*). Micrograph of a tungsten cold field-emitting tip. The radius of curvature of the tip is about 100 nm. (By courtesy of Hitachi Ltd.) (*b*) Schematic diagram (not to scale) of electrode arrangement in a field emission gun. The anode would normally be connected to the body of the microscope, at earth (zero) potential.

(b)

There are several variants of field emission source, which are illustrated schematically in the energy-level diagrams of Figure A4.9. Application of the very strong electric field causes the surface energy barrier to deform to such an extent that it becomes physically very narrow (typically about 10^{-9} m). There becomes a finite probability of electrons from the Fermi level penetrating the barrier even at ambient temperature. This is the *cold field emitter*, CFE, used in some electron guns. Heating a tungsten field emission tip of <100> orientation to about 1800 K and applying a strong electric field can result in atomic rearrangement of the tip to an even sharper end; in this case field emission of electrons occurs into a very narrow angle of cone, with high brightness. This type of emission is called *thermally assisted field emission* (TFE). Confusingly, this name is also used by some for the third type of practical field emitter, more generally called a *Schottky emitter* (SE), in which a high electric field is applied to a heated sharp tip whose work-function has been reduced by a surface coating. This emitter may not be as bright as the cold field emitter but it delivers stable high currents and is less demanding in operation.

A commercial Schottky emitting cathode is less sharp than a cold- or thermally assisted field emitter, with a flat end plane of (100) orientation. A monatomic layer of ZrO is formed on this flat tungsten surface, which reduces the work-function from 4.54 to 2.8 eV. At an operating temperature of about 1800 K the effect of the applied electric field results in electron emission of several nanoamperes and a brightness of 10^8 A cm^{-2} sr^{-1}.

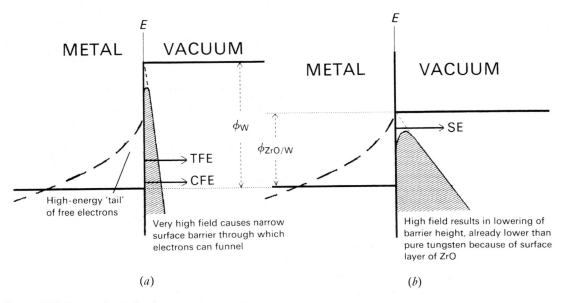

Figure A4.9. Energy-level diagrams showing the difference between the mechanisms of (*a*) cold- and thermally assisted field emission and (*b*) Schottky emission.

Vacuum requirements for field emitters

Cold field emission cathodes are very sensitive to environment. Changes of surface coverage by adsorbed gas result in changes in work-function and very large changes in emission. To operate effectively, the cold field emission cathode requires ultra-high vacuum at the level of 10^{-9}–10^{-10} mbar. Every few hours the tip has to be cleaned by 'flashing' (heating momentarily to drive off adsorbed gas). There is no such problem with the thermally assisted field emitter or a Schottky emitter of the ZrO/W type. These are operated at a temperature of about 1800 K in a vacuum of 10^{-8} mbar and give a highly stable emission current over a lifetime of thousands of hours.

Energy-spread

One property which may govern the selection of emitter type is the energy-spread of the electron beam. The focusing of the finest electron probes can be adversely affected by chromatic aberration. Also, the energy resolution possible in electron energy loss spectroscopy (EELS) is limited by the energy-spread of the primary electron beam.

Electrons from a thermionic triode gun may have an energy-spread of several electron volts, primarily due to interactions within the electron crossover (by the Boersch effect (Boersch, 1953)). Energy-spreads from the field emission sources are much less – 0.2 eV from a cold emitter and 0.3 eV from a Schottky emitter.

The choice between different types of electron gun is therefore dictated by the end use of the beam. The relevant properties of the main contenders are compared in Table A4.1.

Technological advances may result in improved performances being reported, especially in field emission sources. For example, in order to further reduce the source size of a field emission electron gun some workers have sputter-etched the fine tungsten tips to give even smaller wire endings, ultimately to have a single-atom tip. Intense electron emission can be obtained, with brightness exceeding that of the now normal field emitter (e.g. Qian *et al.*, 1993; Müller *et al.*, 1993).

Electron coherence

Just as light from lasers can support interference effects and holograms which are weak, if not impossible, with light from incandescent filament lamps, so electrons from field emitters are more 'coherent' than from a thermionic source. Interference-fringe effects at edges and holes (e.g. Fresnel fringes (p.56)) and the contrast of fine detail are all enhanced. The results of using a pointed filament in a TEM were noted by Hibi & Takahashi (1971), and were illustrated in Figure 2.22.

Table A4.1. *Comparative table of electron sources*

	Thermionic		Field emission	
	W	LaB$_6$	Cold	ZrO/W Schottky
Vacuum requirement, mbar	$<10^{-4}$	$<10^{-6}$	$<10^{-10}$	$<10^{-8}$
Tip flashing	not needed	not needed	every 6–8 h	not needed
Beam Noise, %	1	1	6–10	1
Brightness max, A cm^{-2} sr^{-1}	10^6 (100 kV)	2×10^6 (20 kV)	10^9	10^8
Probe current range	1 pA – 1 µA	1 pA – 1 µA	1 pA – 300 pA	1 pA – 5 nA
Cathode lifetime, h	100	600	>1000	>2000
Current stability/h, %	<1	<1	<5	<1
Energy spread, eV	1–2	0.5–2	0.2–0.4	0.3
Source diameter	20–50 µm	10–20 µm	5–10 nm	15 nm
Work function ϕ, eV	4.54	2.66	4.5	2.8
Operating temp., K	2800	1400–2000	300	1800

Suggested further reading

The electron-optical design of electron guns in microscopes will be found described in Reimer (1993), Hall (1983), Lauer (1982), and Delong (1993), whilst thermionic emission is the subject of chapter 19 of Kohl (1951). The theory and practice of field emission systems is discussed by Kasper (1982).

Bibliography

References

Agar, A W & Revell, R S M (1956). A study of the Formvar replica process. *Brit. J. Appl. Phys.*, **7**, 17–25

Ahn, C C & Krivanek, O K (1983). *EELS Atlas*. Gatan Inc.: Warrendale, PA.

Alani, R & Swann, P R (1993). Chemically assisted ion-beam etching (CAIBE) in a low-angle ion mill. *Proc MSA*, eds. G W Bailey & C L Rieder. San Francisco Press Inc.: San Francisco, pp. 718–9.

Anderson, T F (1951). Techniques for the preservation of three dimensional structure in preparing specimens for the electron microscope. *Trans. NY Acad. Sci.* Ser. II, **13**, 130–4

Andrews, K W, Dyson, D J & Keown, S R (1971) *Interpretation of electron diffraction patterns*, 2nd edition. Hilger: London.

Appleton, T C (1977). Ultrathin frozen sections for x-ray microanalysis of diffusible elements. pp. 247–68 in Meek & Elder (1977)

Armbruster, B L, Carlemalm, E, Chiovetti, R, Garavito, R M, Hobot, J A, Kellenberger, E & Villiger, W (1982). Specimen preparation for electron microscopy using low temperature embedding resins. *J. Microscopy*, **126**, 77–85.

Autrata, R & Hejna, J (1991). Detectors for low voltage scanning electron microscopy. *Scanning*, 13(4), 275–87

Bahr, G F & Zeitler, E (1965a). The determination of the dry mass in populations of isolated particles. *Laboratory Investigation*, **14**, 955–77

Bahr, G F & Zeitler, E (Eds.) (1965b). Quantitative electron microscopy. Symposium proceedings. Vol. 14, No. 6, Pt 2, pp. 739–1340 of *Laboratory Investigation*, Williams and Wilkins: Baltimore.

Baker, J R J (1989). *Autoradiography: A comprehensive overview*. RMS Handbook No. 18. Oxford University Press: Oxford.

Balossier, G, Thomas, X, Michel, J, Wagner, D, Bonhomme, P, Puchelle, E, Ploton, D, Bonhomme, A & Pinon, J M (1991). Parallel EELS elemental mapping in scanning transmission electron microscopy: use of the difference methods. *Microscopy, Microanalysis, Microstructures*, 2(5), 531–46

Barber, D J (1970). Thin foils of non-metals made for electron microscopy by sputter-etching. *J. Mater. Sci.*, **5**, 1–8

Barber, D J (1993). Radiation damage in ion-milled specimens: characteristics, effects and methods of damage limitation. *Ultramicroscopy*, **52**, 101–25

Barer, R & Cosslett, V E (Eds.) (1966). *Advances in optical and electron microscopy*. Vol. 1, 1966; Vol. 4, 1971; Vol. 6, 1975. Academic Press: London.

Battistella, F, Berger, S & Mackintosh, A (1987). Scanning optical microscopy via a scanning electron microscope. *J. Electron Microscopy Technique,* 6(4), 377–84

Bauer, E (1994). Low energy electron microscopy. *Rep. Prog. Phys.*, **57**(9), 895–938

Baumeister, W & Vogell, W (Eds.) (1980). *Electron microscopy at molecular dimensions*. Springer-Verlag: Berlin. (In English)

Beaman, D R & Isasi, J A (1972). *Electron beam microanalysis*. ASTM Special Technical Publication No 506. American Society for Testing and Materials: Philadelphia, PA.

Beesley, J E (1989). *Colloidal gold: A new perspective for cytochemical marking*. RMS Microscopy Handbook No. 17, Oxford University Press: Oxford.

Beeston, B E P, Horne, R W & Markham, R (1973). *Electron diffraction and optical diffraction techniques*. Vol. 1 Pt. II in the Glauert series. North Holland: Amsterdam.

Beutelspacher, H & Van der Marel, H W (1968). *Atlas of electron microscopy of clay minerals and their admixture; a picture atlas.* Elsevier: Amsterdam, London & New York.

Bigelow, W C (1994). *Vacuum methods in electron microscopy.* Vol. 15 in the Glauert series. Portland Press: London.

Binnig, G & Rohrer, H (1982). Scanning tunneling microscopy. *Helv. Phys. Acta*, **55**(6), 726–35

Birks, L S (1969). *X-ray spectrochemical analysis*, 2nd edition. Wiley-Interscience: New York.

Birks, L S (1971). *Electron probe microanalysis*, 2nd edition. Wiley: New York.

Bishop, H E (1966). Some electron backscattering measurements for solid targets. *Proc. 4th Intl. Conf. on x-ray optics and microanalysis, Paris 1965,* eds. Castaing, Deschamps & Philbert, pp. 153–8. Hermann: Paris.

Bleaker, A J & Kruit, P (1991). A condenser objective lens with asymmetric polepieces to facilitate the extraction of secondary and Auger electrons. *Rev. Sci. Instrum.*, **62**(2), 350–6

Boersch, H (1953). Energy distribution of thermal electrons from electron beam producers. *Naturwiss.*, 40(9), 267–8. In German

Bollmann, W (1956). Interference effects in the electron microscopy of thin crystal foils. *Phys. Rev.*, **103**, 1588–9

Böngeler, R, Golla, U, Kassens, M, Reimer, L, Schindler, B, Senkel, R & Spranck, M (1993). Electron–specimen interactions in low-voltage scanning electron microscopy. *Scanning*, **15**(1), 1–18

Bonnell, D A (1993). *Scanning tunneling microscopy and spectroscopy. Theory, techniques and applications.* VCH: Weinheim.

Bostanjoglo, O, Kornitzky, J & Tornow, R P (1991). High-speed electron microscopy of laser-induced vaporization of thin films. *J. Appl. Phys.*, **69**(4), 2581–3

Boyde, A (1973). Quantitative photogrammetric analysis and qualitative stereoscopic analysis of SEM images. *J. Microscopy*, **98**, 452–71

Boyde, A (1994). Bibliography on confocal microscopy. *Scanning*, **16**(1), 33–56

Boyde, A & Ross, H F (1975). Photogrammetry and the SEM. *Photogrammetric Record* 8(46), 408–57

Bradley, D E (1954a). Evaporated carbon films for use in electron microscopy. *Brit. J. Appl. Phys.*, **5**, 65–6

Bradley, D E (1954b). An evaporated carbon replica technique for use with the electron microscope and its application to the study of photographic grains. *Brit. J. Appl. Phys.*, **5**, 96–8

Bradley, D E (1959). High resolution shadowcasting technique for the electron microscope, using the simultaneous evaporation of platinum and carbon. *Brit. J. Appl. Phys.*, **10**(5), 198–203

Bradley, D E (1961). Simple methods of preparing pointed filaments for the electron microscope. *Nature (London)*, **189**, 298

Briggs, A (1985). *Introduction to scanning acoustic microscopy.* RMS Handbook No. 12. Oxford University Press: Oxford.

Briggs, D & Seah, M P (1990). *Practical surface analysis.* Vol. 1, Auger and x-ray photoelectron spectroscopy, 2nd edition. Wiley: Chichester.

Briggs, D & Seah, M P (1992). *Practical surface analysis.* Vol. 2, Ion and neutral spectroscopy. Wiley: Chichester.

de Broglie, L (1924). A tentative theory of light quanta. *Phil. Mag.*, **47**, 446–58

Budd, P M & Goodhew, P J (1988). *Light-element analysis in the transmission electron microscope.* RMS Handbook No.16. Oxford University Press: Oxford.

Bullock, G R (1984). The current status of fixation for electron microscopy; a review. *J. Microscopy*, **133**, 1–15

Bullock, G R & Petrusz, P (Eds.) (1982–9). *Techniques in immunocytochemistry.* Vol. 1, 1982; Vol. 2, 1983; Vol. 3, 1985; Vol. 4, 1989. Academic Press: London.

Burder, D & Whitehouse, P (1992). *Photographing in 3-D*, 3rd edition. The Stereoscopic Society: Chessington, Surrey.

Burmester, C, Braun, H G & Schröder, R R (1994). A new quasi-confocal image plate scanner with improved spatial resolution and ideal detection efficiency. *Ultramicroscopy*, **55**(1), 55–65

Busch, H (1926). Calculation of trajectory of cathode rays in axially symmetric electromagnetic fields. *Ann. Physik*, **81**, ser. 4, 974–93

Buseck, P R, Cowley, J M, & Eyring, L (Eds.) (1988). *High resolution transmission electron microscopy and associated techniques*. Oxford University Press: New York and Oxford.

Busing, W M & Otten, M T (1991). Cryo-observation of biological materials in the TEM. *Microscopy & Analysis*, **24**, 45

Castaing, R (1951). *Application des sondes électroniques a une méthode d'analyse ponctuelle chimique et crystallographique*. Unpublished Ph.D. thesis, University of Paris.

Castaing, R & Henry, L (1962). Magnetic filtering of speeds in electron microscopy. *Comptes Rendus Acad. Sci., Paris*, **255**(1), 76–8. (In French)

Castaing, R & Slodzian, G (1962). Microanalyse par émission sécondaire. *J. de Microscopie*, **1**, 395–410

Chapman, S K (1986). *Maintaining and monitoring the transmission electron microscope*. RMS Handbook No. 8. Oxford University Press: Oxford.

Chatfield, E J, More, J & Nielsen, V H (1974). Stereoscopic scanning electron microscopy at TV scan rates. *Scanning Microscopy, 1974*, 117–24, IITRI: Chicago.

Chen, C J (1993). *Introduction to scanning tunneling microscopy*. Oxford University Press: New York.

Chescoe, D & Goodhew, P J (1990). *Operation of transmission and scanning electron microscopes*. RMS Handbook No.20. Oxford University Press: Oxford.

Cliff, G & Lorimer, G W (1975). The quantitative analysis of thin specimens. *J. Microscopy*, **103**, 203–7

Coates, D G (1967). Kikuchi-like reflection patterns observed in the scanning electron microscope. *Phil. Mag.*, **16**, 1179–84

Cobbold, A J & Gilmour, R E (1971). Direct "Quantimet"–electron microscope linkage for the assessment of particle size distribution of polymer latices. *Brit. Polym. J.*, **3**, 249–58

Cobbold, A J & Mendelson, A E (1971). Ultrathin sectioning of polymers at low temperatures for electron microscopic investigation. *Science Tools*, **18**, 10–12

Colby, J W (1969). Backscattered and secondary electron emission as ancillary techniques in electron probe analysis. In *Advances in electronics and electron physics*, supplement 6, eds. Tousimis & Marton, pp. 177–96. Academic Press: New York.

Cooke, C J & Duncumb, P (1969). Performance analysis of a combined electron microscope and electron probe microanalyser, EMMA. *Proc. 5th Intl. Congr. on x-ray optics and analysis, Tubingen, 1968*, eds. Möllenstedt & Gaulkier, pp. 245–7. Springer: Berlin.

Cornish, D C & Watt, I M (1964). An application of the technique of selected-area replication to the study of the mechanism of glass polishing. *Electron Microscopy 1964*, Prague, vol. A, 429–30

Cosslett, V E (1956). Specimen thickness and image resolution in electron microscopy. *Brit. J. Appl. Phys.*, **7**, 10–13

Cosslett, V E, Camps, R A, Saxton, W O, Smith, D J, Nixon, W C, Ahmed, H, Catto, C J D, Cleaver, J R A, Smith, K C A, Timbs, A E, Turner, P W & Ross, P M (1979). Atomic resolution with a 600 kV electron microscope. *Nature (London)*, **281** (No. 5726), 49–51

Cosslett, V E & Duncumb, P (1956). Microanalysis by a flying spot x-ray method. *Nature (London)*, **177**, 1172–3

Cosslett, V E & Nixon, W C (1960). *X-ray microscopy*. Cambridge University Press: Cambridge & New York.

Costello, M J & McIntosh, T J (1981). Freeze fracture and negative stain analysis of multilamellar lipid phases containing N-alkanes. *Biophys. J.*, **33**, A160

Cowley, J M (Ed.) (1992, 1993). *Electron diffraction techniques.* Vol. 1, 1992; vol. 2, 1993. Oxford University Press: Oxford.

Crang, R F E & Klomparens, K L (Eds.) (1988). *Artifacts in biological electron microscopy.* Plenum: New York & London.

Crewe, A V (1971). High resolution scanning microscopy of biological specimens. *Phil. Trans. Roy. Soc. London B,* **261**, 61–70

Curry, C (1953). *Geometrical optics.* Edward Arnold: London.

Czyrska-Filemonowicz, A, Spiradek, K & Gorczyca, S (1991). Extraction double replica technique for electron microscopy studies of precipitates. *Sonderbände der Praktischen Metallographie,* **22**, 217–26. Dr Riederer-Verlag: Stuttgart.

Danilatos, G D (1991). Review and outline of environmental SEM at present. *J. Microscopy,* **162**, 391–402

Davidson, S M & Booker, G R (1970). Decollimation of a parallel electron beam by thin surface films and its effect on SEM electron channelling patterns. *Electron Microscopy 1970,* Vol. 1, 235–6. Société Française de Microscopie Électronique: Paris.

Davisson, C & Germer, L H (1927). Diffraction of electrons by a crystal of nickel. *Phys. Rev.,* **30**, 705–40

Deininger, C and Mayer, J (1992). Omega energy filtered convergent beam electron diffraction. *Electron Microscopy 92,* Vol. 1, 181–2

Delong, A (1993). Electron sources for electron microscopes. *Microscopy & Analysis,* **38**, 19–21

Delong, A, Hladil, K & Kolařík, V (1994). A low voltage transmission electron microscope. *Microscopy & Analysis,* **39**, 7–9

Dinardo, N J (1994). *Nanoscale characterization of surfaces and interfaces.* VCH: Weinheim.

Dingley, D J (1981). A comparison of diffraction techniques for the SEM. *Scanning Electron Microscopy 1981/IV,* 273–86. SEM Inc.:Chicago.

Dingley, D J & Randle, V (1992). Microtexture determination by electron backscatter diffraction. *J. Mater. Sci.,* **27**(17), 4545–66

Dingley, D J, Baba-Kishi, K Z & Randle, V (1994). *Atlas of backscattering Kikuchi diffraction patterns.* Inst. of Physics Publishing: Bristol & Philadelphia.

van Dorsten, A C, Nieuwdorp, H & Verheoff, A (1950). The Philips 100 kV electron microscope. *Philips Tech. Rev.,* **12**, 33–51

van Dorsten, A C & le Poole, J B (1955). The EM 75kV, an electron microscope of simplified construction. *Philips Tech. Rev.,* **17**, 47–59

Dowell, L G & Rinfret, A P (1960). Low temperature forms of ice as studied by x-ray diffraction. *Nature (London),* **188**, 1144–8

Drummond, I W (1992). Imaging XPS: the different methods explained. *Microscopy & Analysis,* **28**, 29–32

Duke, P J & Michette, A G (Eds.) (1990). *Modern microscopies: Techniques and applications.* Plenum: New York & London.

Duncumb, P (1967). Electron probe microanalysis. *Science Progress,* 55, 511–28

Dupouy, G (1973). Performance and applications of the Toulouse 3 million volt electron microscope. *J. Microscopy,* **97**(1/2), 3–28

Durkin, R & Shah, J S (1993). Amplification and noise in high pressure scanning electron microscopy. *J. Microscopy,* **169**(1), 33–51

Dykstra, M J (1993). *A manual of applied techniques for biological electron microscopy.* Plenum: New York & London.

Echlin, P (1992). *Low temperature microscopy and analysis.* Plenum: New York.

Erre, D, Mouze, D, Thomas, X & Cazaux, J (1993). Progress in digital x-ray projection microscopy. *Inst. Phys. Conf. Ser. No. 130,* Chapter 7, 567–70. Inst. of Physics: Bristol.

van Essen, C G & Schulson, E M (1969). Selected area channelling patterns in the SEM. *J. Mater. Sci.*, **4**, 336–9

Everhart, T E & Thornley, R F M (1960). Wide-band detector for micro-microampere low-energy electron currents. *J. Sci. Instrum.*, **37**, 246–8

Fan, G Y & Ellisman M H (1994). Stereoscopy by tilted illumination in transmission electron microscopy. *Ultramicroscopy*, **55**(2), 155–64

Farnell, G C & Flint, R B (1975). Exposure level and image quality in electron micrographs. *J. Microscopy*, **103**, 319–32

Favard, P & Carasso, N (1973). The preparation and observation of thick biological sections in the high voltage electron microscope. *J. Microscopy*, **97**, 59–81

Fernandez-Moran, H (1953). A diamond knife for ultrathin sectioning. *Exper. Cell Research*, **5**, 255–6

Ferrier, R P (1977). Microanalysis of biological materials using EELS. pp. 193–211 in Meek & Elder (1977)

Ferwerda, J G (1982). *The world of 3–D. A practical guide to stereo photography*. Nederlandse Vereniging voor Stereofotografie: GW Borger.

Fitzgerald, R, Keil, K & Heinrich, K F J (1968). Solid-state energy dispersion spectrometer for electron microprobe x-ray analysis. *Science*, **159**, 528–30

Fitzpatrick, E A (1993). *Soil microscopy and micromorphology*. John Wiley: Chichester.

Forwood, C T & Clarebrough, L M (1991). *Electron microscopy of interfaces in metals and alloys*. Inst. of Physics Publishing: Bristol & Philadelphia.

Franks, J (1980). Ion beam sputter coating for electron microscopy. *Electron Microscopy 1980*, Vol. 1, 538–9. Seventh European Congress on Electron Microscopy Foundation: Leiden.

Fryer, J R (1979). *The chemical applications of transmission electron microscopy*. Academic Press: London.

Fujita, T, Tokunaga, J & Inone, H (1971). *Atlas of scanning electron microscopy in medicine*. Elsevier: Amsterdam.

Gard, J A (Ed.) (1971). *The electron optical investigation of clays*. The Mineralogical Society: London.

Gilkey, J C (Ed.) (1993). *Freeze substitution*. Thematic issue of *Microscopy Research & Technique*, **24**(5/6).

Glauert, A M (1975). *Fixation, dehydration and embedding of biological specimens*. Vol. 3, Pt 1 in the Glauert series. Elsevier: Amsterdam.

Glauert, A M (1991). Epoxy resins: an update on their selection and use. *Microscopy & Analysis*, **25**, 15–20

Goldstein, J I, Newbury, D E, Echlin, P, Joy, D C, Romig, A D Jr, Lyman, C E, Fiori, C & Lifshin, E (1992). *Scanning electron microscopy and x-ray microanalysis*, 2nd edition. Plenum: New York.

Goodhew, P J (1973). *Specimen preparation in materials science*. Vol. 1, Pt 1 in the Glauert series. Elsevier: Amsterdam.

Goodhew, P J (1985). *Thin foil preparation for electron microscopy*. Vol. 11 in the Glauert series. Elsevier: Amsterdam.

Griffith III, E M & Danilatos, G D (Eds.) (1993). Environmental scanning electron microscopy. *Microscopy Research & Technique*, **25**(5/6), 353–534

Grigson, C W B (1965). Improved scanning electron diffraction system. *Rev. Sci. Instrum.*, **36**, 1587–93

Gunning, B E S, & Steer, M W (1974). *Plant cell biology: an ultrastructural approach*. Edward Arnold: Sevenoaks.

Gupta, B L & Roomans, G M (Eds.) (1993). *X-ray microprobe analysis of elements in biology. From origins to current practice*. Scanning Microscopy International: Chicago.

Gütter, E & Menzel, M (1978). An external photographic system for electron microscopes. *Electron Microscopy 1978*, Vol. 1, pp. 92–3. Microscopical Society of Canada: Toronto.

Hall, C E (1983). *Introduction to electron microscopy*, 3rd edition. McGraw-Hill: New York.

Hall, J L (Ed.) (1978). *Electron microscopy and the cytochemistry of plant cells*. Elsevier–North Holland, Biomedical Press: Amsterdam.

Hall, T A (1979). Biological x-ray microanalysis. *J. Microscopy*, **117**, 145–63

Halliday, J S (1962). Wear experiments in the electron microscope. *Electron Microscopy 1962*, Vol. 1, Paper H–6.

Hama, K & Porter, K R (1969). An application of high voltage electron microscopy to the study of biological materials. *J. de Microscopie*, **8**, 149–58

Hammond, C (1992). *Introduction to crystallography*, 2nd edition. RMS Handbook No. 19. Oxford University Press: Oxford.

Hanszen, K J (1971). The optical transfer theory of the electron microscope: fundamental principles and applications. pp. 1–84 in Barer & Cosslett (1971)

Harada, K, Nagata, H & Shimizu, R. (1991). <310> single crystal LaB_6 as thermal field emitter of high brightness electron source. *J. Electron Microscopy*, **40**(1), 1–4

Harland, C J, Akhter, P & Venables, J A (1981). Microcrystallography at high spatial resolution using electron backscattering patterns in a field-emission gun scanning electron microscope. *J. Phys. (E)*, **14**, 175–82

Harris, J R (Ed.) (1981–7). *Electron microscopy of proteins*. In six volumes: Vol. 1, 1981; Vols. 2,3, 1982; Vol. 4, 1983; Vol. 5 (Harris, J R & Horne, R W eds.), 1986; Vol. 6 (Harris, J R & Horne, R W eds.), 1987. Academic Press: London.

Harris, J R & Horne, R W (1994). Negative staining: a brief assessment of current technical benefits, limitations and future possibilities. *Micron*, **25**(1), 5–13

Harris, N S (1989). *Modern vacuum practice*. McGraw-Hill: New York.

Hawkes, P W (Ed.) (1982). *Magnetic electron lenses*. Springer-Verlag: Berlin, Heidelberg, New York.

Hawkes, P W (Ed.) (1985). The beginnings of electron microscopy. Supplement 16 to *Advances in electronics and electron physics*. Academic Press: London.

Hawkes, P W and Kasper, E (1989). *Principles of electron optics*. 3 volumes: Vol. 1 Basic geometrical optics; Vol. 2 Applied geometrical optics; Vol. 3 (1994) Wave Optics. Academic Press: London & San Diego.

Hawkes, P W & Valdrè, U (Eds.) (1990). *Biological electron microscopy. Basic concepts and modern techniques*. Academic Press: London & San Diego.

Hayat. M A (1970). *Principles and techniques of electron microscopy – biological applications*. Vol. 1 by Hayat, Vol. 2 ed. Hayat. Van Nostrand Reinhold Company: New York.

Hayat. M A (1972). *Basic electron microscopy techniques*. Van Nostrand Reinhold Company: New York.

Hayat, M A (Ed.) (1974). *Principles and techniques of scanning electron microscopy – biological applications*. Vols. 1–3. Van Nostrand Reinhold Company: New York.

Hayat, M A (1989a). *Principles and techniques of electron microscopy. Biological applications*. MacMillan Press: London.

Hayat, M A (Ed.) (1989b). *Colloidal gold: Principles, methods and applications*. Vols. 1, 2 (1989); Vol. 3 (1991). Academic Press: San Diego.

Hayat, M A (1992). Quantitation of immunogold labeling. *Micron & Microsc. Acta*, **23**(1/2), 1–16

Heinrich, K F J (1975). Scanning electron probe microanalysis. pp. 275–301 in Barer & Cosslett (1975).

Hermann, R & Müller, M (1991). High resolution biological scanning electron microscopy: a comparative study of low temperature metal coating techniques. *J. Electron Microscopy Technique*, **18**(4), 440–9

Hibi, T & Takahashi, S (1971). Relation between coherence of electron beam and contrast of electron images of biological substances. *J. Electron Microscopy (Japan)*, **20**, 17–22

Hillier, J & Vance, A W (1941). Recent developments in the electron microscope. *Proc. Inst. Radio Engrs.*, **29**, 167–76

Hirsch, P B, Howie, A, Nicholson, R B, Pashley, D W & Whelan, M J (1965). *Electron microscopy of thin crystals*. Butterworths: London.

Hirsch, P B & Humphreys, C J (1970). Theory of SEM channelling patterns for normal and tilted specimens. *Electron Microscopy 1970*, Vol. 1, 229–30

Hockje, S J & Hoflund, G B (1991). Surface characterisation study of InP(100) substrates using ISS, AES and ESCA. I. Comparison of substrate cleaning techniques. *Thin Solid Films*, **197**(1/2), 367–80

Holland, L (1956). *Vacuum deposition of thin films*. Chapman and Hall: London.

Holt, D B & Joy, D C (Eds.) (1989). *SEM microcharacterization of semiconductors*. Vol. 12 of *Techniques of physics*. Academic Press: London.

Horne, R W (1977). Optical diffraction analysis of periodically repeating biological structures. pp. 29–53 in Meek & Elder (1977)

Horne, R W & Markham, R (1973). Application of optical diffraction and image reconstruction techniques to electron micrographs. In Part 2 (pp. 327–435) of *Electron diffraction and optical diffraction techniques*, eds. Beeston, Horne & Markham. Vol. 1 in the Glauert Series. Elsevier: Amsterdam.

Hren, J J, Goldstein, J I & Joy, D C (Eds.) (1979). *Introduction to analytical electron microscopy*. Plenum: New York.

Hsu, T (1992). Technique of reflection electron microscopy. *Micros. Res. & Tech.*, **20**(4), 318–32

Hsu, T & Peng, L-M (1993). Bibliography of REM. *Ultramicroscopy*, **48**(4), 489–90

Hucknall, D J (1991). *Vacuum technology and applications*. Butterworth Heinemann: Oxford.

Hudson, B & Makin, M J (1970). The optimum tilt angle for electron stereomicroscopy. *J. Phys. (E)*, **3**, 311

Hunter, E (1993). *Practical electron microscopy. A beginner's illustrated guide*, 2nd edition. Cambridge University Press: New York.

Jackson, A G (1991). *Handbook of crystallography for electron microscopists and others*. Springer-Verlag: Berlin & New York.

JCPDS (1994). *Powder diffraction file*, 44th set. International Center for Diffraction Data: 12 Campus Boulevard, Newtown Square, PA 19073, USA.

Jeanguillaume, C & Colliex, C (1989). Spectrum image: The next step in EELS digital acquisition and processing. *Ultramicroscopy*, **28**, 252–7

Johannessen, J V (Ed.) (1978). *Electron microscopy in human medicine: vol 2; cellular pathobiology*. McGraw Hill: New York.

Johansson, T (1933). New focusing x-ray spectrometer. *Zeits. f. Physik*, **82**, 507–28

Johnson, C G R & White, E W (1970). *X-ray emission wavelengths and keV tables for non-diffractive analysis*. ASTM Data Series DS46: Philadelphia.

Jongebloed, W L (1990). Trends in biological scanning electron microscopy. *Micron & Microsc. Acta*, **21**, 229–32

Jouffrey, B, Kihn, Y, Perez, J Ph, Sevely, J & Zanchi, G (1980). On some aspects of energy losses studies in electron microscopy up to HVEM. *Micron*, **11**, 253–8

Joyce Loebl (Ed.) (1989). *Image analysis: Principles and practice*. Joyce Loebl: Gateshead.

Joy, D C (1979). The basic principles of electron energy loss spectroscopy. Chapter 7 (pp. 223–44) in *Introduction to analytical electron microscopy*, eds. Hren, Goldstein & Joy. Plenum: New York.

Joy, D C (1991). Contrast in high-resolution scanning electron microscope images. *J. Microscopy*, **161**(2), 343–55

Joy, D C (1995). *Monte Carlo modeling for electron microscopy and microanalysis*. Oxford University Press: Oxford.

Joy, D C, Newbury, D E & Davidson, D L (1982). Electron channeling patterns in the scanning electron microscope. *J. Appl. Phys.*, **53**, R81–R122

Joy, D C, Romig, A D Jr & Goldstein, J I (Eds.) (1986). *Principles of analytical electron microscopy*. Plenum: New York.

Kasper, E (1982). Field electron emission systems. pp. 207–60 in *Advances in optical and electron microscopy*, vol. 8, eds. Barer & Cosslett. Academic Press: London.

Kaufmann, R, Schurmann, M, Hillencamp, F, Wechsung, R & Heinen, H J (1979). Laser microprobe mass analysis: achievements and aspects. *Scanning Electron Microscopy 1979*, ed. Johari, O. pp. 279–90. SEM Inc.: Chicago.

Kay, D H (Ed.) (1965). *Techniques for electron microscopy*, 2nd edition. Blackwell: Oxford.

Kessel, R G & Shih, C Y (1974). *Scanning electron microscopy in biology – a student's atlas on biological organisation*. Springer-Verlag: Berlin.

Knoll, M (1935). Static potential and secondary emission of bodies under electron irradiation. *Z. Tech. Phys.*, **16**, 467–75

Kohl, W H (1951). *Materials technology for electron tubes*. Reinhold: New York.

Köhler, J K (Ed.) (1973). *Advanced techniques in biological electron microscopy*. Springer-Verlag: Berlin.

Kok, L P & Boon, M E (1992). *Microwave cookbook for microscopists*, 3rd edition. Coulomb Press: Leyden.

Kossel, W & Möllenstedt, G (1938). Elektroneninterferenzen im Konvergenten Bundel. *Ann. der Phys.*, **26**, 660–1

Krivanek, O L (1992). Developments in electron detectors and recording systems. *Electron Microscopy 92*, Vol. 1, 83–7

Kruit, P, Bleeker, A J & Pijper, F J (1988). High resolution Auger analysis in STEM. *EUREM 88, Inst. Phys. Conf. Ser. No. 93*, Vol. 1, 249–54

Kunath, W, Riecke, W D & Ruska, E (1966). Spherical aberration of saturated strong objective lenses. In *Electron Microscopy 1966*, vol. 1, pp. 139–40

Lane, G S (1969). The application of stereographic techniques to the SEM. *J. Phys. (E)*, **2**, 565–9

Langmuir, D B (1937). *Proc. Inst. Radio Engrs.*, **25**, 977–91

Larson, P K & Dobson, P J (Eds.) (1988). *Reflection high-energy electron diffraction and reflection electron imaging of surfaces*. Plenum: New York & London.

Lauer, R (1982). Characteristics of triode electron guns. pp. 137–206 in *Advances in optical and electron microscopy*, vol. 8, eds. Barer & Cosslett. Academic Press: London.

Leapman, R D (1992). EELS quantitative analysis. p. 47 in *Transmission electron energy loss spectrometry in material science*, eds. Disko, Ahn & Fultz. TMS: Warrendale, Pa.

Lee, R E (1993). *Scanning electron microscopy and x-ray microanalysis*. Prentice Hall: New Jersey.

van de Leemput, L E C & van Kempen, H (1992). Scanning tunneling microscopy. *Rept. Prog. Phys.*, **55**(8), 1165–240

Lewis, P R & Knight, D P (Eds.) (1992). *Cytochemical staining methods for electron microscopy*, 2nd edition. Vol. 14 in the Glauert series. Elsevier: Amsterdam & New York

Lide, D R (Ed.) (1990). *Handbook of Chemistry and Physics*, 71st edition. CRC Press Inc.: Boca Raton, FL.

Liebl, H (1975). Ion probe microanalysis (review article). *J. Phys. (E)*, **8**, 797–808

Livesey, S A, del Campo, A A, McDowell, A A & Stasny, J T (1991). Cryofixation and ultra-low temperature freeze-drying as a preparation technique for TEM. *J. Microscopy*, **161**, 205–16

Loretto, M H (1993). *Electron beam analysis of materials*, 2nd edition. Chapman & Hall: London.

Loretto, M H & Smallman, R E (1975). *Defect analysis in electron microscopy*. Chapman and Hall: London.

Love, G & Scott, V D (1981). Updating correction procedures in quantitative electron-probe micro-analysis. *Scanning* **4**, 111–30

Lyman, C E, Newbury, D E, Goldstein, J I, Williams, D B, Romig A D Jr, Armstrong, J, Echlin, P, Fiori, C D, Joy, D C, Lifshin, E & Peters, K R (1990). *Scanning electron microscopy, x-ray microanaly-sis and analytical electron microscopy. A laboratory workbook*. Plenum: New York.

Lyman, C E, Goldstein, J I, Williams, D B, Ackland, D W, Von Harrach, S, Nichols, A W & Statham, P J (1994). High-performance x-ray detection in a new analytical electron microscope. *J. Microscopy*, **176**, 85–98

Mackenzie, A P (1993). Recent advances in electron probe microanalysis. *Rept. Prog. Phys.*, **56**(4), 557–604

Mackenzie, R (1992). The position sensitive atom probe. *Microscopy & Analysis*, **27**, 17–19

Maher, E F (1985). The SOMSEM – an SEM-based scanning optical microscope. *Scanning*, **7**, 61–5

Mandal, A K & Wenzl, J E (1979). *Electron microscopy of the kidney in renal disease and hypertension*. Plenum: New York.

Mansfield, J (Ed.) (1984). *Convergent-beam electron diffraction of alloy phases*. By the Bristol Group of J Steeds, compiled by J Mansfield. Inst. of Physics Publishing: Bristol & Philadelphia

Marks, L D & Smith, D J (1983). HREM and STEM of defects in multiply-twinned particles. *J. Microscopy*, **130**, 249–61

Marraud A & Bonnet M (1980). Images en relief et films gauffrés. *Science et Avenir 1980*.

Marti, O & Amrein, M (1993). *STM and SFM in biology*. Academic Press: London.

van der Mast, K D, Barth, J D & le Poole, J B (1974). A continuously renewed laser-heated tip in a T F emission gun. *Electron Microscopy 1974*, vol. 1, 120–1.

Matsuyama, H & Koike, K (1994). Twenty-nm resolution spin-polarized scanning electron microscopy. *J. Electron Microsc.*, **43**(3), 157–63

McCabe, A R & Smith, G D W (1983). The oxidation of rhodium–platinum – a study by field ion microscopy and imaging atom probe techniques. *Platinum Met. Rev.*, **27**, 19–25

McCrone, W C & Delly, J G (1973). *The particle atlas*. Four volumes, 2nd edition. Ann Arbor Science Publishers Inc.: Ann Arbor, MI.

McLaren, A C (1991). *Transmission electron microscopy of minerals and rocks*. Cambridge University Press: Cambridge.

McMullan, D (1995). Scanning electron microscopy 1928–1965. *Scanning*, **17**(3), 175–85

Meek, G A (1976). *Practical electron microscopy for biologists*, 2nd edition. Wiley: London, New York, Sydney & Toronto.

Meek, G A & Elder H Y (Eds.) (1977). *Analytical and quantitative methods in microscopy*. Cambridge University Press: Cambridge.

Menter, J W (1956). The direct study by electron microscopy of crystal lattices and their imperfections. *Proc. Roy. Soc. A*, **236**, 119–35

Miller, M K and Smith, G D W (1989). *Atom probe microanalysis: principles and applications to materi-als problems*. Materials Research Society: Pittsburgh.

Misell, D L & Brown, E B (1987). *Electron diffraction: an introduction for biologists*. Vol. 12 of the Glauert series. Elsevier: Amsterdam.

Moor, H (1970). High resolution shadowcasting by the use of an electron gun. *Electron Microscopy 1970*, Vol. 1, 413–14. Société Francaise de Microscopie Électronique: Paris.

Moor, H (1987). Theory and practice of high pressure freezing. Chapter 8, pp. 175–91, in *Cryotechniques in biological electron microscopy*, eds. Steinbrecht & Zierold. Springer-Verlag: Berlin.

Moor, H & Mühlethaler, K (1963). Fine structure in frozen etched yeast cells. *J. Cell Biol.*, **17**, 609–28

Morel, G (1993). *Hybridization techniques in electron microscopy*. CRC Press: Boca Raton, FL.

Mori, N, Oikawa, T, Harada, Y & Miyahara, J (1990). Development of the Imaging Plate for the TEM and its characteristics. *J. Electron Microscopy*, **39**(6), 433–6

Müller, E W (1937). Electron microscope observation of field cathodes. Zeits. f. Physik, **106**, 541–50. (In German)

Müller, E W (1951). The field ion microscope. *Z. Phys.*, **131**, 136–42, (In German)

Müller, E W, Panitz, J A & McLane, S B (1968). The atom-probe field ion microscope. *Rev. Sci. Instrum.*, **39**, 83–6

Müller, E W & Tsong, T T (1969). *Field ion microscopy – principles and applications*. Elsevier: New York.

Müller, H U, Völkel, B, Hofmann, M, Wöll, Ch & Grunze, M (1993). Emission properties of electron point sources. *Ultramicroscopy*, **50**(1), 57–64.

Müllerová, I & Frank, L (1993). Very low energy microscopy in commercial SEMs. *Scanning*, **15**(4), 193–201.

Mulvey, T (Ed.) (1996). *The Growth of Electron Microscopy*. Vol. 96 of *Advances in imaging and electron physics*, Ed. Hawkes, P W. Academic Press: San Diego

Murphy, J A (1978). Non-coating techniques to render biological specimens conductive. *Scanning Microscopy*, **II**, 175–95

Murr, L E (1991). *Electron and ion microscopy and microanalysis*, 2nd edition. Marcel Dekker Inc: New York, Basel & Hong Kong.

Murray, J W (1971). *An atlas of British recent foraminiferids*. Heinemann: London.

Nermut, M V (1994). Electron microscopy of human immunodeficiency virus. *Microscopy & Analysis*, **42**, 11–13

Neubauer, G & Schnitger, A (1970). Ein neues Betrachtungsgerät für Stereobildpaare aller Formate. *Beitr. elektronemikrosk. Directabbild. Oberflächen*, **3**, 411–14

Newbury, D E & Myklebust, R L (1979). Monte Carlo electron trajectory simulation of beam spreading in thin foil targets. *Ultramicroscopy*, **3**, 391–5.

Newbury, D E, Joy, D C, Echlin, P, Fiori, C E & Goldstein, J I (1986). *Advanced scanning electron microscopy and x-ray microanalysis*. Plenum: New York.

Nicolosi, J A, Groven J P, Marlo, D & Jenkins, R (1986). Layered synthetic microstructures for long wavelength x-ray spectrometry. *Optical Engineering*, **25**, 964–9

Oatley, C W (1972). *The scanning microscope, Pt 1, The instrument*. Cambridge University Press: Cambridge.

Oatley, C W (1975). The tungsten filament gun in the scanning electron microscope. *J. Phys. (E)*, **8**, 1037–41

Oatley, C W, Nixon, W C & Pease, R F W (1965). Scanning electron microscopy, in *Advances in electronics and electron physics*, **21**, 181–247

Orloff, J (1993). High-resolution focused ion beams. *Rev. Sci. Instrum.*, **64**(5), 1105–26

Oster, C F Jr & Skillman, D C (1962). Determination and control of electron microscope magnification. *Electron Microscopy 1962*, Vol. 1, paper EE–3. Academic Press: New York.

Page, T F (1993). The modern SEM: an increasingly versatile and integrated tool for the microstructural and microchemical analysis of engineering ceramics, coatings and composites. *Inst. Phys. Conf. Ser. No. 138*, Section 7, pp. 295–300. Inst. of Physics: Bristol.

Pålsgård, E, Lindh, U, & Roomans, G M (1994). Comparative study of freeze-substitution techniques for x-ray microanalysis of biological tissue. *Micros. Res. & Tech.*, **28**(3), 254–8

Pedler, C M H & Tilly, R (1966). A new method of serial reconstruction from electron micrographs. *J. Roy. Micr. Soc.*, **86**, 189–97

Pelc, R & Žižka, Z (1993). Electron microprobe studies of the effect of long-term vacuum storage of freeze-dried cryosections on distribution of elements in cells. *J. Microscopy*, **170**(2), 167–71

Pfefferkorn, G & Boyde, A (1974). Review of replica techniques for scanning electron microscopy. In *Proc. 7th annual SEM symposium*, Eds. Johari & Corvin, pp. 75–82. IITRI: Chicago.

Phillipp, F, Höschen, R, Osaki, M, Möbus, G and Rühle, M (1994). New high-voltage atomic resolution microscope approaching 1 Å point resolution installed in Stuttgart. *Ultramicroscopy*, **56**(1–3), 1–10

Polak, J M & Varndel, I M (Eds.) (1984). *Immunolabelling for electron microscopy*. Elsevier: Amsterdam.

le Poole, J B (1964). Miniature lenses. *Electron Microscopy 1964*, Vol. A, 439–40

le Poole, J B, Bok, A B & Rus, P J (1970). A compact 1 MV electron microscope. *Electron Microscopy 1970*, Vol. 1, p. 13. Société Francaise de Microscopie Électronique: Paris.

Pozsgai, I (1982). Energy dispersive x-ray fluorescence analysis in the scanning electron microscope. *Electron Microscopy 1982*, pp. 781–2. Deutsche Gesellschaft für Elektronenmikroskopie e.V.:Frankfurt.

Pozsgai, I (1985). On the atomic number dependence of the x-ray fluorescence analysis in the electron microscope. *Proc. EMAG '85; Inst. Phys. Conf. Ser. No. 78*, Chapter 7, pp. 201–4. Inst. of Physics: Bristol.

Pozsgai, I (1991). X-ray microfluorescence analysis inside and outside the electron microscope. *X-ray spectrometry*, **20**, 215–23

Pozsgai, I (1993). Trace element analysis in the SEM. *Microscopy & Analysis*, **38**, 11–13

Praprotnik, B, Driesel, W, Dietzsch, C & Niedrig, H (1994). HV-TEM *in-situ* investigation of the tip shape of indium liquid metal ion emitter. *Surf. Science*, **314**(3), 353–64

Prewett, P D & Mair, G L R (1991). *Focused ion beams from liquid metal ion sources*. Research Station Press Ltd: Taunton.

Qian, W, Scheinfein, M R & Spence, J C H (1993). Brightness measurements of nanometer-sized field-emission-electron sources. *J. Appl. Phys.*, **73**(11), 7041–5

Quintana, C (1994). Cryofixation, cryosubstitution, cryoembedding for ultrastructural, immunocytochemical and microanalytical studies. *Micron*, **25**(1), 63–99

Reed, S J B (1993). *Electron microprobe analysis*, 2nd edition. Cambridge University Press: Cambridge.

Reid, N & Beesley, J E (1990). *Sectioning and cryo-sectioning for electron microscopy*. Vol. 13 in the Glauert series. Elsevier: Amsterdam.

Reimer, L (1985). *Scanning electron microscopy*. Springer-Verlag: Berlin & New York.

Reimer, L (1993). *Transmission electron microscopy: physics of image formation and microanalysis*, 3rd edition. Springer-Verlag: Berlin & New York.

Reimer, L (1993b). *Image formation in low-voltage scanning electron microscopy*. SPIE: Bellingham, WA.

Reimer, L & Pfefferkorn, G (1977). *Raster Elektronenmikroskopie*, 2nd edition. Springer-Verlag: Berlin.

Reiss, G (1992). Micromanipulation inside the scanning electron microscope. *Microscopy & Analysis*, **31**, 5–7

Richards, B P & Footner, P K (1992). *Role of microscopy in semiconductor failure analysis*. RMS Handbook No. 25. Oxford University Press: Oxford.

Riecke, W D (1961). The exactness of agreement between selected and diffracting areas in le Poole's selected area diffraction technique. *Optik*, **18** , 278–93. (In German)

van Riessen, A & Terry, K W (1982). X-ray induced x-ray fluorescence in a JSM-35c scanning microscope. *JEOL News*, **20E**, 19–23

Rivière, J C (1990). *Surface analytical techniques*. Oxford University Press: Oxford.

Robards, A W & Sleytr, U B (1985). *Low temperature methods in biological electron microscopy*. Vol. 10 in the Glauert series. Elsevier: Amsterdam.

Robards, A W & Wilson, A J (Eds.) (1993). *Procedures in electron microscopy*. Wiley: Chichester.

Roberts, S G & Page, T F (1981). A microcomputer-based system for stereogrammetric analysis. *J. Microscopy*, **124**, 77–88

Robinson, V N E (1973). A reappraisal of the complete electron emission spectrum in scanning electron microscopy. *J. Phys. (D)*, **6**, L105–7

Rochow, T G & Tucker, P A (1994). *Introduction to microscopy by means of light, electrons, x-rays or acoustics*, 2nd edition. Plenum: New York & London.

Roomans, G M & Shelburne, J D (Eds.) (1983). *Basic methods in biological x-ray microanalysis*. Scanning Microscopy International: Chicago.

Roos, N & Morgan, A J (1990). *Cryopreservation of thin biological specimens for electron microscopy*. RMS Handbook No. 21. Oxford University Press: Oxford.

Ruska, E (1980). *The early development of electron lenses and electron microscopy*. Translated into English by T Mulvey. Hirzel: Stuttgart.

Russ, J C (1990). *Computer assisted microscopy: The measurement and analysis of images*. Plenum: New York & London.

Rutherford, G K (Ed.) (1974). *Soil microscopy*. Proc. 4th Intl working meeting on soil micromorphology, Kingston, Ontario, 1973. The Limestone Press: Kingston, Ontario.

Saparin, G V & Obyden, S K (1988). Color display of video information in scanning electron microscopy: principles and application to physics, geology, soil science, biology and medicine. *Scanning*, **10**, 87–106

Saxton, W O (1978). Computer techniques for image processing in electron microscopy. In *Advances in electronics and electron physics*, Supplement 10, pp. 1–289. Academic Press: New York & London.

Saxton, W O & Chang M (1988). On-line image analysis and microscope control in EM. *EUREM 88; Inst. Phys. Conf. Ser. No. 93*, Vol. 1, 59–64

Scanga, F (1964). *Atlas of electron microscopy; biological applications*. Elsevier: Amsterdam.

Schlüter, G, Hund, A & Thaer, A (1979). Kombination LM/REM. *Labor-Praxis*, **3**, 22–5

Scott, V D, Love, G & Reed, S J B (1994). *Quantitative electron-probe microanalysis*, 2nd edition. Ellis Horwood: London.

Sekine, T, Nagasawa, Y, Kudoh, M, Sakai, Y, Parkes, A S, Geller, J D, Moganic, A & Hirata, K (1982). *Handbook of Auger Spectroscopy*. JEOL Ltd: Tokyo.

Septier, A (1966). The struggle to overcome spherical aberration in electron optics. pp. 204–74 in Barer & Cosslett (Eds.) (1966)

Serra, J (1984, 1988). *Image analysis and mathematical morphology*. Vol. 1, 1984; Vol. 2, *Theoretical advances* (Ed. Serra), (1988). Academic Press: San Diego.

Shaw, B A, Evans, J T & Page, T F (1994). Scanning electron acoustic microscopy imaging of subsurface microcracks produced in gear fatigue. *J. Mats. Sci. Lett.*, **13**(21), 1551–4

Shechtman, D & Brandon, D G (1970). Experiments with a conical opaque stop. *Electron Microscopy 1970*, Vol. 1, 41–2. Société Francaise de Microscopie Électronique: Paris.

Shindo, D, Hiraga, K, Oikawa, T & Mori, N (1990). Quantification of electron diffraction with Imaging Plate. *J. Electron Microscopy*, **39**(6), 449–53

Siegbahn, K, Nordling, C, Fahlman, A, Nordberg, R, Hamrin, K, Hedman, J, Johansson, G, Bergmark, T, Karlsson, S E, Lindgren, I & Lindberg, B (1967). *ESCA – Atomic, molecular and solid state structure studied by means of electron spectroscopy*. Almqvist & Wiksells: Uppsala.

Sigee, D C, Morgan, A J, Sumner, A T & Warley, A (1993). *X-ray microanalysis in biology – experimental techniques and applications*. Cambridge University Press: Cambridge.

Sjöstrand, F S (1958). Ultrastructures of retinal rod synapses of the guinea-pig eye as revealed by three-dimensional reconstructions from serial sections. *J. Ultrastructure Res.*, **2**, 122–70

Slayter, E M & Slayter H S (1992). *Light and electron microscopy*. Cambridge University Press: New York and Cambridge.

Smart, P & Tovey N K (1981). *Electron microscopy of soils and sediments: examples*. Clarendon Press: Oxford.

Smart, P & Tovey, N K (1982). *Electron microscopy of soils and sediments : techniques*. Clarendon Press: Oxford.

Smith, B A & Nutting, J (1956). Direct carbon replicas from metal surfaces. *Brit. J. Appl. Phys.*, 7, 214–17

Smith, G C (1994). *Surface analysis by electron spectroscopy*. Plenum Press: New York.

Sommerville, J & Scheer, U (Eds.) (1987). *Electron microscopy in molecular biology – a practical approach*. IRL Press: Oxford & Washington.

Spence, J C H & Zuo, J M (1992). *Electron Microdiffraction*. Plenum: New York.

Steeds, J W (1979). *Convergent beam electron diffraction*. Chapter 15, pp. 387–422 in Hren *et al.* (1979).

Steere, R L (1957). Electron microscopy of structural detail in frozen biological specimens. *J. Biophys. & Biochem. Cytology*, 3, 45–60

Steere, R L (1981). Preparation of freeze-fracture, freeze-etch, freeze-dry, and frozen surface replica specimens for electron microscopy in the Denton DFE–2 and DFE-3 freeze-etch units. Chapter 5, pp. 131–81 in *Current trends in morphological techniques*, vol. II, ed. J E Johnson Jr., CRC Press Inc.: Boca Raton, FL.

Steere, R L, Erbe, E F & Moseley, J M (1977). A resistance monitor with power cut-off for automatic regulation of shadow and support film thickness in freeeze-etching and related techniques. *J. Microscopy*, 111, 313–28

Steinbrecht, R A & Zierold, K (Eds.) (1987). *Cryotechniques in biological electron microscopy*. Springer-Verlag: Berlin & Heidelberg.

Stewart, A D G & Snelling, M A (1964). A new scanning electron microscope. *Electron Microscopy 1964*, Vol. 1, 55–6. Czech. Academy of Science: Prague.

Swann, P R, Humphreys, C J & Goringe, M J (Eds.) (1974). *High voltage electron microscopy: Proc. 3rd Intl. Conf.*, Oxford 1973. Academic Press: London & New York.

Sung, C & Williams, D B (1991). Principles and applications of convergent beam electron diffraction: a bibliography (1938–1990). *J. Electron Microscopy Technique*, 17, 95–118

Tanaka, M & Terauchi, M (1985). *Convergent-beam electron diffraction*. JEOL: Tokyo.

Tanaka, M, Terauchi, M & Kaneyama, T (1988). *Convergent-beam electron diffraction II*. JEOL: Tokyo.

Tanaka, M, Terauchi, M & Tsuda, K (1994). *Convergent-bean electron diffraction III*. JEOL: Tokyo.

Thellier, M, Ripoll, C & Berry, J-P (1991). Biological applications of secondary ion mass spectrometry. *Microscopy & Analysis*, 23, 13–15

Thibaut, M (1992). Laser microprobe for analysis in microbiology. *Microscopy & Analysis*, 28, 35–7

Thomas, G & Goringe, M J (1979). *Transmission electron microscopy of materials*. Wiley: New York.

Thompson, M N (1981). Materials science applications of hybrid scanning modes. *Electron Microscopy and Analysis 1981, (EMAG 81)*, pp. 83–6. Inst. of Physics: Bristol

Tochigi, H, Sonehara, K & Tanaoka, I (1962). An application of oxide-cored cathode to telefocal electron gun for electron microscope. *Electron Microscopy 1962*, Vol. 1, paper KK–13. Academic Press: New York.

Tomokiyo, Y (1992). Application of convergent beam electron diffraction to extract quantitative information in materials science. *J. Electron Microscopy*, 41(6), 403–13

Tonomura, A (1994). *Electron holography*. Springer–Verlag: Berlin.

Tribe, M A, Erant, M R & Snook, R K (1975). *Electron microscopy and cell structure*. Cambridge University Press: Cambridge.

Tromp, R M, Hamers, R J & Demuth, J E (1986). Quantum states and atomic structure of silicon surfaces. *Science*, 234, 304–9

Valentine, R C (1966). The response of photographic emulsions to electrons. In *Advances in optical and electron microscopy*, Vol. 1, pp. 180–203, Eds. Barer & Cosslett. Academic Press: London.

Vainshtein, B K (1964). *Structure analysis by electron diffraction*. Translated into English and edited by Feigl, E & Spink, J A. Pergamon Press: Oxford

Valdrè, U, Robinson, E A, Pashley, D W, Stowell, M J & Law, T J (1970). An ultra-high vacuum electron microscope specimen chamber for vapour deposition studies. *J. Phys. (E)*, 3, 501–6

Vale, S H & Statham, P J (1986). STEM image stabilisation for high resolution microanalysis. *Electron Microscopy 1986*, vol. 1, 573–4. Japanese Society of Electron Microscopy: Tokyo

Vaughan, J G (1979). *Food microscopy*. Academic Press: London.

Venables, J A, Hembree G G, Liu J, Harland C J, & Huang, M (1994). Advances in Auger electron spectroscopy and imaging. *Electron Microscopy 1994*, Vol. 1, 759–62

Veneklasen, L H (1992). The continuing development of low energy electron microscopy for characterizing surfaces. *Rev. Sci. Instrum.*, 63(12), 5513–32

Verhoeven, J D & Gibson, E D (1976). Evaluation of LaB$_6$ cathode electron gun. *J. Phys. (E)*, 9, 65–9

Villiger, W & Bremer, A (1990). Ultramicrotomy of biological objects: from the beginning to the present. *J. Struct. Biol.*, 104(1–3), 178–88

Wall, J S & Hainfield, J F (1986). Mass mapping with the scanning transmission electron microscope. *Ann. Rev. Biophys. Biophys. Chem.* 15, 355–76

Walther, P, Wehrli, E, Hermann, R, & Müller, M, (1995). Double-layer coating for high-resolution low-temperature scanning electron microscopy. *J. Microscopy*, 179(3), 229–37

Watt, I M (1964). A simple method for the single or repeated replication of selected areas of surfaces in electron microscopy. *J. Sci. Instrum.*, 41, 107–8

Watt, I M (1974). Reduction in specimen-level heating during carbon depositions by the Bradley technique. *Electron Microscopy 1974*, Vol. 1, 402–3. Australian Academy of Science: Canberra.

Watt, I M (1978). A comparison of gold and platinum sputtered coatings for scanning electron microscopy. *Electron Microscopy 1978*, Vol. II, 94–5. Microscopical Society of Canada: Toronto.

Watt, I M (1991). Carbon – the electron microscopist's most useful element. *Microscopy & Analysis*, 26, 21–3

Watt, I M & Wraight, N A (1971). A new magnification test specimen for scanning electron microscopes. *Metron*, 3, 153–6

Watts, J F (1990). *Introduction to surface analysis by electron spectroscopy*. RMS Handbook No. 22. Oxford University Press: Oxford.

Weakley, B S (1981). *A beginner's handbook in biological electron microscopy*, 2nd edition. Churchill Livingston: Edinburgh.

Weavers, B A (1971). Combined high resolution microscopy and electron probe x-ray microanlaysis and its applications to medicine and biology. *Micron*, 2, 390–404

Weibel, E R (1969). Techniques for quantitative evaluation of structure. 'Morphometry'. *International Review of Cytometry*, 26, 235–302

Wells, O C, Boyde, A, Lifshin, E & Rezanowich, A (1974). *Scanning electron microscopy*. McGraw-Hill: New York.

Wenk, H R (1976). *Electron microscopy in mineralogy*. Springer-Verlag: Berlin.

Whalley, W B, (Ed.) (1978). *Scanning electron microscopy in the study of sediments*. (Papers presented in a symposium at Swansea, Sept. 1977). Geo. Abstracts Ltd, University of East Anglia: Norwich.

White, J G, Amos, W B & Fordham, M (1987). An evaluation of confocal versus conventional imaging of biological structures by fluorescence light microscopy. *J. Cell Biol.*, 105, 41–8

Williams, D B, Pelton, A R & Gronsky, R (Eds.) (1991). *Images of materials*. Oxford University Press Inc.: New York.

Williams, M A (1977). *Quantitative methods in biology*. Vol. 6, Pt II in the Glauert series. Elsevier: Amsterdam.

Wilson, A J & Robards, A W (1984). *An atlas of low temperature scanning electron microscopy*. University of York: York.

Wilson, A R (1993). The accuracy, precision and stability of a (scanning) transmission electron microscope with a parallel electron-energy-loss spectrometer assessed by measurements of the plasmon energy of Al. *Phil. Mag. B*, **67**(2), 181–92

Wilson, T (1990). *Theory and practice of confocal microscopy*. Academic Press: San Diego.

Wischnitzer, S (1981). *Introduction to electron microscopy*, 3rd edition. Pergamon: Oxford.

Woldseth, R (1973). *X-ray spectrometry*. Kevex Corporation: Burlingame, CA.

Woodruff, D P & Delchar T A (1994). *Modern techniques of surface science*, 2nd edition. Cambridge University Press: Cambridge.

Yagi, K (1993). RHEED & REM. Chapter in *Electron diffraction techniques* (Ed. Cowley J M), Vol. 2. Oxford University Press: Oxford.

Yakobi, B G & Holt, D B (1990). *Cathodoluminescence microscopy of inorganic solids*. Plenum: New York.

Yamamoto, N, Spence, J C H & Fathy, D (1984). Cathodoluminescence and polarization studies from individual dislocations in diamond. *Phil. Mag. B*, **49**(6), 609–29

Zanchi, G, Kihn, Y, Sevely, J & Jouffrey, B (1983). On the influence of high voltage in electron energy loss spectroscopy. *Proc. 7th Intl. Conf. on HVEM*, Berkeley, 1983, Eds. Fisher, Gronsky & Westmacott, pp. 85–8. Lawrence Berkeley Lab. Report LBL–16031, UC–25, CONF–830819: Berkeley, CA.

Ziebold, T O (1967). Precision and sensitivity in electron microprobe analysis. *Anal. Chem.*, **39**, 859–61

Zworykin, V R, Hillier, J & Snyder, R L (1942). A scanning electron microscope. In *ASTM Bulletin No. 117*, 15–23. ASTM: Philadelphia.

Additional literature on electron microscopy

Laboratory handbooks on electron microscopy

Practical methods in electron microscopy. Edited by Audrey M Glauert.
A series of practical handbooks for electron microscopists, written by recognised authorities. Published by Elsevier, Amsterdam until volume 14, after which the series was continued by Portland Press, London and Chapel Hill.

Vol. 1   Part I *Specimen preparation in materials science* by P J Goodhew (1972)
Part II *Electron Diffraction and optical diffraction techniques* by
B E P Beeston, R W Horne and R Markham

Vol. 2   *Principles and practice of electron microscope operation* by A W Agar, R H Alderson and Dawn Chescoe (1974).

Vol. 3   Part I *Fixation, dehydration and embedding of biological specimens* by Audrey M Glauert (1974). Part II *Ultramicrotomy* by Norma Reid

Vol. 4   *Design of the electron microscope laboratory* by R H Alderson (1975)

Vol. 5   Pt I *Staining methods for sectioned material* by P R Lewis and D P Knight (1977) Pt II *X-ray microanalysis in the electron microscope* by J A Chandler.

Vol. 6   Pt I *Autoradiography and immunocytochemistry* by M A Williams (1977) Pt
II *Quantitative methods in biology* by M A Williams

Vol. 7   *Image analysis, enhancement and interpretation* by D L Misell (1978)

Vol. 8   *Replica, shadowing and freeze-etching techniques* by J H M Willison and A J
Rowe (1980).

Vol. 9   *Dynamic experiments in the electron microscope* by E P Butler and K F Hale
(1981)

Vol. 10   *Low temperature methods in biological electron microscopy* by A W Robards
and U B Sleytr (1985)

Vol. 11   *Thin foil preparation for electron microscopy* by P J Goodhew (1985). 1st
revision of Vol. 1 Pt I.

Vol. 12   *Electron diffraction: An introduction for biologists* by D L Misell and E B
Brown (1987)

Vol. 13   *Sectioning and cryosectioning for electron microscopy* by Norma Reid and J
E Beesley (1990)

Vol. 14   *Cytochemical staining methods for electron microscopy* by P R Lewis and D P
Knight (1992). Revision of Vol. 5 Pt I.

Vol. 15   *Vacuum methods in electron microscopy* by W C Bigelow (1994)

Procedures in Electron Microscopy, principal editors A W Robards & A J Wilson.
This is a loose-leaf compilation of descriptions and formulae covering most aspects
of electron microscopy. Published by John Wiley, Chichester. Originally published
in 1993 it is planned to be augmented by biannual updates. It is intended for practical use at the workbench. The main chapters are headed:

1.   Setting up an electron microscope facility
2.   Biomedical specimens
3.   Materials specimens
4.   General preparation techniques for TEM
5.   Basic biological preparation techniques for TEM
6.   Basic materials specimen preparation techniques for TEM
7.   Specialized specimen-dependent preparation techniques for TEM
8.   Immunocytochemical techniques for TEM and SEM
9.   General techniques for TEM observation
10.   General preparation techniques for SEM
11.   Basic biological preparation techniques for SEM
12.   Basic materials preparation techniques for SEM
13.   Specialized specimen-dependent preparation techniques for SEM
14.   General techniques for SEM observation
15.   Microanalysis
16.   Low-temperature methods for TEM and SEM
17.   Image analysis and quantitation

Royal Microscopical Society Microscopy Handbooks.
This is a series of clear and practical guides to many aspects of light and electron microscopy. Initially published by Oxford University Press, Oxford; from No. 27 onwards they have been published by Bios Scientific Publishers, Oxford.
The complete series of titles is:

1. *An introduction to the optical microscope* by Savile Bradbury (1984); Revised edition 1989.
2. *The operation of the transmission electron microscope* by D Chescoe & P J Goodhew (1984)
3. *Specimen preparation for transmission electron microscopy of materials* by P J Goodhew (1984)
4. *Histochemical protein staining methods* by J James & J Tas (1984)
5. *X-ray microanalysis in electron microscopy for biologists* by A J Morgan (1985)
6. *Lipid histochemistry* by O Bayliss High (1984)
7. *The light microscopy of synthetic polymers* by D A Hemsley (1984)
8. *Maintaining and monitoring the transmission electron microscope* by S K Chapman (1986)
9. *Qualitative polarized-light microscopy* by P C Robinson & S Bradbury (1992)
10. *Introduction to fluorescence microscopy* by J S Ploem & H J Tanke (1987)
11. *An introduction to immunocytochemistry: current techniques and problems* by J M Polak & S van Noorden (1984); revised edition 1987
12. *An introduction to scanning acoustic microscopy* by A Briggs (1985)
13. *An introduction to photomicrography* by D J Thomson & S Bradbury (1987)
14. *Enzyme histochemistry* by J D Bancroft & N M Hand
15. *RMS Dictionary of light microscopy* by S Bradbury, P J Evennett, H Haselmann & H Piller (1989)
16. *Light-element analysis in the transmission electron microscope* by P M Budd & P J Goodhew (1988)
17. *Colloidal gold – a new perspective for cytochemical marking* by J E Beesley (1989)
18. *Autoradiography – a comprehensive overview* by J R J Baker (1989)
19. *Introduction to crystallography* by C Hammond (1990); Revised edition 1992
20. *The operation of transmission and scanning electron microscopes* by D Chescoe & P J Goodhew (1990)
21. *Cryopreparation of thin biological specimens for electron microscopy – methods and applications* by N Roos & A J Morgan (1990)
22. *An introduction to surface analysis by electron spectroscopy* by J F Watts (1990)
23. *Basic measurement techniques for light microscopy* by S Bradbury (1991)
24. *The preparation of thin sections of rocks, minerals and ceramics* by D W Humphries (1992)

25. *The role of microscopy in semiconductor failure analysis* by B P Richards & P K Footner (1992)
26. *Enzyme histochemistry – a laboratory manual of current methods* by C J F van Noorden & W M Frederiks (1992)
27. *In-situ hybridization* by A R Leitch, T Schwarzacher, D Jackson & I J Leitch (1994)
28. *Biological microtechnique* by J Sanderson (1994)
29. *Flow cytometry* by M G Ormerod (1994)
30. *Food microscopy* by O Flint (1994)
31. *Scientific PhotoMACROgraphy* by B Bracegirdle (1994)
32. *Microscopy of textile fibres* by P H Greaves & B P Saville (1995)
33. *Modern PhotoMICROgraphy* by B Bracegirdle & S Bradbury (1995)

Proceedings of international conferences on electron microscopy

Much useful state-of-the-art information is contained in the published proceedings of the biennial international and regional conferences on electron microscopy, and a list is given below of the volumes available to date. The quality of these papers may be a little variable and (as with any published work) the results and conclusions may have been subject to later revision or withdrawal, but in many cases the short presentation is the only public airing of the work, which may have formed part of an otherwise confidential project. Unfortunately not all the abstracting journals cover these conference proceedings, and it is prudent to test the available abstracts systems with a known paper before relying on any one of them for total coverage of the current literature.

The proceedings of the annual SEM symposia originally organised by the Illinois Institute of Technology Research Institute (IITRI), and now by Scanning Microscopy International, are a valuable source of information on techniques and applications of the SEM as well as containing very comprehensive bibliographies. Further details are given later.

The IFSEM conferences

Beginning with a meeting at Delft in 1949 a series of periodic international gatherings have been held to discuss electron microscopy, under the auspices of what is now the International Federation of Societies for Electron Microscopy, (IFSEM). Initially, the contributed papers were edited and published as the Conference Proceedings, often several years after the conference. Starting in 1962 the published *Proceedings* have taken the form of volumes of illustrated two-page extended abstracts of each contribution, produced as a handbook for use by the participants at the conference, and entitled *Electron Microscopy, 19...*. The number of papers is frequently very large and more recently the invited review papers have been printed in addition to the two-page abstracts. The Proceedings of the 1994 International

Congress in Paris totalled 3821 pages, issued in five parts. The two main series of conferences are the International and the Asia Pacific and European Regional Meetings, held alternately at two-year intervals. The venue, date and title of the International and European Regional meetings, with the title, editors and publishers of their Proceedings are listed below:

Delft 1949. *Proceedings of the Conference on Electron Microscopy*, eds. A L Houwink, J B le Poole & W A le Rütte. Hoogland: Delft (1950).

Paris 1950. *Comptes Rendus du Premier Congrès International de Microscopie Électronique.* Editions de la revue d'optique théorique et instrumentale: Paris (1953)

Gent 1954. *Rapport Europees Congrès Toegepaste Electronenmicroscopie.* Ed. and publ. by G Vandermeerssche: Uccle-Bruxelles

London 1954. *The Proceedings of the Third International Conference on Electron Microscopy, London 1954*, ed. R Ross. Royal Microscopical Society: London (1956)

Stockholm 1956. *Electron Microscopy, Proceedings of the Stockholm Conference*, eds. F J Sjöstrand & J Rhodin. Almqvist & Wiksells: Stockholm (1957)

Berlin 1958. *Vierter Internationaler Kongress für Elektronenmikroskopie*, eds. W Bargmann, *et al.* Springer: Berlin, Göttingen & Heidelberg (1960)

Delft 1960. *The Proceedings of the European Regional Conference on Electron Microscopy.* Two vols., eds. A L Houwink & B J Spit. Nederlandse vereniging voor electronen-microscopie: Delft (1961)

Philadelphia 1962. Fifth International Congress for Electron Microscopy. *Electron Microscopy 1962.* Two vols., ed. S S Breese. Academic Press: New York.

Prague 1964. Third European Conference on Electron Microscopy. *Electron Microscopy 1964.* Two vols., ed. M Titlbach. Czechoslovak Academy of Science: Prague

Kyoto 1966. Sixth International Congress for Electron Microscopy. *Electron Microscopy 1966.* Two vols., ed. R Uyeda. Maruzen: Tokyo.

Rome 1968. Fourth European Conference on Electron Microscopy. *Electron Microscopy 1968.* Two vols., ed. D S Bocciarelli. Tipographia Poliglotta Vaticana: Rome.

Grenoble 1970. Seventh International Conference on Electron Microscopy. *Electron Microscopy 1970* Three vols., ed. P Favard. Société Francaise de Microscopie Electronique: Paris.

Manchester 1972. Fifth European Congress on Electron Microscopy. *Electron Microscopy 1972.* Institute of Physics: London.

Canberra 1974. Eighth International Congress for Electron Microscopy.

Electron Microscopy 1974. Two vols., eds. J V Sanders & D J Goodchild. Australian Academy of Science: Canberra.

Jerusalem 1976. Sixth European Congress for Electron Microscopy. *Electron Microscopy 1976.* Two vols.: Vol. I ed. D G Brandon, Vol. II eds. D G Brandon & Y Ben-Shaul. Tal International Publishing Co: Jerusalem.

Toronto 1978. Ninth International Congress for Electron Microscopy. *Electron Microscopy 1978.* Three vols., ed. J M Sturgess. Microscopical Society of Canada: Toronto.

The Hague 1980. Seventh European Congress on Electron Microscopy, EUREM 80, with 9th International Conference on X-ray Optics and Microanalysis, IX ICXOM. *Electron Microscopy 1980.* Four vols.: Vol. I eds. P Brederoo & G Boom, Vol. II eds. P Brederoo & W de Priester, Vol. III eds. P Brederoo & V E Cosslett, Vol. IV eds. P Brederoo & J van Landuyt. Seventh European Congress on Electron Microscopy Foundation: Leiden.

Hamburg 1982. Tenth International Congress on Electron Microscopy, *Electron Microscopy 1982.* Three vols., ed. Congress Organising Committee. German Society for Electron Microscopy: Frankfurt/Main

Budapest 1984. Eighth European Congress on Electron Microscopy, EUREM 84. *Electron Microscopy 1984.* Three vols., eds. A Csanády, P Röhlich & D Szabó. The Programme Committee of the Eighth EUREM: Budapest

Kyoto 1986. Eleventh International Congress on Electron Microscopy, XI ICEM. *Electron Microscopy 1986.* Four vols., eds. T Imura, S Marusa & T Suzuki. The Japanese Society of Electron Microscopy: Tokyo.

York 1988. 9th European Congress on Electron Microscopy, EUREM 88. *EUREM 88.* Three vols., eds. P J Goodhew & H G Dickinson. *Inst. Phys. Conf. Ser. No. 93.* Institute of Physics: Bristol.

Seattle 1990. XIIth International Congress for Electron Microscopy, XII ICEM. *Electron Microscopy 1990.* Four vols., eds. L D Peachey & D B Williams. San Francisco Press Inc.: San Francisco.

Granada 1992. 10th European Congress on Electron Microscopy, EUREM 92. *Electron Microscopy 1992.* Three vols., Vol. I eds. A Ríos, J M Arias, L Megías-Megías & A López-Galindo; Vol. II eds. A López-Galindo & M I Rodríguez-García; Vol. III eds. L Megías-Megías, M I Rodríguez-García, A Ríos & J M Arias. University of Granada: Granada.

Paris 1994. 13th International Congress on Electron Microscopy, 13 ICEM.

Electron Microscopy 1994. Three vols., eds. B Jouffrey & C Colliex. Les Éditions de Physique: Paris.

Dublin 1996. 11th European Congress on Electron Microscopy, EUREM 96. Proceedings published as bound volumes and CD-ROM.

Other national and international conferences

Similar volumes of *Proceedings* are sometimes published following annual or irregular meetings of some of the larger electron microscope societies. For example, *Proceedings of the Annual Meetings of the Microscopy Society of America* (MSA, formerly the Electron Microscope Society of America), are published by San Francisco Press Inc.: San Francisco, CA.

EMAG 75 was a meeting on 'Developments in Electron Microscopy and Analysis' held by the Electron Microscopy and Analysis Group, EMAG, of the Institute of Physics. The proceedings, edited by J A Venables were published by Academic Press: London (1976). Further conferences with the same theme were held in subsequent alternate years, resulting in *EMAG 77, 79, 81, 83, ...* volumes published in their *Conference Series* by the Institute of Physics: Bristol and London.

Proceedings of the annual German meetings on Scanning Electron Microscopy, published under the title 'Beitrage zur elektronenmikroskopischen Direktabbildungen von Oberflächen' and edited by G Pfefferkorn are published by Remy, PO Box 6622, D-4400 Münster, Germany. Volume Lit. 6–8 contains 1940 references from the literature of 1972–75.

SEM symposia and journals

A series of annual symposia on scanning electron microscopy was started in 1968. These symposia were initially known as the *IITRI Symposia*, organised by Dr Om Johari at the Illinois Institute of Technology Research Institute, Chicago.

In 1978 the organisation was put on a full-time basis through Scanning Electron Microscopy, Inc. At the same time the published *Proceedings* of the Symposia were incorporated into a journal, *Scanning Electron Microscopy*, which also contained other contributed material.

From 1987 the journal became a soft-bound quarterly, *Scanning Microscopy*, whose range of contents was broadened to include scanned images using laser and ion beams, and scanning light microscopy.

From 1982 an additional series of annual conferences on basic topics in scanning microscopy and related techniques has been dedicated to Professor G E Pfefferkorn of Münster, Germany. *Scanning Microscopy* Supplements are issued based on these conferences. The topics covered so far in this 'Pfefferkorn Conference' series are:

1st (1982)   Electron beam interactions with solids for microscopy, microanalysis and microlithography. (Eds. D F Kyser, D E Newbury, H Niedrig & Shimizu).

2nd (1983)   The science of biological specimen preparation for microscopy and microanalysis. (Eds. J -P Revel, T Barnard & G H Haggis).

3rd (1984)   Electron optical systems for microscopy, microanalysis and microlithography. (Eds. J J Hren, F A Lenz, E Munro & P B Sewell).

4th (1985)   The science of biological specimen preparation for microscopy and microanalysis 1985. (Eds. M Müller, R P Becker, A Boyde & J J Wolosewick).

5th (1986)   Physical aspects of microscopic characterization of materials. (Eds. J Kirschner, K Murata & J A Venables).

6th (1987)   Image- and signal processing in electron microscopy. (Eds. P W Hawkes, F P Ottensmeyer, W O Saxton & A Rosenfeld).

7th (1988)   The science of biological specimen preparations for microscopy and microanalysis 1988. (Eds. R M Albrecht & R L Ornberg).

8th (1989)   Fundamental electron and ion beam interactions with solids for microscopy, microanalysis and microlithography. (Eds. J Schou, P Kruit & D E Newbury).

9th (1990)   The science of biological specimen preparations for microscopy and microanalysis 1990. (Eds. L Edelmann & G M Roomans).

10th (1991)   Signal- and image processing in microscopy and microanalysis. (Ed. P W Hawkes).

11th (1992)   Physics of generation and detection of signals used for microcharacterization. (Eds. G Remond, R H H Gijbels, L L Levenson & R Shimizu).

12th (1993)   Science of biological x-ray microanalysis. (Eds. B L Gupta & G M Roomans).

13th (1994)   Cathodoluminescence.

14th (1995)   The science of biological specimen preparation.

Details of the annual Scanning Microscopy symposia, the 'Pfefferkorn Conferences', and the quarterly journals *Scanning Microscopy;* and *Cells and Materials,* can be obtained from Scanning Microscopy International, PO Box 66507, Chicago (A M F O'Hare), IL 60666–0507, USA.

Other books on electron microscopy and its applications

The literature on microscopes and their applications is extensive and may be distributed in a technical library under a variety of classifications.

In the widely used *Dewey Decimal Classification* (20th edition, 1989 : Forest Press: Albany, NY, USA) the *Instrumentation and Techniques of Microscopy* have the class number 502.82. Interdisciplinary works in microscopy, such as microtomy, share the same classification. *Electron Microscopes* are classified as 502.825, but the *Manufacturing Technology of Electron Microscopes* is 681.413.

Microscopy applied in different fields is classified as a sub-section of that branch of science. Thus:

Microscopy	natural sciences	502.82
	biology	578
	medical diagnosis	616.0758
Other relevant classifications are:		
	Electron optics (physics)	537.56
	Electron metallography	669.950 282
	Microstructure in materials science	620.112 99
	Medical sciences	610
	Chemistry	540
	applied	660

The above classification numbers refer to the stated version of the *Dewey Decimal Classification* and are given as a guide only; the classification used in an individual library may differ from this, and the library's index of classification should always be consulted.

Periodicals dealing with electron microscopy

Keeping up with the contemporary literature on the developments and applications of electron microscopy is a formidable task for the individual. Assistance can be purchased from international abstracting organisations who will, for a subscription, supply lists of current articles with certain chosen keywords in their titles. By the use of computer storage and sorting of data the appearance of an article in any one of thousands of journals can be reported within weeks of its publication. Linked with this service can be the rapid supply of a copy of the actual article. Three well-known sources of this alerting service are Inspec, Chemical Abstracts (CA) and ISI (Institute for Scientific Information).

For those who prefer to keep their own tabs on current literature there are nine journals which are almost certain to contain relevant articles:

Journal of Electron Microscopy. The official organ of the Japanese Society of Electron Microscopy. Published by JEM, c/o Center for Academic publications Japan, 4–16 Yayoi 2-chome, Bunkyo-ku, Tokyo, 113 Japan. Bimonthly, in English.

Journal of Microscopy. (Originally the *Journal of the Royal Microscopical Society.*) Published by Blackwell Scientific Publications, Osney Mead, Oxford OX2 0EL. Monthly, in English.

Micron. (Originally published as *Micron & Microscopica Acta.*) Published by Pergamon Press (Elsevier Science Ltd.), The Boulevard, Langford Lane, Kidlington, Oxford OX5 1GB. Six issues per year, in English.

Microscopy and Analysis. Published in UK, European and USA editions by RGC, 1 Gable Cottage, Post House Lane, Bookham, Surrey, KT23 3EA. Controlled-circulation bimonthly, in English.

Microscopy, Microanalysis, Microstructures. (Before 1990 this was the *Journal de Microscopie et de Spectroscopie Electroniques.*) Official organ of the French Society for Electron Microscopy. Published by Les Editions de Physique, Ave du Hoggar, Z.I. de Courtaboeuf, B.P. 112, 91944 Les Ulis Cedex A, France. Bimonthly, in English or French.

Microscopy Research and Technique. (Before 1992 this was called *Journal of Electron Microscopy Technique.*) Published by Wiley-Liss Inc, 605 Third Avenue, New York, NY 10158, USA. Monthly, twice in February, April, June, August, October & December, in English.

Optik. Official organ of the German Society for Electron Microscopy. Publisher: Wissenschaftliche Verlagsgesellschaft mbH, Birkenwaldstrasse 44, PO Box 10 10 61, D-70009 Stuttgart, Germany. Monthly, in German or English.

Scanning. Published by FAMS Inc., Box 832, Mahwah NJ 07430, USA. Bimonthly, six issues plus Supplement annually, in English.

Ultramicroscopy. Published by Elsevier Science BV, Molenwerf 1, PO Box 211, 1000 AE Amsterdam, The Netherlands. Monthly, in English.

Articles on the development of devices for microscopes may be found in *Measurement Science & Technology* (the *Journal of Physics* (E), formerly *J. Sci. Instrum.*), *Journal of Physics (D)* (formerly *Brit. J. Appl. Phys.*) and the corresponding American journals *Review of Scientific Instruments* and *Journal of Applied Physics.* The journals published by Scanning Microscopy International were mentioned in an earlier section (page 465).

The contents lists of most of the above list are reprinted (together with a large number of other periodicals) in the weekly issues of '*Current Contents*®', published by ISI in Philadelphia. The Life Sciences and Physical, Chemical and Earth Sciences editions are the relevant ones.

There are numerous abstracting publications which include electron microscopy topics: the most widely known are probably *Chemical Abstracts* (published by the American Chemical Society) and *Physics Abstracts* (published by Inspec). Their annual indexes are valuable for searching into earlier publications.

One of the most powerful literature compilations is the *Science Citation Index*® from ISI. This is published bimonthly in four volumes: *Permuterm subject index, Source index*, and *Citation index A-K* and *K-Z.* Every article from over 5000 publications, including all the microscopy ones listed above, can be traced through its title (in the *Permuterm subject index*) or its author (*Source index*). Publications from individual laboratories or university departments can be tracked. In addition, the *Citation index* enables the reader to go forward in time from an earlier literature reference to locate all the occasions on which this reference has been quoted in later work.

An annual publication which selects several important subjects for detailed review is *Advances in Optical and Electron Microscopy*, originally edited by R Barer and V E Cosslett, subsequently by T Mulvey and C J R Sheppard, and published by Academic Press. In 1995 this publication was amalgamated with another Academic Press annual, *Advances in Electronics and Electron Physics*, edited by P W Hawkes, to form *Advances in Imaging and Electron Physics*.

Names and addresses of electron microscope manufacturers and their agents

Readers particularly interested in obtaining the most up-to-date information on commercial electron microscopes should apply to the manufacturers or their local agents. The relevant addresses are given below for Britain, USA and country of origin (if different). They are thought to be correct at the time of going to print.

AMRAY
Amray Inc., 160 Middlesex Turnpike, Bedford, MA 01730, USA
CAMECA
Cameca Ltd, PO Box 88, Wilmslow, Cheshire, SK9 5BE, UK
Cameca, 103 bd Saint-Denis/BP6, 92403 Courbevoie Cedex, France
Cameca Instruments Inc., 204 Spring Hill Road, Trumbull, CT 06611–1356, USA
ELECTROSCAN
Nikon UK Ltd, Instruments Division, Nikon House, 380 Richmond Road, Kingston, Surrey, KT2 5PR, UK
ElectroScan Corporation, 66 Concord Street, Wilmington, MA 01887, USA
GRESHAM–CAMSCAN
Gresham–CamScan, CamScan House, Pembroke Avenue, Waterbeach, Cambridge CB5 9PY, UK
CamScan USA, Inc., 508 Thomson Park Drive, Cranberry Township, PA 160666–6425, USA
HITACHI
Hitachi Scientific Instruments, Nissei Sangyo Co. Ltd, 7 Ivanhoe Road, Hogwood Lane Industrial Estate, Finchampstead, Wokingham, Berkshire, RG40 4QQ, UK
Nissei Sangyo America Ltd, 460E Middlefield Road, Mountain View, CA 94043, USA
Hitachi Limited, CPO Box 1316, Tokyo 100–91, Japan
JEOL
JEOL (UK) LTD, JEOL House, Silver Court, Watchmead, Welwyn Garden City, Hertfordshire, AL7 1LT, UK
JEOL (USA) INC., 11 Dearborn Road, Peabody, MA 01960, USA

JEOL LTD, 1–2 Musashino 3-chome, Akishima, Tokyo 196, Japan
RJ LEE
RJ Lee Instruments EUROPE, Bennett House, 1 High Street, Edgware, Middlesex, HA8 7HR, UK
RJ Lee Instruments Ltd, 515 Pleasant Valley Road, Trafford, PA 15085, USA
LEO
LEO Electron Microscopy Ltd, Clifton Road, Cambridge, CB1 3QH, UK
LEO Electron Microscopy Inc., One Zeiss Drive, Thornwood, NY 10594, USA
LEO Elektronenmikroskopie GmbH, D-73446 Oberkochen, Germany
PHILIPS
Philips Electron Optics, York Street, Cambridge, CB1 2QU, UK
Philips Electronic Instruments Company, Electron Optics Group, 85 McKee Drive, Mahwah, NJ 07430, USA
Philips Electron Optics, Building AAE, PO Box 218, 5600 MD Eindhoven, The Netherlands
TESLA
DSR Customer Services Ltd, 93 Wellington Street, Howley, Warrington, Cheshire, WA1 2DA, UK
Tesla Elmi a.s, Purkyňova 99, 61245 Brno, Czech Republic
TOPCON
ISS Group, Pellowe House, Francis Road, Withington, Manchester, M20 9XP, UK
Topcon Technologies Inc., 37 West Century Road, Paramus, NJ 07652, USA
Topcon Corporation, 75–1, Hasunuma-cho, Itabashi-ku, Tokyo, 174 Japan
VG
Fisons Instruments/VG Microscopes Ltd, Birches Industrial Estate, Imberhorne Lane, East Grinstead, West Sussex, RH19 1UB, UK
Fisons Instruments Inc, 55 Cherry Hill Drive, Beverly, MA 01915, USA

Name index

The name mentioned is cited on the page(s) indicated.

Subject index